International E-Conference on Cutting Edges in Mechanical Engineering (CEME 2020)

IOP Conference Series: Materials Science and Engineering
Volume 1168

Greater Noida, India
20 February 2021

Editors:

Avinash Ravi Raja
Mohd Zaheer Khan Yusufzai
Sudesh Singh
Mahipal Singh Rawat
Gagan Varshney

ISBN: 978-1-7138-3432-8
ISSN: 1757-8981

Printed from e-media with permission by:

Curran Associates, Inc.
57 Morehouse Lane
Red Hook, NY 12571

Some format issues inherent in the e-media version may also appear in this print version.

Institute of Physics
Dirac House, Temple Back
Bristol BS1 6BE UK

Phone: 44 1 17 929 7481
Fax: 44 1 17 920 0979

techtracking@iop.org

Additional copies of this publication are available from:

Curran Associates, Inc.
57 Morehouse Lane
Red Hook, NY 12571 USA
Phone: 845-758-0400
Fax: 845-758-2633
Email: curran@proceedings.com
Web: www.proceedings.com

TABLE OF CONTENTS

IOP Conf. Series: Materials Science and Engineering 1168 (2021) 011001 doi:10.1088/1757-899X/1168/1/011001

International E- Conference on

"Cutting Edges in Mechanical Engineering" (CEME-2020)

(20th February, 2021)

Editors
Dr. Avinash Ravi Raja,
Assistant Professor, Greater Noida Institute of Technology, Gr. Noida, INDIA
Dr. Mohd Zaheer Khan Yusufzai,
Associate Professor, Indian Institute of Technology (BHU), Varanasi, INDIA
Dr. Sudesh Singh,
Assistant Professor, Sharda University, Gr. Noida, INDIA
Dr. Mahipal Singh Rawat,
Professor, Greater Noida Institute of Technology, Gr. Noida, INDIA
Mr. Gagan Varshney,
Assistant Professor, Greater Noida Institute of Technology, Gr. Noida, INDIA

The conference was aimed to bring together leading academic scientists, researchers and research scholars to exchange and share their experiences and research results and discuss the latest advancement and cutting edges in Mechanical Engineering. The conference has provided a leading interdisciplinary platform for researchers, practitioners and educators to present and discuss the most recent innovations, trends, and concerns as well as practical challenges encountered and solutions adopted in the fields of advances in Mechanical Engineering.

Objectives of the conference:

- To create awareness about current advancement in the area of Mechanical Engineering in order to promote R&D activities for real life applications.
- The conference will create a unique environment for participants to establish professional networks in their respective fields, particularly with representatives from academia.
- To improve the common man's life by developing new innovative engineering ideas, technical tools or models or products of their needs

Participants was requested to submit papers broadly in the areas of Industrial Engineering, Non-conventional Energy, Nano Technology, Materials Engineering, Smart Materials, Thermal Engineering, Production Technology, IT enabled manufacturing and design, Simulation & Design, Applied Mechanics, Automobile Engineering, Application of Artificial Intelligence Machine Learning, IoT and Manufacturing, Process improvement through software development, Environment Engineering, Industry 4.0, Cryogenics, Heat and Mass Transfer, Bio-Fuels, Mechatronics, Robotics and Solar Energy.

Conference was proceed and enriched with Inaugural Chief Guest Speaker, Professor (Dr.) Yogendra Joshi, Georgia Tech University, USA, Guest of Honour, Dr. Kamendra Kumar, Director, TCIL, Delhi, Special invitee, Dr. S. Biswas, Ex Director (R&D), BHEL, Delhi, Inaugural keynote speaker Professor (Dr.) Ashok G. Matani,, Government College of Engineering, Amarawati (AP).

The organizing committee addresses the thanks to esteemed associates of the conference, eminent speakers, the members of the technical committee, session chairs, reviewers, participants, and all the members who have supported the conference.

Dr. Dr. M P S Rawat, **Program Coordinator (CEME-2020)**

Dr. Avinash Ravi Raja, **Organizing Secretary (CEME-2020)**

Mr. Vaibhav Gangwar, **Organizing Secretary (CEME-2020)**

International Advisory Committee

Prof.(Dr.) Yoginder K Joshi , Gorgia Tech, USA

Dr. Shishpal Rawat, Board President, SCI, USA

Dr. Rituparna Datt, Research Scientist, University of South Alabama, Alabama

Dr. Prakash Duraisamy, University of South Alabama, Alabama

Dr. Satyandra K. Gupta, University of Southern California, California

Dr. Hendrik Simon Cornelis Metselaar, University of Malaya, Kuala Lumpur, Malaysia

National Advisory Committee

Dr. Pradeep Kumar, Professor, IIT Roorkee.

Dr.Hari Singh, Professor, NIT Kurukshetra

Prof. R. K. Saini, National VP, CSI

Dr. R S Walia, Professor, PEC Chandigarh

Dr.D K Dviwedi, Professor, IIT Roorkee

Dr.Sanjeev Kumar, Professor, YMCAUST Faridabad

Dr. N K Singh, GM, BHEL, R & D Hyderabad

IOP Conf. Series: Materials Science and Engineering 1168 (2021) 011001 doi:10.1088/1757-899X/1168/1/011001

Dr. Pardeep Gupta, Professor, SLIET Punjab

Dr. H S Pali, Professor NIT Jammu

Dr. A K Sharma, Professor IIT Roorkee

Dr. R K Saxena, Professor SLIET Punjab

Dr. Arvind Jayant, Professor SLIET Punjab

Dr. K S Sangwan, Professor BITS Pilani

Dr. Dinesh Saini, Professor, NIT Jamshedpur

Dr. T U Siddiqie, Professor, IET Bareilly

Dr. Nilamber Singh, Professor, NIT Patna

Dr. Poonam Kumari, Professor, IIT Guwahati

Dr. Kamal Jangra, Professor, PEC Chandigarh

Dr. Jaiveer Singh, Professor, IIT Jodhpur

Dr. Satish Shekhawat, Powergrid, Gurugram

Sh. Sandeep Puri, MD, HydeEngg & Consulting

Sh. Sanjay Negi, CEO, Market Insight Consultants

Dr. M. Z. Khan Yushufzai, Associate. Prof. IIT BHU

Dr. Meghanshu Vashista, Associate. Prof. IIT BHU

Conference Adviser

Dr. Dhiraj Gupta, Director, Greater Noida Institute of Technology, Gr. Noida, INDIA

Dr. Anuranjan Mishra, Dean R&D, Greater Noida Institute of Technology, Gr. Noida, INDIA

Conference Chair

Dr. Alok Manas Dubey, Professor & Head, Greater Noida Institute of Technology, Gr. Noida, INDIA

IOP Conf. Series: Materials Science and Engineering 1168 (2021) 011001 doi:10.1088/1757-899X/1168/1/011001

Technical Chair

Mr. Gagan Varshney, Assistant Professor, Greater Noida Institute of Technology, Gr. Noida, INDIA

Associate Professor, Vellore Institute of Technology, Vellore, IN

Program Coordinator

Dr. M P S Rawat,
Professor, Greater Noida Institute of Technology, Gr. Noida, INDIA

Organizing Secretary

Dr. Avinash Ravi Raja
Assistant Professor, Greater Noida Institute of Technology, Gr. Noida, INDIA

Mr. Vaibhav Gangwar
Assistant Professor, Greater Noida Institute of Technology, Gr. Noida, INDIA

Host

Mr. Girendra Bhati
Assistant Professor, Greater Noida Institute of Technology, Gr. Noida, INDIA
Mr. S.Q. Husain
Assistant Professor, Greater Noida Institute of Technology, Gr. Noida, INDIA
Mr. Anuj Dixit
Assistant Professor, Greater Noida Institute of Technology, Gr. Noida, INDIA

Session Chairs

Dr. Radha Raman Mishra, BITS Pilani, INDIA

Dr. S.L. Verma, Noida Institute of Technology, Greater Noida, INDIA

Dr. T.U. Siddiqui, M.J.P Rohilkhand University, Bareilly, INDIA

Dr. Anil Kumar, KNIT Sultanpur, INDIA

Dr Sanjeev Kumar, JC Bose University of Science, YMCA, Faridabad, INDIA

Dr. Sudesh Singh, Sharda University, Greater Noida, INDIA

Paper Review Committee

Miss. Pooja Verma	Mechanical Engineering, Indian Institute of Technology (BHU) Varanasi

IOP Conf. Series: Materials Science and Engineering 1168 (2021) 011001 doi:10.1088/1757-899X/1168/1/011001

Dr. Preeti Verma	National Research Laboratory for Conservation of Cultural Property, Lucknow
Mr. Devashish Rajpoot	Mechanical and Aerospace Engineering Department, Nanyang Technical University
Dr. Sanjay Kumar Gupta	Mechanical engineering, Indian Institute of Technology (BHU) Varanasi
Dr. Pawan Kumar Yadav	Mechanical Engineering, IIT (BHU)
Dr. Anup kumar	Department of Mechanical Engineering, Maryland Institute of Technology and Management Jamshedpur
Dr. Amrendra kumar	mechanical engineering, NIT Durgapur
Dr. Rabindra Prasad	Mechanical engineering department, IIT(BHU) varanasi
Dr. Manish Deo	Mechanical engineering department, IIT(BHU) varanasi
Mr. Mithlesh kumar Mahto	Mechanical engineering department, IIT(BHU) varanasi
Dr. Radha raman mishra	Mechanical engineering department,Asst. Prof. BITS PILANI
Mr. Vishnu tiwari	Mechanical engineering department, JSS NOIDA
Dr. Sakendra kumar	Mechanical engineering department, DIET SHYAMPUR rishikesh
Mr. Ashwani Sharma	Mechanical Engineering, IIT BHU Varanasi
Mr. Abhimanyu Chaudhari	Mechanical Engineering, IIT BHU
Dr. K R Ramkumar	Postdoctoral researcher, IIT Madras
Dr. Yogesh Kumar	NIT, Patna
Dr. Rohit Kumar Singh Gautam	MECHANICAL DEPARTMENT, KCIT, Banda
Dr. Akhand Rai	MECHANICAL DEPARTMENT, KCIT, Banda
Dr Navneet Chaudhary	Galgotia Engineering college
Dr Rahul Saini	ABES engg college, Ghaziabad, up
Dr Asim Mishra	ITS engineering college
Dr. Omji Shukla	NIT Patna
Dr. Nitin Kotkunde	BITS, Hyderabad
Dr. Purushottam Kumar	BIT, Sindri
Dr. Kandpal Bhaskar	IPEC Ghaziabad
Dr. Nitin Jauhari	Graphic Era University, Dehradoon
Dr. T V V L N Rao	SRM Institute of Science and Technology, Kattankultahur

Dr. Shubrajit Bhaumik	SRM Institute of Science and Technology, Kattankultahur
Dr. Jitendra Kumar Katiyar	SRM Institute of Science and Technology, Kattankultahur
Dr. Vinod Kumar Sharma	VIT University, Vellore
Dr. Shushant Singh	UIT, Dehradun
Dr. Parshant Kumar	GEC, Jammu
Mr. Shuhaib Mushtaq	Islamic University of Science and Technology Awantipora
Mr. Pankaj Kumar	Mechanical engineering department, IIT(BHU) varanasi
Dr. Sumit Kumar	Mechanical engineering department, IIT(BHU) varanasi
Dr. Randhawa	PEC, chandigarh
Dr. Avinash Ravi Raja	Greater Noida Institute of Technology, Greater Noida
Mr. Adarsh Kumar	Mechanical engineering department, IIT(BHU) varanasi
Mr. Syed Qaisar Husain	Greater Noida Institute of Technology, Greater Noida
Mr. Sandeep Patidar	Greater Noida Institute of Technology, Greater Noida
Mr. GIRENDRA BHATI	Greater Noida Institute of Technology, Greater Noida
Mr. R N Yerrawar	M E S College of Engineering, Pune, India
Dr. Suresh Prasad	Greater Noida Institute of Technology, Greater Noida
Mr. Rajesh Kumar	Greater Noida Institute of Technology, Greater Noida
Dr Tauseef Uddin Siddiqui	M.J.P Rohilkhand University, Bareilly
Mr. Prabhakar Singh	Greater Noida Institute of Technology, Greater Noida
Mr. Anuj Dixit	Greater Noida Institute of Technology, Greater Noida
Dr. Shyam Lal Verma	Noida Institute of Engineering & Technology, Greater Noida
Mr. Sandeep Chauhan	GL Bajaj Institute of Technology, Greater Noida
Mr. Manish Maurya	Accurate Institute of Management & Technology, Greater Noida
Mr. Shivansh Mahrotra	IIT delhi.
Mr. Jitendra Kumar Tripathi	Greater Noida Institute of Technology, Greater Noida
Dr. Srujan Manohar	Pondicherry Engineering College
Mr. Pranav Salhan	Delhi Technological University
Dr. Yatin Agarwal	Greater Noida Institute of Technology, Greater Noida
Mr. Navin Kumar	Greater Noida Institute of Technology, Greater Noida
Mr. Kapil Kumar	Greater Noida Institute of Technology, Greater Noida
Mr. Varun Yadav	Motheson

IOP Conf. Series: Materials Science and Engineering 1168 (2021) 011001 doi:10.1088/1757-899X/1168/1/011001

Dr. Ajeet Kumar	Mechanical engineering department, IIT(BHU) varanasi
Mr. Syed Mohd Faisal	Integral university, lucknow
Dr. Hemat Nautiyal	PEC, chandigarh
Mr. Viveksheel Rajput	PEC, chandigarh
Dr. Mahipal Singh Rawat	Greater Noida Institute of Technology, Greater Noida
Mr. Hitesh Dhiman	BHEL, Delhi
Mr. Trinath Mahala	Greater Noida Institute of Technology, Greater Noida

Peer review declaration

All papers published in this volume of *IOP Conference Series: Materials Science and Engineering* have been peer reviewed through processes administered by the Editors. Reviews were conducted by expert referees to the professional and scientific standards expected of a proceedings journal published by IOP Publishing.

- **Type of peer review: Single-blind**
- **Describe criteria used by Reviewers when accepting/declining papers. Was there the opportunity to resubmit articles after revisions? YES**

Type of peer review:

To ensure quality and professional ethics a peer review mechanism has been adopted. The reviewers (at least two for each paper) received the manuscript with all the details of the authors and a form, prepared for the conference, with certain fields for comments and recommendations. The authors, at the next step, received anonymous reviews with the following possible recommendations: accept as is/minor revisions/major revisions/reject. The papers were revised by the authors accordingly and then forwarded to the same reviewers for final recommendation. The final decision was taken by the editors.

- **Conference submission management system: EASY CHAIR**
- **Number of submissions received: 111**
- **Number of submissions sent for review: 97**
- **Number of submissions accepted: 30**
- **Acceptance Rate (Number of Submissions Accepted / Number of Submissions Received X 100): 27.03%**
- **Average number of reviews per paper: 2**
- **Total number of reviewers involved: 72**
- **Any additional info on review process (ie plagiarism check system): HAS BEEN CHECKED THROUGH TURNITIN**
- **Contact person for queries:**

DR. AVINASH RAVI RAJA,
Assistant Professor, greater noida institute of technology , Greater noida, Uttar Pradesh, INDIA,
E-mail: avinash129@gmail.com, M: +91 9598809859

A Comparative Study Between Different NCAPs and The Future of BNVSAP

Pranav Salhan[1], **Prasoon Jain**[2]

[1,2] Department of Mechanical Engineering
[1,2] Delhi Technological University, India

E-mail: pranav22107@gmail.com

Abstract. India is the third largest road network which is used by one percent of the world's total vehicle population. On this road network six percent of the total of global road traffic accidents occur. India is the sixth-largest car market in the world but is still the only country amongst the global top 10 car markets to not have a testing program. To reduce the number of road traffic accidents, many countries have come up with safety programmes such as Euro NCAP (New Car Assessment Program), US NCAP, Latin NCAP, etc. In 2014, Indian government proposed a similar safety assessment program called as Bharat New Vehicle Safety AssessmentProgram (BNVSAP) but it is since delayed in its implementation due to a number of reasons. The aim of this study is to compare existing NCAPS and the new BNVSAP and propose some improvements in it by studying the experiences gained by the older NCAPS and observing the difference in the situation of road traffic in India and in other countries by analysing real worldaccident data.

1. Introduction

India being a developing country doesn't have a strong infrastructure to enforce traffic rules (such as speed cameras, traffic monitors, etc.). There were about three million new car registrations in the country with the customers not knowing about the safety ratings of their vehicles. The consumers aren't aware nor are educated about the importance of safety ratings for cars. Therefore, the start of a new crash assessment programme as soon as possible is a must. In 2019 almost 154 thousand people died in roadaccidents with over-speeding being one of the main causes. To bring the cost down automakers oftendon't include airbags in their low budget offerings or just one in front of the driver which is the reasonwhy most Indian cars tested abroad failed to pass the crash tests. A successful evaluation programmemust be able to adapt to changes and new developments. Experiences from older programmes such asUS NCAP and Euro NCAP should be studied and used to improve existing plans. Different NCAPs have different circumstances which should be taken into consideration. Said circumstances include real- world accident data, financial status of area, consumer market, etc.

Dummy readings are used to evaluate risk factors that help replicate real-world accidents. An effective evaluation programme must be easily accessible to all consumers. Any customer in the marketlooking for a new vehicle must have access to authentic ratings. Social awareness through publicationmediums (such as magazines and newspapers) and on the internet.

2. Literature Review

A reliable tool for depicting crash tests on automobiles are computer software. They have a great effect on crashworthiness of a car as they decrease the time and cost spent in development of the automobile. Simulations are done on computer software to decrease the number of tests and to optimize the design of the automobile. This kind of simulation is done by Ambati et al. [1] in their research paper. They took a model of Chevrolet C1500 pick-up truck and simulated it for frontal impact crash tests. The computer software they used was LS-DYNA. They validated the results from the actual test performed by National Crash Analysis Centre (NCAC). They found that almost half of the energy of the impact is absorbed by the engine, bumper and rails in 0.04 sec after the initiation of the crash. They also noticed that there is minimum intrusion of components in the cabin and minimum deformation of the cabin.

The physical tests are done to validate the performance of an automobile in terms of its crashworthiness. In crashworthiness engineering, there are useful technique called as state of art technique which use FEA (finite element analysis) and lumped parameter models or full car and component tests. U.N. Gandhi [2] put forward an analytical model which consists of two parts: a transfer function part which consists of ARMA (Autoregressive moving average) of white noise and the second part is a differential equation which consists of stiffness, damping characteristics and mass. The lumped parameters which are present in the model are time varying and they corelate with the structural characteristics involved during crash test. The important of this technique is seen by its ability to predict structural parameters from frontal and side impact test data.

Similar to the gasoline vehicles, the crash testing of electric vehicles is also very interesting. They are also tested by the NCAP. Therefore, in their study Z. Zhang et al. [3] differentiated the safety tests of electric automobiles with that of gasoline automobiles. They researched the crash test procedure, crash standards of electric vehicle. They collected the data from both sets of tests and differentiated the gasoline automobiles with that of electric automobiles.

One of the most fatal injuries involved in car crashes are head injuries. Therefore, it becomes very important to study the biomechanical limits of human organisms. Erika et al. [4] researched the same and prepared an article in which they proposed a material which protect solid parts of the construction from damaging from collisions and simultaneously it should be light in weight. They found that DIN 1.4301 steel was the optimum material as it can resist a collision at 24 km/h and the strain member completes RCAR requirements. Not only the strength and stiffness of the car was improved but it was light in weight and more fuel efficient.

One of the emerging topics in the field of safety in traffic accidents is high voltage safety in electric vehicle. This field includes the study of design of high and low voltage cables in automobiles, integrity of high voltage components of the vehicle after its crash, high voltage cut-off safety, crashworthiness of automobiles. Lian et al. [5] provided a solution to improvise the HV safety performance in a rear impact type collision. They also made possible that the high voltage safety requirements of the vehicle are met according to Chinese regulations and FMVSS (US Federal Motor Vehicle Safety Standard) 305 for a rear impact at a speed of 80 km/h.

Nowadays, many safety systems are equipped with automobiles to decrease the number of crashesand personal injuries, more importantly fatal injuries. One such popular safety system is ESC or ESP (Electronic Stability Control). The major production of this feature started in 1998. The role of this system is to hold on the automobile in control in difficult maneuvering conditions such over-steer or under-steer through braking wheels separately and/or by regulating engine power [6]. Tingvall et al. [7]took the data of car crashes happening in Sweden in between 1998 to 2004 and showed the positive impacts of this system on real life crashes. They did a statistical analysis on the data and found that theeffectiveness of ESC on all kinds of crashes was 16.7 ± 9.3% and the effectiveness of crashes with injured occupants was 23.0 ± 9.2%. This shows how effective it is in reducing car crashes and fatal injuries.

3. Analysis of road accident scenario at national level

According to a report in 2018 by WHO, India comes first in terms of the number of road accidents occurring in a year. Even the most populous country, China, is behind India in this situation. Current situation of road traffic accidents in India is very concerning as it is worsening each year instead of improving which is generally the case in most countries. According to the latest data of the year 2019 provided by NCRB (National Crime Records Bureau) [8], a total of 4,37,396 road accidents were reported in 2019 in which 4,39,262 were injured and 1,54,732 people died. As a result, the rate of deaths per thousand vehicles has increased from 0.6 in 2018 to 0.61 in 2019 (See Table 1). A large percentageof about 13.7% of the total road accidents deaths were due to car accidents.

The state wise distribution reveals that the highest share of the total fatalities was of Uttar Pradeshwith 15.8% (3,355 out of 21,196). According to the distribution by month, the highest road accidentsoccurred in the month of January with 40,548 and least in the month of September with 31,441. By distributing the total road accidents time-wise it is observed that between 18:00 hrs to 21:00 hrs, 19.0%of the total number of road accidents occurred (83,097 out of 4,37,396) which is highest followed by 17.4% during 15:00 hrs to 18:00 hrs and 15.3% during 12:00 hrs to 15:00 hrs. When distributed cause-wise, it shows that one of the major causes of road accidents deaths is due to defect in mechanical condition of motor vehicle and it accounts for 1.3% (2045 out of 1,54,732) [9].

Table 1. Growth in number of vehicles and road accidents in India (2015-2019) [8].

SI. No.	Year	Road accidents (In thousands)	Persons Injured (in thousand)	Persons Killed (in Nos.)	No. of Vehicles (In thousand) #	Rate of Deaths per thousand Vehicles (COl.7/Col.9)
1	2015	464.7	482.4	1,48,707	2,10,023	0.71
2	2016	473.0	485.5	1,51,801	2,30,031	0.66
3	2017	445.7	456.2	1,50,093	2,53,311	0.59
4	2018	445.5	446.5	1,52,780	2,53,311	0.60
5	2019	437.4	439.2	1,54,732	2,53,311	0.61

Table 2. Annual summary of road accidents in Great Britain [10].

Year	Fatality	Serious injury	Slight injury	Total injury
1980	5,953			323,000
1990	5,217	60,000	275,000	336,000
1997	3,599			324,000
1998	3,421	41,000	281,000	322,000
1999	3,423	39,000	278,000	317,000
2000	3,409	38,000	279,000	317,000
2002	3,431	36,000	263,000	299,000
2004	3,221	31,000	245,000	278,000
2006	3,172	29,000	226,559	255,000
2008	2,538	26,000	202,333	228,000
2010	1,857	20,803	185,995	206,798
2012	1,754	23,039	170,930	193,969
2014	1,775	22,807	169,895	194,477
2016	1,792	24,101	155,491	179,592
2018	1,782	25,484	133,112	160,378
2019	1,870	27,820	129,820	157,630

Table 2. shows the gradual decrease in fatalities caused by road accidents after the formation of Euro NCAP in 1996. There are various other factors affecting the statistics such as stricter traffic rules, technological advancements, etc. but gives great hope.

4. Comparison between different NCAPs

4.1. US NCAP

National Highway Traffic Safety administration (NHTSA) first carried out safety evaluating crash tests in 1979. Initially only frontal crash was evaluated but with time new tests were introduced including the side crash test in 1997. The tested cars are awarded with ratings between one star to five stars to illustrate the results conveniently to the consumers. US NCAP assessments comprises of four broad areas [11]:

- Adult Occupant Protection (for the front occupants);
- Child Occupant Protection;
- Pedestrian Protection
- Safety Assist, which evaluates driver-assistance and crash-avoidance technologies.

Our focus will be on the adult occupant protection which include the following tests:

4.1.1. Frontal Crash Test

The test vehicle is driven into a rigid barrier with full overlap with a test speed of 56 km/h. The car has an average-sized male dummy in the driver seat and a female dummy of small size in passenger seat with both dummies being secured with seat belts. Car structures need to be stiff to reduce lower leg and head injuries. This test also assesses the restraint system of the car, such as airbags and seatbelts.

4.1.2. Side Barrier Crash Test

To represent an intersection type accident a 3015 lb. (1367 kg) moving barrier at a speed of 38.5 mph(62 km/h) is crashed into the standing test vehicle. The aim of this test is to evaluate the injuries to thehead, abdomen, chest and pelvis.

4.1.3. Side Pole

This test involves launching the vehicle at 75-degree angle at 32km/h towards a rigid and narrow pole in the vicinity of the driver. Head protection is of utmost importance in this test. This test is formalized to test the car's capability to guard the driver's head, chest and pelvis on impact.

4.2. EURO NCAP

Europe's car safety assessment programme also known as Euro NCAP was formed in 1996 for safer cars. It is a voluntary non-profit to provide consumers with apt safety information of new cars available in the market. This programme includes various tests like frontal and side impact crash test to evaluate the car and give a star rating out of five. It is not compulsory for manufacturers to get their vehicles tested. The vehicle models chosen for testing are either selected by Euro NCAP or sponsored by manufacturer. Euro NCAP also assess the potential risks to vulnerable pedestrians on collision with the vehicle at city speeds. Head, lower leg and upper leg impact on the bumper and bonnet are tests using dummy. They give extra points for effective pedestrian collision prevention [12]. Adult occupant protection tests are:

4.2.1. Mobile progressive deformable barrier

The test vehicle is launched at 50 km/h towards a deformable moving barrier mounted on an oncoming trolley of weight 1400kg, also moving at 50 km/h with a 50% overlap. The aim of the test is to evaluate the crumple zones and the compatibility of the test vehicle.

4.2.2. *Side mobile impact barrier*
Same setup as US but speed is again reduced from 62 km/h to 50km/h. Since there is little energy absorbing space on the side of the vehicle, this test has encouraged stronger B-pillars and side airbags.

4.3. *LATIN NCAP*
Latin New Car Assessment Program was founded in 2010 to assess the safety levels of new cars for the Latin America and Caribbean. Latin NCAP awards vehicles with a safety rating between 0 and 5 stars. This programme provides safety ratings based on the assessment of adult occupant protection, child occupant safety, pedestrian and liable road user's protection and safety assistance systems offered by the various vehicle models. The adult occupant protection includes the following 3 major crash tests:

4.3.1. *Frontal Impact Test*
The frontal impact test is performed at 64 km/h where the car is crashed into a deformable barrier with 40% of its width front on the driver side (offset). Results obtained from the crash test dummies are used to assess safety provided to adult occupants in the front seat. This test encourages manufacturers for prevent excessive Intrusion in the cabin and to install advanced restraint system.

4.3.2. *Side Impact Test*
Vehicles are tested with the same setup as the Euro NCAP's Side mobile impact barrier test.

4.3.3. *Side Pole Impact Test*
This test involves launching the vehicle laterally at 29 km/h towards a rigid and narrow pole. Head protection is of utmost importance in this test.

Additionally, new technologies include the Autonomous Emergency Braking (AEB) low-speed system which uses forward-looking radar, cameras or optical sensors or a combination of these to help quickly and accurately detect vehicles, pedestrians or other potential obstacles and the apply the brakes in low-speed situations.

5. The proposed NCAP for Indian market: BNVSAP
Bharat New Vehicle Safety Assessment Programme (BNVSAP) is a proposed NCAP for the Indian market which was initially expected to begin mid-2014 but is since postponed. The tested cars will receive star ratings based on their performance in various tests with credits given for inclusion of features such as ABS, ESC, EBD, etc. [13]. This will also help boost the number of cars exported to foreign markets. The proposed crash tests are:

5.1. *Frontal Impact Test*
The frontal impact test is performed at 56km/h where the car is crashed into a deformable barrier with 40% of its width front on the driver side (offset). Results obtained from the crash test dummies are used to assess protection given to adult occupants in the front seat. This test encourages manufacturers to prevent excessive Intrusion in the cabin and to install advanced restraint systems.

5.2. *Side Impact Test*
A barrier which is deformable is attached to a trolley which is propelled with a speed 50 km/h towards the static test vehicle at right angles. This has driven the development of stronger B-pillars with energy-absorbing structures in door panels and the addition of curtain airbags in cars.

Gradually cars will be required to meet tougher regulations and as time progresses more features willbe made compulsory like airbags and anti-lock braking system. Some new safety features include automatic emergency braking at low speed, hill descent assist, side crash avoidance, etc. but can currently only be found in expensive options. Global NCAP is a charity project, which serves as a harmonization tool between various NCAPs. Today's car market is globalized, where companies sell the same vehicles and vehicle platforms in different parts of the world [14]. Global NCAP aims to bring technical knowledge and support to emerging car markets. Their aim also includes promotion of information exchange between various NCAPs and social awareness [15].

6. Conclusion

A successful evaluation programme must be able to adapt to changes and new developments. Experiences from older programmes such as US NCAP and Euro NCAP should be studied and used to improve existing plans. Due to the late arrival of BNVSAP, the addition of new tests (such as side pole test) and technologies as quickly as possible is a must to maintain a global standard. Different NCAPs have different circumstances which should be taken into consideration. Said circumstances include real-world accident data, financial status of area, consumer market, etc. These tests encourage manufacturers to build safer vehicles for passengers and lead to development of new technologies to avoid accidents. Companies boast their car's ratings if they have successful crash tests through advertisements which encourages other competing companies to introduce safer alternatives. Nowadays manufacturers use a combination of materials in their chassis such as steel, aluminium and carbon fiber to find the required combination of strength, rigidity and lightness. Low Young's modulus and energy absorbing materials should be selected for B-pillars, etc. Steel and aluminium are the most common materials used in commercial vehicles today.

7. References

[1] U.N. Gandhi, S.J. Hu, Data-based approach in modeling automobile crash, International Journal of Impact Engineering, Volume 16, Issue 1, 1995, Pages 95-118.

[2] Z. Zhang, S. He and K. Wang, "Research on the Crash Test's Safety in NCAP of Electrical Vehicle," 2017 9th International Conference on Measuring Technology and Mechatronics Automation (ICMTMA), Changsha, 2017, pp. 310-315.

[3] Erika Fechová, Jozef Kmec, Alena Vagaská, Dražan Kozak, Material Properties and Safety of Cars at Crash Tests, Procedia Engineering, Volume 149, 2016, Pages 263-268.

[4] Lian, Y., Zeng, D., Ye, S., Zhao, B., & Wei, H. (2018). High-Voltage Safety Improvement Design for Electric Vehicle in Rear Impact. Automotive Innovation. 10.1007/s42154-018-0030-z.

[5] Sferco R, Page Y, Le Coz J-Y, Fay P. (2001) Potential effectiveness of electronic stability programme (ESP)—What European field studies tell us. Paper no. 2001-S2-O-327. Proceedings of the 17th International Technical Conference on the Enhanced Safety of Vehicles (CD-ROM). National Highway Traffic Safety Administration, Washington, DC.

[6] Lie, A., Tingvall, C., Krafft, M., & Kullgren, A. (2006). The Effectiveness of Electronic Stability Control (ESC) in Reducing Real Life Crashes and Injuries. Traffic Injury Prevention, 7(1), 38–43.

[7] Traffic accidents, ncrb.gov.in, Retrieved from https://ncrb.gov.in/sites/default/files/Chapter-1A-Traffic-Accidents_2019.pdf

[8] Singh, S. K. (2017). Road Traffic Accidents in India: Issues and Challenges. Transportation Research Procedia, 25, 4708–4719.

[9] Reported Road Casualties Great Britain, wikipedia.org, Retrieved from https://en.wikipedia.org/wiki/Reported_Road_Casualties_Great_Britain

[10] Seeck, A., Coxon, C., & Wani, K. (2003). Global NCAP harmonisation - initial position, requirements and prospects. International Journal of Crashworthiness, 8(1), 5–15.

[11] The Ratings Explained, euroncap.com, Retrieved from https://www.euroncap.com/en/vehicle-safety/the-ratings-explained/

[12] Department and Laboratories, araiindia.com, Retrieved from https://www.araiindia.com/services/department-and-laboratories/psl.

[13] Seeck, A., Coxon, C. & Wani, K. Global NCAP harmonisation - initial position, requirements and prospects. International Journal of Crashworthiness 8, 5–15 (2003).

[14] Global NCAP, Retrieved from http://www.globalncap.org/about/

IOP Conf. Series: Materials Science and Engineering 1168 (2021) 012002 doi:10.1088/1757-899X/1168/1/012002

Multi-Response Optimization of Weld-process variables for Micro-Friction Stir Welding of Al6061(T6) and Cu101 sheets using TOPSIS method

M V N Srujan Manohar[1*] and K Mahadevan[2]

[1] Research Scholar, Department of Mechanical Engineering, Pondicherry Engineering College, Puducherry 605014, INDIA
[2] Professor, Department of Mechanical Engineering, Pondicherry Engineering College, Puducherry 605014, INDIA

*Corresponding Author: srujansujith8990@pec.edu

Abstract. This work is an attempt to identify the optimum weld-process variables for micro-friction stir welding of Al6061(T6) and Cu101 sheets using multi-response optimization. The experiments were conducted according to L9 orthogonal array with selected input variables namely tool-rotational speed and tool-travel speed. The response variables considered in this study are ultimate tensile strength, micro-hardness and surface-roughness which determine the strength of the welded-joint. The multi-response optimization approach namely Technique for Order Preference by Similarity to Ideal Solution(TOPSIS) method is used to determine the optimum process variables that provides the best value among all response variables. Finally, the optimal solution shows that multi-responses of micro-friction stir welds can be improved through TOPSIS method.

1. Introduction

Joining of aluminium and copper materials has gained significant scope in electrical and electronic industries due to their high corrosion-resistance and best conductivity properties. Among various joining techniques, Friction stir welding (FSW) has wide scope in joining dissimilar-materials without extending their melting zones [1]. The studies on FSW technique in present days is slowly focusing on thin sectioned materials joining with thickness of 1000 μm(1 mm) or low. This micro-level joining paved the way for the development of micro-friction stir welding (μFSW). Most of the research studies on FSW process were implemented on joining thick dimensioned plates in similar/dissimilar form and the work intended for joining thin dimensioned dissimilar materials in micro-level approach was identified to be less [2],[3].

Considering cost and time aspects, industries prefer easy and reliable operations to grasp optimization problems. In fact, Taguchi-approach is the basic designing concept generally used to optimize operational variables. But this approach focuses on single response criterion and does not give sufficient data about the remaining response variables involved. Generally performance of the product while manufacturing is examined by several responses. Hence Taguchi-approach turns out to be time consumption approach [4].

By overcoming the limitations observed by Taguchi-approach in solving multi-response problems, several techniques have been implemented to solve the multi-criteria decision making problems so that quality responses are simultaneously optimized and the solutions are evaluated for best levels. Several research studies were carried out to develop a mathematical relation between process variables and mechanical responses of welded-joint with different experimental design procedures. In general, multi-criteria decision making problems are usually subdivided into continuous and discrete types. In continuous type, response variables are determined in continuous domain having more number of alternative choices whereas in discrete type, response variables are determined randomly with limited number of pre-specified alternatives [5].

More research studies are focused on GRA-(Grey relational analysis) technique for solving multi-response optimization problems. This technique depends upon grey-system theory fitted for working out problems having intricate relationship between factors and levels. In similar manner, there is another method called Technique for order of preference by similarity to ideal solution (TOPSIS), which is simple and effective for solving multi-criteria decision making problems. Both GRA and TOPSIS methods are examined for several applications for selecting optimum parameters of process variables on multi-responses, hence those studies concludes that the solution achieved by TOPSIS method somewhat identified as better one when compared with GRA method [6],[7]. Very few works were carried out for optimizing multi-response problems in micro-FSW applications using TOPSIS method. With this intention, the present study has been reported to carry out TOPSIS optimization for weld-process variables effect on output responses such as Ultimate Tensile strength (UTS), Micro-Hardness (MH) and Surface Roughness (SR).

2. Design of Experiments

For conducting design of experiments, Taguchi-approach is selected to probe the effect of weld-process variables on output responses with limited experimental runs. In present study, two main weld-process variables such as tool-rotational speed and tool-travel speed were chosen based on several literature studies. Weld-process variables along with their levels are shown in Table 1. In Taguchi-method, relevant orthogonal array is selected based on degrees of freedom(DF). The total DF of process-parameters should be less than the DF of relevant orthogonal array. As per Table 1, total DF is calculated as four for two factors and three levels. Hence L9 orthogonal array was selected and this design is suitable for studying the effect of input process variables on output mechanical responses such as UTS, MH and SR.

Table 1. Weld-process variables and levels.

Weld-process variable	Unit	Level			DF
		1	2	3	
Tool-rotational speed	rpm	1100	1400	1800	2
Tool-travel speed	mm/min	40	60	80	2
Total DF					4

3. Materials and Methodology

Thin-gauge Al6061(T6) and Cu101 sheets of 0.8 mm thickness are selected for welding using μFSW technique. The complete welding process is operated on a semi-automated vertical milling machine using pinless tungsten-carbide tool. Weld-process variables are compiled for different tool-rotational speeds and tool-travel speeds by preserving three constant parameters such as tool plunge depth of 0.3 mm, tool tilt angle of 0.50 and tool dwell time of 5 seconds for all the levels [8]. After processing of all the weld-joints, quality of the weld-joint is determined by examining mechanical behaviours such as tensile test, micro-hardness test and surface roughness test.

Before proceeding to mechanical behaviour analysis, the weld-samples are cut as per ASTM E8M standard dimensions using Wire EDM technique as shown in Figure 1. Total three specimens were cut from each weld-sample. Finally nine samples were tested for tensile-behaviour for examining higher values at different levels using high precision computer control universal testing machine. Similarly micro-hardness test was carried out on all nine weld-samples using Vickers digital hardness machine for examining the intermetallic layers of Al/Cu mixture. In similar manner, surface roughness test was carried out on all nine weld-samples using Digital surface roughness tester for examining better surface finish at different levels.

IOP Conf. Series: Materials Science and Engineering 1168 (2021) 012002 doi:10.1088/1757-899X/1168/1/012002

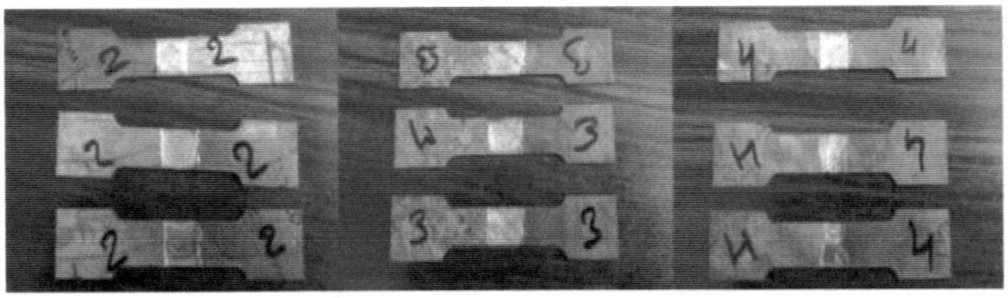

Figure 1. Machines samples as per ASTM E8M standards

The measured responses for all the weld-samples are tabulated in Table 2 and their effects are plotted on Figure 2.

Table 2. L_9 orthogonal array for measured responses

Weld-Sample No. (WS) ⇩	Tool rotational speed – TRS	Tool traverse speed – TTS	Ultimate tensile strength –UTS	Micro Hardness – MH	Surface Roughness – SR
	Weld-process variables		Response variables		
Units⇨	rpm	mm/min	MPa	HV	μm
WS-1	1100	40	83	80	3.364
WS-2	1100	60	72	71	3.452
WS-3	1100	80	80	73	3.380
WS-4	1400	40	138	55	3.108
WS-5	1400	60	162	64	3.024
WS-6	1400	80	157	59	3.112
WS-7	1800	40	175	54	1.841
WS-8	1800	60	145	58	1.432
WS-9	1800	80	191	60	1.172

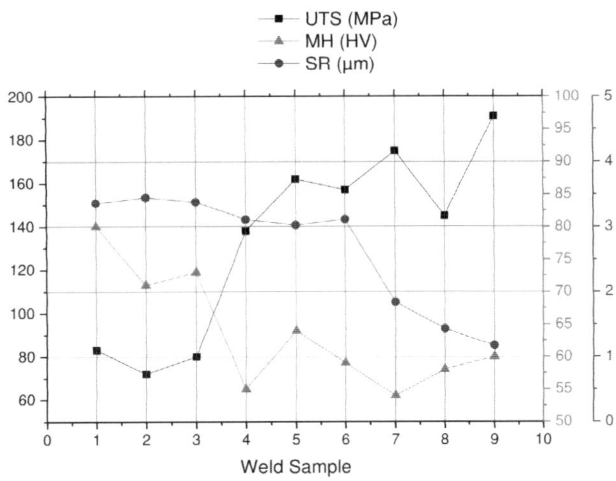

Figure 2. Graphical plots for measured response variables

4. Results and Discussions based on TOPSIS-method

Technique for order of preference by similarity to ideal solution (TOPSIS) method is a multi-response optimization tool for solving complex decision making problems in industrial applications. Initially this method was developed by Ching-Lai-Hwang & Yoon in 1981 [9] with further evolutions by K. Yoon in 1987 [10] and Hwang, Lai & Liu in 1993 [11]. The basic approach of this method is for choosing alternatives that have shortest distance from positive-ideal solution(PIS) and longest distance from negative-ideal solution(NIS). Finally ranks are created for alternatives based on the solutions obtained from multi-response problems. The classical TOPSIS algorithm [12] based on the experiment data is calculated in following steps,

Step 1: The initial step is to construct the decision-matrix which consists of 3 attributes (response variables) and 9 alternatives (experimental-runs) by removing all units. Now the decision matrix is normalized as per equation (1) and the computed values are shown in Table 3.

$$v_{ij} = (x_{ij}) / \sqrt{\sum_{i=1}^{9} x_{ij}^2} \tag{1}$$

for i = 1,2..., 9 and j = 1,2 and 3

Where v_{ij} = normalized value and x_{ij} = measure of jth attribute to ith alternative

Table 3. Normalization

Weld-Sample No.	Normalized Response values		
	UTS	MH	SR
WS-1	0.19754	0.41449	0.40219
WS-2	0.17136	0.36786	0.41271
WS-3	0.19040	0.37822	0.40410
WS-4	0.32844	0.28496	0.37158
WS-5	0.38556	0.33159	0.36154
WS-6	0.37366	0.30568	0.37206
WS-7	0.41650	0.27978	0.22010
WS-8	0.34510	0.30050	0.17120
WS-9	0.45458	0.31086	0.14012

Step 2: In second step, relative-weights should be assigned to each of the attribute. In present study, all response variables are given equal significance and hence the relative-weight of 0.333 is considered. The weighted normalization is deliberated as per equation (2) and the computed values are shown in Table 4.

$$V_{ij}= w_j \text{ x } v_{ij} \qquad (2$$

for i = 1,2..., 9 and j = 1,2 and 3.

Where V_{ij} = weight-normalized value and w_j = weight of the jth attribute

Table 4. Weighted-Normalization

Weld-Sample No.	Weight-Normalized Response values		
	UTS	MH	SR
WS-1	0.06578	0.13802	0.13392
WS-2	0.05706	0.12249	0.13743
WS-3	0.06340	0.12594	0.13456
WS-4	0.10937	0.09489	0.12373
WS-5	0.12839	0.11042	0.12039
WS-6	0.12442	0.10179	0.12389
WS-7	0.13869	0.09316	0.07329
WS-8	0.11491	0.10006	0.05701
WS-9	0.15137	0.10351	0.04666

Step 3: In third step, the best (positive-ideal) and worst (negative-ideal) solutions are deliberated as per equation (3) and equation (4). The computed values are shown in Table 5.

$$V^+ = \{(\max v_{ij} \mid j\epsilon O),(\min vij \mid j\epsilon P \mid i=1,2,\ldots..9)\} \quad (3)$$

$$V^- = \{(\min v_{ij} \mid j\epsilon O),(\max vij \mid j\epsilon P \mid i=1,2,\ldots..9)\} \quad (4)$$

Where O = beneficial attribute, and P = non-beneficial attribute (if applicable)

Table 5. Best (PIS) and Worst (NIS) values

Ideal-solution	Response values		
	UTS	MH	SR
V^+	0.15137	0.13802	0.04666
V^-	0.05706	0.09316	0.13743

Step 4: In fourth step, the separation measure for each alternative from PIS and NIS is deliberated as per equation (5) and equation (6). The computed values are shown in Table 6.

$$S_i^+ = \sqrt{\sum_{i=1}^{9}(v_{ij} - V_j^+)^2} \quad (5)$$

$$S_i^- = \sqrt{\sum_{i=1}^{9}(v_{ij} - V_j^-)^2} \quad (6)$$

for i = 1,2..., 9 and j = 1,2 and 3.

Step 5: In fifth step, closeness coefficient (CC) value for each alternative from the separation measure is deliberated as per equation (7) and the computed values are shown in Table 6.

$$CC_i = \frac{S_i^-}{S_i^- + S_i^+} \quad (7)$$

for i =1,2 ..., 9; $0 \leq CC_i \leq 1$

Step 6: In sixth step, ranking is given in preference order as shown in Table 6 that depends upon CC-values arranged in descending order,

WS-9>WS-7>WS-8>WS-5>WS-6>WS-4>WS-1>WS-3>WS-2

Based on the obtained ranks, WS-9 is considered as best-optimal solution and WS-2 is considered as worst-optimal solution among all alternatives(experimental-runs).

Table 6. Separation Measure and Closeness Coefficient Values

Weld-Sample No.	Separation Measure		Closeness Coefficient – CC	Rank
	S^+	S^-		
WS-1	0.12223	0.04583	0.27269	7
WS-2	0.13181	0.02933	0.18201	9
WS-3	0.12494	0.03351	0.21148	8
WS-4	0.09780	0.05409	0.35613	6
WS-5	0.08201	0.07533	0.47877	4
WS-6	0.08946	0.06925	0.43631	5
WS-7	0.05368	0.10381	0.65912	2
WS-8	0.05363	0.09931	0.64930	3
WS-9	0.03450	0.13130	0.79189	1

Finally, the average-CC value of each weld-process variable for all levels is deliberated and the computed values are shown in Table 7. By means of tabular data, a graph was plotted between average-CC value and levels of weld-process variables as shown in Figure 3.

Table 7. Average Closeness Coefficient Value (average-CC)

Level	Weld-process parameter	
	TRS	TTS
1	0.22206	0.42932
2	0.42374	0.43669
3	**0.70011**	**0.47989**

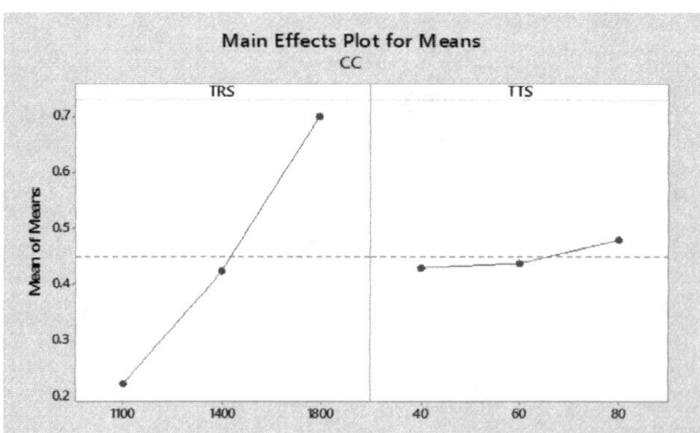

Figure 3. Effect of weld-process variables on average-CC

5. Conclusion

Thin gauge Al6061(T6) and Cu101 sheets of 0.8 mm thickness were successfully welded using micro-friction stir welding approach. Weld-process variables such as tool-rotational speed and tool-travel speed were significantly influenced the output responses such as UTS, MH and SR.

Based on the Taguchi L9 orthogonal array design concept and Technique for order of preference by similarity to ideal solution (TOPSIS) approach, the optimum weld combination is observed as TRS3TTS3 (tool-rotational speed of 1800 rpm and tool-travel speed of 80 mm/min) that yields high closeness coefficient value of 0.79189 for producing high quality welded-joint.

6. Future Scope

Micro-Friction stir welding of Aluminium alloy and copper alloy sheets has been demonstrated successfully in the present work, but further work is required to improve the tool characteristics, and also to demonstrate the mechanical properties of the welds at different zones.

7. References

[1] Jiahu Ouyang, Eswar and Radovan Kovacevic 2006 *J. Mater. Process. Technol.* **172** 110–122
[2] Scialpi A, Giorgi, Filippis, Nobile and Panella 2007 *Mater. & Des.* **29** 928–936
[3] Farazila Yusof, Firdaus, Jamaludin and Hamdia 2013 *Proc. of the 1st int. joint symposium on joining and welding*, 219–224
[4] Parameshwaranpillai T and Lakshminarayanan 2011 *Adv. Compos. Mater.* 2011 **20** 37–41
[5] Vyas Gayatri S and Misal Chetan S 2013 *Int. J. Adv. Comput. Theory Eng.* **2** 2319 – 2526
[6] Sudhagar S, Sakthivel, Prince J Mathew and Ajith Arul Daniel 2017 *Measurement* **108** 1-8
[7] Mohammad Asjad and Faisal Talib 2018 *Int. J. Adv. Oper. Manage.* **10** 109 – 129
[8] Manohar M V N S and Mahadevan K 2020 *Mater. Today Proc.* (Elsevier BV: Amsterdam)
[9] Ching-Lai Hwang and Yoon K 1981 *Lecture Notes in Economics and Mathematical Systems* (Springer) **186** 58-191
[10] Yoon K 1987 *J. Oper. Res. Soc.* **38** 277-286
[11] Ching-Lai Hwang, Lai and Liu 1993 *Comput. Oper. Res.* **20** 889-899
[12] Ching-Lai Hwang and Yoon K 1995 (SAGE publications, California)

Acknowledgement

We would like to acknowledge Microlab, Chennai and Kalasalingam Academy of Research and Education, Krishnankoil.

Author Contributions

M V N Srujan Manohar: Conceptualization, Methodology, Resources, Validation and Drafting
Dr K Mahadevan: Supervision, Investigation, Writing – Review & Editing.

Comparative Study of Electric Tractor and Diesel Tractor.

R Dhond[1], U Srivastav[1], Dr. B T Patil[1] and H Vaishnav[1]

[1]Department of Production Engineering, Fr. Conceicao Rodrigues College of Engineering, Mumbai-400104

Abstract. The increase in population day-by-day causes a direct and indirect effect on country's pollution and resources. These have adverse effects on the environment and contributes directly to climate change. A transition from an internal combustion engine to an electric vehicle is vital, since with a lot of motor and battery technology available globally. This paper focuses on different types of motors and batteries that can be sourced in India, for the conversion of a diesel tractor to an electric tractor and the running costs related to the same is compared.

1. Introduction

To desperately fight against all the factors that hamper the environment, various methods were adopted by many countries, pacts and treaties were signed to abate the effect of this global cause: pollution, which is a major factor leading to climate change cannot be ignored at all. Some important agreements include The Kyoto Agreement of 1997 and The Paris Agreement of 2015. Focusing on mobility, it is one important aspect with which a country's GDP might increase considerably, but it also responsible for approximately 40% of pollution caused in India. The major pollutants released by burning fuels are Nitrogen oxides (NOx), Ozone (O3), Carbon Monoxide (CO) [1]. Thus, the transition from moving from conventional energy sources to clean energy sources directly effects on the country's growth and development, and the environment. We can observe the transition in 2-wheelers and small passenger vehicles. These modes of transports are very frugal and have a running cost of approximately Rs.2/km which is drastically less when compared to IC engine vehicles. Climate change is the most concerning factor and is one of the reasons to move towards electric mobility. India has also promised to reduce harmful greenhouse gases by approximately 35% to that in 2005 by 2030.

The reason why the transition towards EV is expanding day by day is due to:

- Advancements and Rapid growth in renewable energy technology have also facilitated in the growth of EV, making it parsimonious and highly efficient wind turbines and solar cells are less expensive, cleaner and have less load on the grid when compared to the technology used a few years back, these systems are getting smarter and more efficient day-by-day. This is also possible because of data analysis and digital revolution by using GPS to reduce the cost and achieve high efficiency. Due to the abundance of skilled manpower and technology in manufacturing and the IT sector is also responsible for this growth.
- Betterment in battery technology has improved battery efficiency, high energy density, quick recharge and improved better battery health. The improvement in motor-technology makes the complete system economical and improves in performance.
- Energy security is higher in electricity when compared to diesel and petrol, the system requires extensive labor and supply chain cost and also prone to frequent fluctuation in cost. The cost of this is also affected by weather, geopolitical factors and other miscellaneous expenses.

This concept is in development for commercial vehicles which will reduce carbon emissions hence resulting in pollution reduction. It is observed that commercial vehicles (light) are majorly bound in the city limits and have higher run time when compared to private vehicles, electrifying commercial vehicles will help in reducing pollution drastically. The tractor is the most used and essential part of the agriculture sector, usage of electric vehicles in the farming sector is an untouched domain, expanding usage and innovations in this sector will help in reducing the overall development cost for farmers and in an agricultural country like India where the agriculture sector contributes 17.7% of the GDP and reduction in working capital will be profitable for the farmers and consumers as the probability of fluctuation cost of the commodity is reduces [2]. Based on available tractor models, motors and batteries, the suitable battery pack capacity is calculated and by combining different types of batteries and motors. The basic cost for the battery pack along with the running cost of the diesel and electric tractor is compared to prove the usage of electric tractor is more frugal. This study aims to compare an internal combustion engine tractor with a lithium-ion based electric tractor.

2. Literature Review

According to Anil Khurana, V. V. Ravi Kumar and Manish Sidhpuria, to control the increasing amount of CO2 emissions the urgency and need to shift to EV is inevitable. The transport sector is the highest emitter of this greenhouse gas. The author has approached using two parts in a questionnaire, first part focuses on the demographics and the second part on model variables. The author has used his own sample and variable to do the modelling. For the survey the author has considered attributes like environmental concern, economic benefits, social influence, self-image, attitude and behavioural intention. [3].

The research done by I. H. Ryu, D. C. Kim and K. U. Kim, formulates all the necessary data required for power transmission and efficiency of a four-wheel-drive diesel tractor. The research also points out necessary details regarding international standards for power transmission creating a foundation to support the research. The research relies on intensive parameters, like axle torque, engine speed, velocity,etc., for calculation of efficiency. The values obtained had a range of 56% to 86%, and an estimate mean of 72.5%. These values were also refered in our case to calculate the battery pack size considering 80% efficiency.[4]

An exploratory study by Mr.Bajrang Lal & Dr.Ajmer Singh stated that in a diesel tractor, CO2 emissions are causing maximum environmental effects. Over 68.86% of the rural population was influenced by agriculture giving 54% of tractor sales in India which is highest in the world. This lead to an increase in the demand for high fuel efficiency with less maintenance [5].

Study on the Development of the Electric Tractor by Yuko UEKA, Jun YAMASHITA, Kazunobu Sato, Yoshinori DOI the study here resulted in remodeling the tractor that using and adding the weight have very less effect on the performance of the tractor. Thus, electric equipment for travelling and tillage reduced 70% of energy consumption. Further, CO2 emissions were reduced to 70% from internal combustion to Electric vehicles with a 10KW tractor. The main advantage was that electrification eliminated the direct emission of exhaust gases [6].

An overview of lithium-ion batteries for electric vehicles by Xiaopeng Chen, Weixiang Shen, Thanh Tu Vo, Zhenwei Cao, Ajay Kapoor reviewed the most commonly used lithium-ion batteries by specifying how well it will perform on an electric vehicle. This provided a brief idea about the characteristics of LiFePO.4 (LPO), LiCoO2 (LCO), LiMn2O4 (LMO) and LiNiMnCoO2 (NMC) depending upon cost, safety, operating temperature, thermal stability, etc. Further analysis resulted that LCO is the best option for an EV but manufacturers generally use NMC, LFP and LMO as they are cheap and safe for the environment [7].

Mr. A. Rakesh Kumar and Dr. Sanjeevikumar Padmanaban stated various concepts and technology related to electric vehicles and sub-systems which support the electric vehicle ecosystem. This includes charging infrastructure, battery chemical research and usability, battery management system (BMS) and converters, and electric motors. These have helped us selecting our motor and battery technology. Real-life comparison between diesel and electric car energy consumptions [8].

Sreeram k, Preetha PK and Prabaharan Poornachandran have given deeper insights on the real condition of electric vehicles in India as well as on a global level, including power production, plans, dependency on fossil fuels, and minimizing the dependency on fossil fuels. The paper also highlights different types of motors available and their merits and demerits. Motors like Brushless DC Motor (BLDC), Permanent Magnet Synchronous Motor (PMSM), Induction Motor (IM), Switched Reluctance Motor (SRM) and Reluctance Motor (SYNRM) are included. Also, real-world cases of electric vehicles are used and compared which includes a comparison between Mahindra e-verito and Mahindra e2o. A brief idea of hybrid systems that can be used in the transition phase from IC engine vehicles to electric vehicles as no complete dependency will be there on either of the two technologies. Hybrid technology uses a motor to assist which is powered by either a Lithium-ion cell or a supercapacitor depending on the application. Considering the shortcomings and future capabilities improvement in battery technology, reduction in charging time and better infrastructure of charging can be expected [9].

In a recent case study by Diego Troncon and Luigi Alberti, focuses on non-road vehicles and machinery and electrification of it. The system here in focus was a tractor of 100HP. The idea was the conversion from the standard internal combustion engine to a hybrid powertrain. With experiments carried out in real operating conditions on a traditional tractor, the performance requirements for the hybrid powertrain is analyzed with an electric motor in focus. The author also highlights the point where complete electric tractors would be suitable for light tasks. The study also supports the use of Permanent Magnet motors and analyzing the system considering the thermal-equivalent torques. Further design and analysis were carried out using Finite Element Analysis and the system was validated through thermal analysis of the electric motor. The main purpose for conversion to the hybrid system was to downsize the internal combustion engine and the choice for permanent magnet motor was selected to make the tractor cost-effective in terms of production and ownership [10].

A study on electric vehicles in India by Mohamed M, G Tamil Arasan and G Sivakumar highlights Battery Electric Vehicles (BEV) and Hybrid Electric Vehicle (HEV), the two systems are briefly explained along with the advantages of using EVs. The timeline of all EVs is also included right from 1832-2040. Most importantly the scope and opportunities for EVs in India are explained which includes government initiatives, battery technology, commercial/industrial sector implementation.[11]

IOP Conf. Series: Materials Science and Engineering 1168 (2021) 012003 doi:10.1088/1757-899X/1168/1/012003

3. Methodology

3.1. Diesel Tractor

The process of designing began with selecting and comparing the IC engine tractors available on sale. Some of the top-selling utility tractors in India are from prominent manufacturers which includes Mahindra&Mahindra, Escort, Sonalika, John Deer, etc.

Utility tractors are compact low-cost tractors ranging from 35HP to 70HP. Despite the low cost, these tractors are used for small to medium-sized farms and farmers increasingly adopt them. Utility tractors are capable of carrying out almost all modern farming machinery such as plough, cultivator, thresher, water pump, etc. The diesel tractor which we used for comparison: **Mahindra 575DI (year 2020).**

Specifications:

Table 1.

Name	Values	Symbol
Vehicle Weight	1860kg	
Lifting capacity	1600kg	
Total Weight (*W*)	3460kg	*W*
Height (*h*)	2.1m	*h*
Length (*L*)	3.51m	*L*
Width (*w*)	1.95m	*w*
Frontal Area (A_f) = ($h * L$)	7.37mm^2	*Af*
Coefficient of Drag (C_d)........[28]	0.9	C_d
Air Density	1.2 kg/m^3	
Front Tyre	0.08m	
Rear Tyre	0.17m	
Rolling Resistance (field)	0.1	R_r
Rolling Resistance (Tar road)	0.02	*Rrt*
Top Speed	40kmph	

*Linear wheel Travel= 2*π*(Rear tyre)* = 1.08m

To calculate the power consumed by the tractor depends on parameters of resistance, drag and gradient. The rolling resistance is the resistance offered by the surface of the road to the vehicle. The drag is the resistive force offered by the air to the vehicle and gradient force is the resistance offered by gravity due to the inclination of the surface of the road.

IOP Conf. Series: Materials Science and Engineering 1168 (2021) 012003 doi:10.1088/1757-899X/1168/1/012003

To calculate the power required [Table 2]:

$P_r = F_r + F_{Cd} + F_{gcr}$

P_r is the power required, F_r is the rolling resistance force, F_{Cd} is the Drag force, F_{gcr} is the gradient force.

$F_r = m * g * v * R_r$..... (rolling resistance = 0.1) [12]

$F_r = m * g * v * R_{rt}$...........(rolling resistance = 0.02)

Where m is the mass of the vehicle with attachments, g is the gravitational force, v is the velocity of the vehicle, R_r is the rolling resistance and R_{rt} is the rolling resistance for the tar road

$F_{Cd} = 0.5 * (Air\ density * C_d * A_f * v^3)$

Where C_d is the coefficient of drag, A_f is the frontal area.

$F_{cgr} = m*g*\sin\Theta$

Considering inclination of 10° (Θ).

Where F_{gcr} is the force acting on an inclined plane and Θ is the angle of inclination.

*RPM=(Speed*60)/(linear distance travelled)*

The tractor efficiency is considered as 80% including all the loses.

$Pr = Pr*1.2$

$Torque = (P_r*60)/(2*\pi*RPM)$.

Table 2.

Speed (kmph)	Speed (m/s)	Fr	Fcd	Fgcr	Pr (Watts)	Torque
10	2.78	9418.89	80.99	5764.36	18317.1	1134.19
15	4.17	14128.33	273.33	5764.36	24199.2	998.94
20	5.56	18837.78	647.89	5764.36	30300	938.09
25*	6.94	3767.56	1265.41	5764.36	12956.8	320.91
35*	9.72	5651.33	3472.28	5764.36	17865.6	316.07

3.2. Electric Tractor

The drivetrain of an electric tractor consists of [*Fig. 1*] [8]:

- Motor
- Battery
- Battery Management System
- Power Converters (Drive signal control)

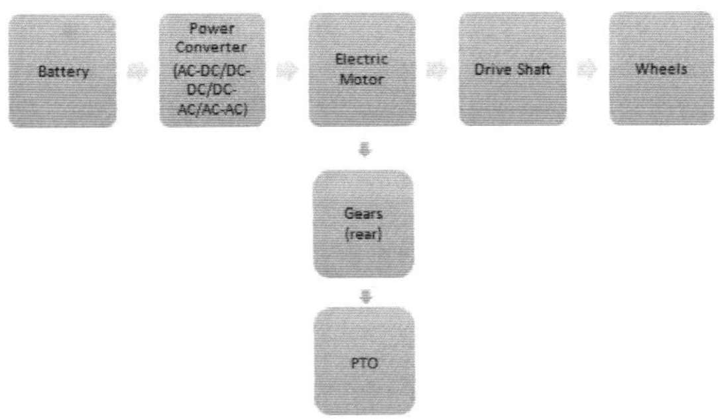

Figure 1. EV Block Diagram

The theoretical power calculations provided above will require an appropriate electric motor as well as several batteries. The motor selection was based on availability as well, some motors selected based on the motor technology are:

- Induction motor
- PMDC motor
- PMAC motor

The motor's choices are PMAC ME1302, PMDC ME1909, PMDC ME1002, Induction Motor 1LE7 IE2. The PMAC ME1302 is a motor with a peak power output of 38kW and continuous power output of 15kW, the voltage and current ratings are 120V and 220A respectively. The motor has a weight of 16kgs. The second choice is using two PMDC ME1909 motor, each motor has a peak power output of 20kW and continuous power of 11.5kW. The motor has a voltage rating of 72V and a current rating of 200A. The weight of the motor is 17kgs. The PMDC ME1002 is the motor with the highest power output, the peak power and continuous power of this motor is 63kW and 26kW respectively. The motor requires a voltage of 144V and 170A of current. The weight of this motor is 87kgs. The 1LE7 IE2 induction motor is the heaviest motor at 274kgs. It produces a peak power of 37kW and requires 415V-420V and 64A of 3-phase current. *[Fig. 4]*

In the single motor setup, the motor replaces the IC engine keeping the gearbox, the four-wheel-drive (4WD) system and power take-off(PTO) intact. In the dual motor setup, the motors are placed on each axle, i.e. front axle and rear axle. This system eliminates the 4WD gearbox/differential and the motor placed on the rear axle is also responsible for the functioning of the PTO.

For EV based on the comparison between different types of battery chemical technologies and the availability of batteries in India from the local vendors the three types of battery technologies selected are:

- $LiFePO_4$ (Lithium-Iron-Phosphate)
- $Li_4Ti_5O_{12}$ (Lithium-titanate)
- NMC (Lithium Mixed Nickel-Manganese-Cobalt Oxide)

Battery:

Table 3.

Battery	Voltage	Current	Grade (C)	Life Cycle	Weight (kg)
NMC	3.7V	2.5Ah	3C	1100	0.045
LPO	3.2V	6Ah	3C	1500	0.14
LTO	2.3V	30Ah	10C	25000	1.23

Calculations for the number of cells:

1. $N_{cs}= V_{bp}/V_{bc}$
2. $N_{cp}= I_{bp}/I_{bc}$

N_{cs} is the number of cells in series, V_{bp} is motor voltage, V_{bc} is the voltage of each battery cell, N_{cp} is the number of cells in parallel, I_{bp} is the current of the motor, I_{bc} is the current of each battery cell.

Required data of one cell:-

Voltage (V), Rated capacity if one battery (Ah), Grade(C).

Power(Wh)= (Ah)(V)*

Discharge rate(Current) = (C)(Ah)*

Run time= 1/(C) ..... in hours

Calculation of energy stored, current and voltage for a set of batteries in series and parallel

Capacity of battery pack= (Current of pack)(Run time)*

Power of the battery pack= (Voltage of battery pack)(Capacity of battery pack)*

In the table given below, the cost shows the amount required for charging a battery once. Since, the cost of electricity varies in different regions in India and also varies on the units consumed, the range of cost is calculated by considering the cost of 1 unit or 1kWh as Rs.6 and Rs.9 for lower and upper limit respectively.

The cost of LTO cells is the highest but the life cycle of these cells is 3 times the life cycle of NMC/LPO.

The calculation of the cells in the battery pack is made based on the 100% efficiency of the motor. The efficiency of the motor, transmissions along with the electronic components is considered to be a minimum of 80%. [13]

NMC cells:

Table 4.

Motor	Cells			Capacity(kWh)	Cost per Charge (INR)
	Series	Parallel	Total		
PMAC	40	88	3520	32.560	195.36-293.04
PMDC	40	94	3760	34.780	206.68-313.02
PMDC(M1002)	36	89	3204	29.637	177.82-266.73
Induction Motor	113	31	3503	32.402	194.41-291.62

LPO cells:

Table 5.

Motor	Cells			Capacity(kWh)	Cost per Charge (INR)
	Series	Parallel	Total		
PMAC	42	40	1680	32.256	193.54-290.30
PMDC	47	38	1786	34.291	205.75-308.62
PMDC(M1002)	46	34	1564	30.028	180.17-270.25
Induction Motor	132	13	1716	32.947	197.68-296.52

LTO cells:

Table 6.

Motor	Cells			Capacity(kWh)	Cost per Charge (INR)
	Series	Parallel	Total		
PMAC	55	8	440	30.360	182.16-273.24
PMDC	66	8	528	36.432	218.59-327.88
PMDC(M1002)	65	7	455	31.395	188.37-282.55
Induction Motor	216	2	432	29.808	178.84-268.27

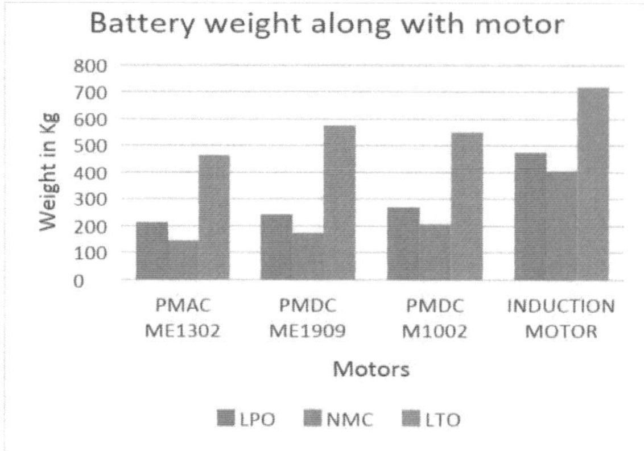

Figure 2. Battery weight comparison

The running cost of a diesel tractor is calculated to compare it with the running cost of an electric tractor. Here the major focus is the Haryana and Punjab region where the average farm size is approx. 5acres which is roughly 22,500 m^2. Therefore, the length and breadth of the field are 150m X 150m. Here the widest accessory used for farming is considered, which is offset 22 disc harrow. The width of the accessory is 2.5m. So to use the disc harrow throughout the field, the tractor needs to cover a distance of 9km is to be covered. Therefore, the average speed of the tractor is considered to be 6kmph, and using this information average time is calculated and considering the average fuel consumption of the tractor, the total fuel of approximately 10 litres is required. The cost of diesel as per August 2020, the running cost for the diesel tractor was calculated to be **Rs.93.5/km.**

The electric tractor's running cost can be calculated by considering the travel factor as 13.335 and with calculated battery pack capacity as 24.199 KW. The range and average speed calculated is 200kms and 15 kmph respectively. The power consumption is calculated by dividing power and speed, the cost for units consumed per km is calculated as 1.613 units/km. The cost per unit in India as per August 2020 is considered and the cost per km for the electric tractor is **Rs.14.15/km**. *[Fig. 3.]*

From the above calculations, it can be deduced that the electric tractor is **84.86%** efficient in terms of cost.

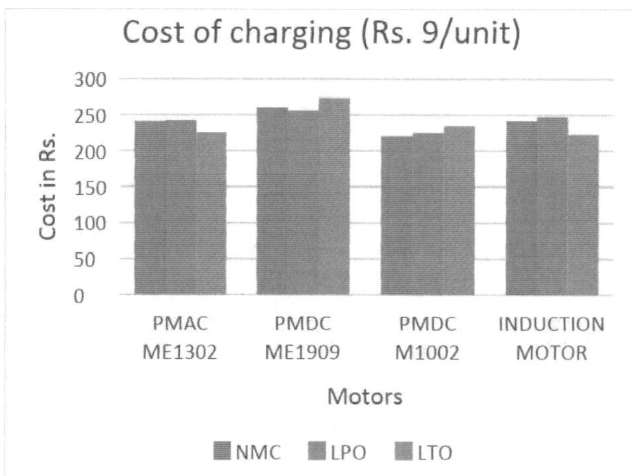

Figure 3. Charging Cost Comparison

4. Conclusion

The paper presented shows an analytical comparison between an electric vehicle and a diesel tractor by studying the vehicle, battery and motor specifications. The tractors with 40-50HP are dominating the tractor market and it is necessary to bring a change by considering the pollution and the increase in the fuel cost. The electric tractor will have fewer environmental effects than the diesel tractor. Furthermore, the running cost of the electric tractor and the diesel tractor is being calculated and the result shows that electric tractors are less expensive in the long-term use which is 84% cheaper than the diesel tractor.

5. References

1. Ramachandra, T. V. "Emissions from India's transport sector: statewise synthesis." *Atmospheric Environment* 43, no. 34 (2009): 5510-5517.
2. https://www.jagranjosh.com/general-knowledge/what-is-the-sectorwise-contribution-in-gdp-of-india-1519797705-1, JAN 3, 2020
3. Khurana, Anil, VV Ravi Kumar, and Manish Sidhpuria. "A Study on the Adoption of Electric Vehicles in India: The Mediating Role of Attitude." *Vision* 24, no. 1 (2020): 23-34.
4. Ryu, I. H., D. C. Kim, and K. U. Kim. "Power efficiency characteristics of a tractor drive train." *Transactions of the ASAE* 46, no. 6 (2003): 1481.
5. Lal, Bajrang, and Ajmer Singh. "Understanding the Tractor Market: an Exploratory Study in Haryana (India)." *Global Journal of Management And Business Research* (2016).
6. Ueka, Yuko, Jun Yamashita, Kazunobu Sato, and Yoshinori Doi. "Study on the development of the electric tractor: specifications and traveling and tilling performance of a prototype electric tractor." *Engineering in agriculture, environment and food* 6, no. 4 (2013): 160-164.
7. Chen, Xiaopeng, Weixiang Shen, Thanh Tu Vo, Zhenwei Cao, and Ajay Kapoor. "An overview of lithium-ion batteries for electric vehicles." In *2012 10th International Power & Energy Conference (IPEC)*, pp. 230-235. IEEE, 2012.
8. Kumar, Mr A. Rakesh, and Dr Sanjeevikumar Padmanaban. "Electric Vehicles for India: Overview and Challenges." *IEEE India Council Newsletter* (2019).
9. Preetha, P. K., and Prabaharan Poornachandran. "Electric Vehicle Scenario in India: Roadmap, Challenges and Opportunities." In *2019 IEEE International Conference on Electrical, Computer and Communication Technologies (ICECCT)*, pp. 1-7. IEEE, 2019.
10. Troncon, Diego, and Luigi Alberti. "Case of Study of the Electrification of a Tractor: Electric Motor Performance Requirements and Design." *Energies* 13, no. 9 (2020): 2197.
11. Aswani, Geetika, Vikas Singh Bhadoria, and Jay Singh. "Electric vehicles in india: Opportunities and challenges." In *2018 International Conference on Automation and Computational Engineering (ICACE)*, pp. 65-71. IEEE, 2018.
12. Grunditz, Emma. *Design and assessment of battery electric vehicle powertrain, with respect to performance, energy consumption and electric motor thermal capability*. Chalmers University of Technology, 2016.
13. Hashemnia, Nasser, and Behzad Asaei. "Comparative study of using different electric motors in the electric vehicles." In *2008 18th International Conference on Electrical Machines*, pp. 1-5. IEEE, 2008.

Solar Power Generation, Utilization and Management for facilitating Sustainable Development in India:Review

Utkarsh Mohan[2,*] ,Priyanka Singh[1], Krishna Mohan Agarwal[2], Sanjeev Kumar Sharma[2], Ashish Yadav[2] , Chinmay Sharma[2]

[1]*Civil Engineering Department, Amity University Uttar Pradesh, Noida, 201303, India*

[2]*Mechanical Engineering Department, Amity University Uttar Pradesh, Noida, 201303, India*

**Corresponding author. Tel - +918840776380; E- mail address: mohanutkarsh99@gmail.com*

ABSTRACT

The approach of the paper is to present a review of solar power generation, utilization and its management for facilitating sustainable development in India. It briefs about the allusive estimation of the solar energy and its utilization. The Solar Energy is a gigantic energy source that can be used for any of the day-to-day requirements, including engineering, the refrigeration, the water heating, the heat of industrial processes, transportation and sustainable production and even for the cleaning of environment. The solar radiation is the purest and cleanest form of the energy where no mass is attached, and it can be converted in to any forms easily. The solar energy that falls on the surface of the earth is sufficient to provide solution to all our requirements on this planet. Among all the sectors of India, power sector is largest one which is important for the growth of Indian economy. There is perpetual and rising demand of power in the all the sectors. So, to meet the increasing demand and to overcome the shortage of power, generation of power by renewable energy source is must for sustainability in this field. In the today's scenario, generation of power from renewable energy source is necessary as compared to non-renewable energy sources. This paper emphasizes on the the scope of generation, usage and its management for sustainable development ofIndia.

Keywords— Solar Power, Structures, Non-renewable Energy Source, Renewable Energy Source, Sustainable Development

1. Introduction

Renewable and non-renewable energy sources are the two primary types of energy sources available on the planet. The renewable energy sources (RES) fulfill one sixth of total energy demand of the globe [1]. The sources of renewable energy include solar, geothermal, wind, marine energies and hydropower. Renewable energy sources are devoid of pollutants and can be replenished. This is one of the biggest advantages of RES [2, 3]. The usages of RES have increased over past decade and are expected to increase by 80 percent in the next century [4, 5]. The Global energy scenario over next three decades has been presented in Table 1. In the current era, the world is moving towards the concept of sustainable development. The environmental impact of human activities is the most important parameters measuring the sustainable development [7]. The literature review shows that with the advancement in technology and rapid industrialization, the usage of fossil fuels has immensely increased, which is responsible for global climatic change [8]. The developed Nations are moving towards sustainable development through measures of reduction in emission of greenhouse gases, conservation of forests, renewable energy consumption etc. [9]. Excessessive use of fossil fuels has alarm the risk of its depletion and there is acute necessity of searching for new energy sources to fulfill human demands [10, 11]. Moreover, use of fossil fuels had lead to greenhouse effect and environment pollution and necessity of cleaner fuel is required [12, 13]. This paper attempts to provide a critical review of the scope of available renewable energy sources, to fulfill the energy demands through its proper utilization and management.

Table 1: Global Solar energy scenario 2040 [14]

Total consumption (million tons oil equivalent)					
Year	2001	2010	2020	2030	2040
Solar thermal energy	4.1	15	66	244	480
Photovoltaic Energy	0.1	2	24	221	784
Solar thermal electricity	0.1	0.4	3	16	68

2. Renewable energy sources

In future, renewable energy sources are indispensible options in the generation of the power and sustainable growth. RES can be recycled infinitely and there is no possibility of its depletion [15,16]. Fig. 1 illustrates the different sources of renewable energy. It can be further sub categorized into solar power, biomass, wind power and hydro power. Non-RES are frequently used for domestic purposes which are the major source of emission of greenhouse gases and air pollutants. [17,18]. Solar home systems, solar dryers, solar cookers, photovoltaic, thermal power generation, and water heaters are some of the most common uses of solar energy.[19,20]

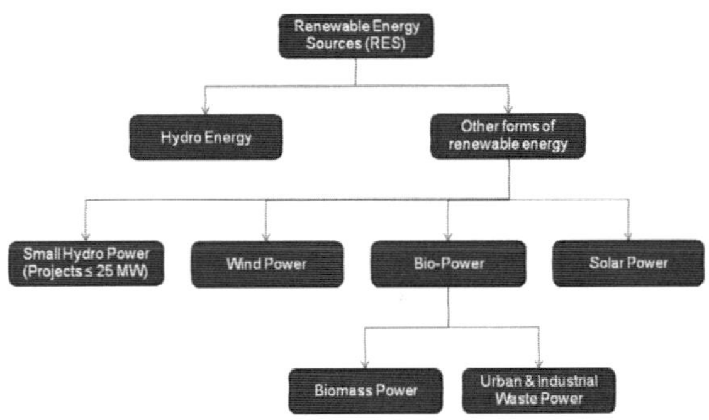

Figure 1: Sources of Renewable Energy [21]

IOP Conf. Series: Materials Science and Engineering 1168 (2021) 012004 doi:10.1088/1757-899X/1168/1/012004

2.1.Solar Energy

Solar energy is abundantly available energy sources. Sun emits enormous amount of energy in a day sufficient to fulfill energy demand of the earth for whole year, but only 50% of Sun's energy is received by Earth and far lesser amount is used to meet energy requirements [22,23]. There is a extensive possibility for utilization of solar energy [24,26]. Generation of power by solar energy saves emission of 39 billion tons of carbon dioxide per year [27,29]. Among the domestic applications solar cooking is one of the most usual applications of solar energy. Solar cookers of different varieties are easily available and very economical. Solar water heater is another prevalent application of solar energy for domestic use and it is estimated that it can extenuate carbon dioxide emissions significantly [30]. Solar drying technology is the most efficient process dry vegetables and fruits clean and hygienic environment with zero energy costs and it results in saving of time energy and has lesser area requirements. It tremendously improves the quality of fruits and vegetables and reduces their deterioration comprehensively [31,32].

In country like India, the energy which is freely available throughout the year is solar energy. Its inherent advantage is its ample availability in remotest corner of our country [33].India receives nearly 5,000 trillion kWh of solar energy per year, which is equivalent to 600 GW, due to its varied geological terrain. This is much more than the country's current total energy consumption. [34, 35].

The technologies used to generate solar energy can be divided into two categories:

- Solar thermal power plants are a type of solar thermal power plant that uses sunlight to generate electricity
- Photovoltaic Electricity Generation System (Solar PV)

2.2. Solar thermal powerplants

Solar thermal power plants convert solar radiation into high-temperature heat using mirrors and reflectors to generate electricity. This energy is used to generate steam by heating a working fluid. The steam is then used to spin a turbine or power an engine, which drives a generator and generates electricity. [36]. Solar collectors are one of the main components of solar thermal system. The solar collectors are a special kind of heat exchangers which convert solar radiation into electrical energy [37]. However, the initial cost for this type of power generation is quite high as compared to other sources of electricity generation but it involves least amount of carbon emission as shown in Table 2.

Table 2 When traditional technologies are compared to solar power generation in terms of cost and emissions is shown below [38].

Electricity Generation technology	Carbon emissions (gC/kWh)	Generation costs (US¢/kWh)
Solar thermal and solar PV systems	0	9–40
Pulverized coal–natural gas turbine	100–230	5–7

2.3. Solar Photovoltaic Electricity Generation System

Solar photovoltaic cells are the source of generation of electricity in which solar energy is converted into electricity by use of photovoltaic cells. Photovoltaic models produce electricity directly from the sunlight without any gas emission and noise pollution. The development of new technology has reduced the cost of solar power generation drastically. The photovoltaic models require very large surface area even for generation of small quantity of energy [39, 40]. The inverter is the primary components of photovoltaic system its main function is to convert DC power to AC power, to meet the voltage and power quality requirements of substations. Due to abundance of silicon, the Silicon solar cells are the most preferred photovoltaic cells especially for space and Terrestrial applications. Photovoltaic system is promising sources of energy generation and it works with least carbon dioxide emission [41,42].

Fig. 2 Represents some thermal technologies which are in general practice.

Figure 2: Solar Thermal Technologies [43]

In the generation of electricity by solar photovoltaic, a minimal amount of the non-renewable energy is being utilized during its installation, manufacturing and transportation of solar PV modules and its components [44]. The energy consumption during the manufacturing, its installation and transportation of the solar PV systems is being is assessed LCA (life cycle assessment). It indicates the efficiency of plant as well as its economic analysis. Presently, India ranks at the sixth position in the generation of the electricity and it is about 4 % of the world's electricity generation per annum[44-46]. The direct conversion of solar energy to electricity is mainly done through photovoltaic cells. PV's effect is determined by photon interactions with energies equal to or greater than the band gap of PV materials.. Basically, there are two classifications of the PV systems, which are basically Flat-plate PV systems and Concentrator PV systems. Flat Plate PV systems typically use flat plate[47].

Flat-plate collectors are a large number of cells mounted at a fixed angle on a rigid flat surface, which collects optimum sunlight throughout the year. Both direct sunlight and diffuse sunlight reflected from clouds, the ground, and objects are used in this system. Concentrator PV systems make use of specialised, effective solar cells and concentrating optics to boost the intensity of sunlight striking the cells, which is usually very low

[48]. It comes in two different configurations: point-focus and line-focus. Reflective parabolic dishes or circular Fresnel lenses (which use saw-tooth grooves in a plastic sheet) are used in point-focus systems to focus sunshine onto a cell mounted at the focal point, while linear parabolic reflectors or linear Fresnel lenses are used in line-focus systems to focus sunlight on cells mounted along the focal line. Concentrator systems, in particular, use two-axis tracking to track the sun and capture as much direct sunlight as possible[49-50].

3. Generation and Utilization of Solar Power in India

Solar PV-based power generation is being given more importance in India in order to quickly increase the share of electricity production from renewable energy. The Indian government expects to produce 175 GW of electricity from renewables by 2022 as part of the Jawaharlal Nehru National Solar Mission [51]. Fig.3. demonstrate the solar radiation zone of India. The suggested goal is to produce five times the current amount of electricity from renewable sources. 100 GW of the 175 GW target will be created solely from solar energy, with the remainder coming from wind, biomass, and small hydro. Out of total solar-based energy, the Indian government seeks to generate 40 GW from rooftop solar PV. India currently has an installed capacity of 8.7 GW of solar PV power plants. [52].

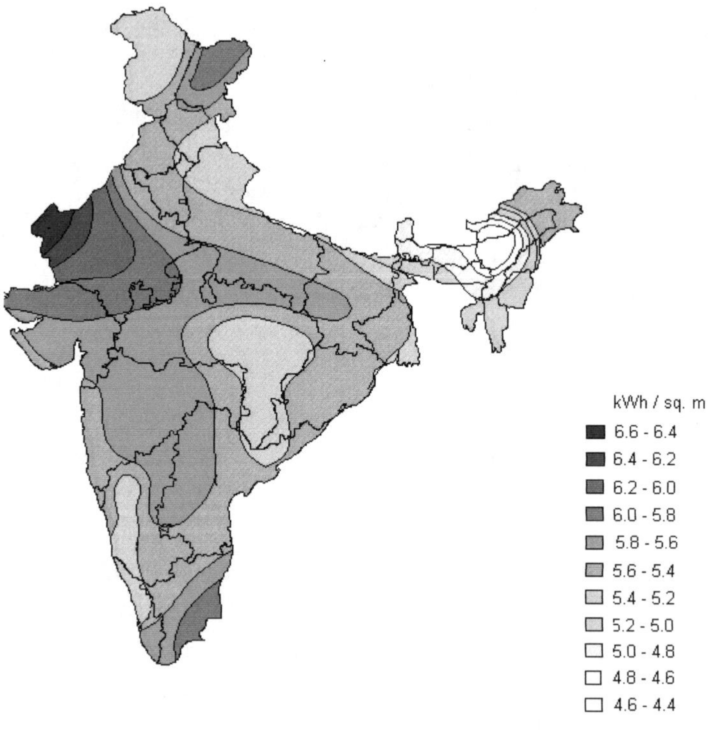

Figure 3: Solar radiation zones as per TERI based on the IMD database [53]

The Ministry of New and Renewable Energy (MNRE) has devised a plan to establish a number of solar parks throughout the country, each with a capacity of solar projects of at least 500 megawatts. The Scheme suggests

IOP Conf. Series: Materials Science and Engineering 1168 (2021) 012004 doi:10.1088/1757-899X/1168/1/012004

that the Government of India provide financial support to construct solar parks with the aim of facilitating the creation of infrastructure required for the establishment of new solar power projects in terms of land allocation, transmission and evacuation lines, access roads, water availability, and other factors in a targeted way [54,55]. SECI, a central public sector company under the Ministry of New and Renewable Energy, has been implementing various schemes to grow the solar sector in India based on the availability of solar radiation.Fig 4.shows the India Solar resource map which shows the areas of maximum and minimum radiation, which is exploited in finalizing solar parks.

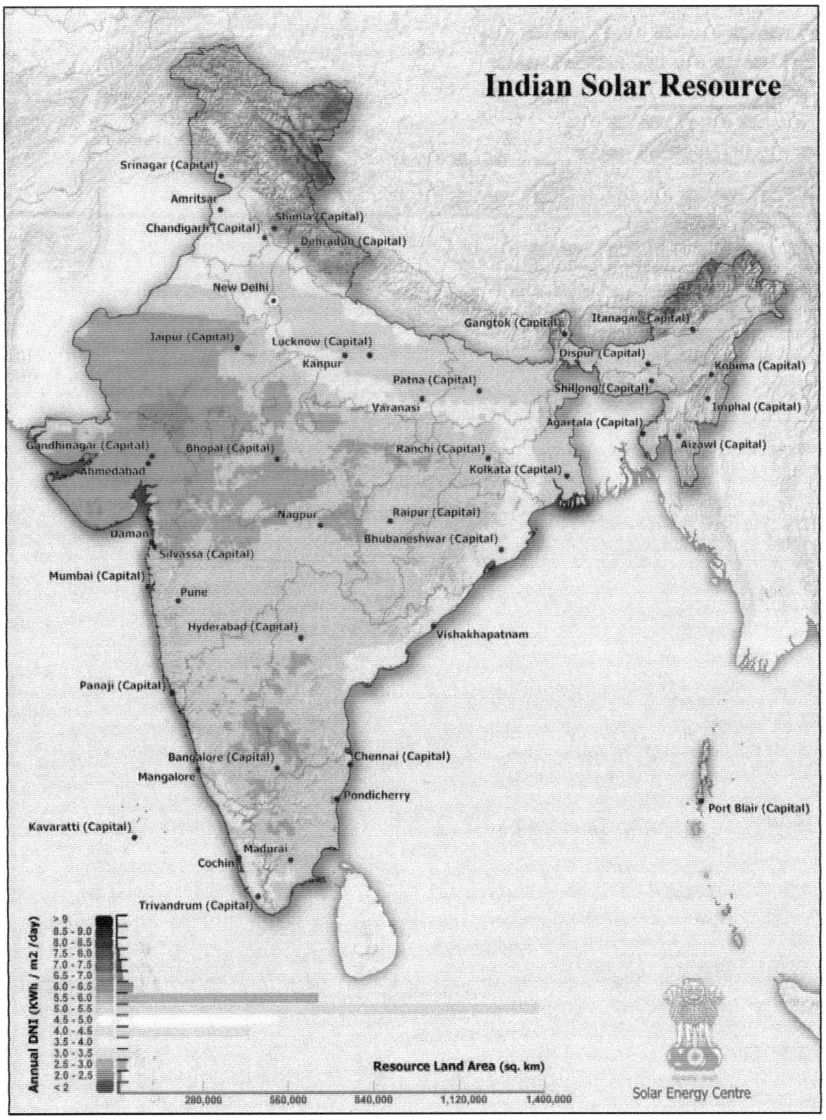

Figure 4: India Solar Resource Map [56]

The state appoints a nodal agency to oversee the solar park's construction. The Solar Park is a focused area for solar power generation project growth.Land needed for the construction of Solar Power Projects with cumulative capacity of 500 MW and above is identified and purchased as part of solar park growth, and various infrastructures such as transmission systems, water, road connectivity, and communication networks are constructed. [57]. By avoiding emissions equal to the solar park's produced capacity, the state will be able to minimise its carbon footprint. Rajasthan and northern Gujarat receive the most annual global radiation, as can be seen. As of September 2015, Rajasthan had the highest share of solar power generation in India, with 28.4 percent, followed by Gujarat with 24.4 percent [58-60]. Table 3 shows current solar power plants in different state of India.

Table 3: Some Solar plants of India [61]

Name of Solar Plants	**Peak Power (MW)**	**Commission Year**
Charanka Solar Park, Charanka village, Patan, Gujarat	224	Apr-12
Welspun Solar MP project, Neemuch, Madhya Pradesh	151	Mar-13
Mahagenco Solar Project, Maharashtra	130	Mar-13
Rajgarh Solar PV ,Rajghar Madhya Pradesh	50	Mar-14
Welspun Energy Rajasthan Solar Project ,Phalodhi, Rajasthan	50	Mar-13
Talcher Kaniha Solar PV , Odisha	10	Mar-14
Unchahar Solar PV, Unchahar, Uttar Pradesh	10	Mar-14
Sharda Construction, Latur, Maharashtra	10	June-15
Jalaun Solar Power Project, Uttar Pradesh	50	Jan-16
Mandsaur Solar Farm, Madhya Pradesh	250	July-18
Rewa Ultra Mega Power Project,Madhya Pradesh	750	July-18
Pavagada Solar Park, Karnataka	2050	Dec-19
Bhadla Solar Park, Rajasthan	2245	Mar-20

3.1. Grid connected roof top solar PV System

Solar PV systems have become more feasible and appealing in recent years. The available roof-top area on the buildings can also be used to set up solar PV power plants, obviating the need for free land, as shown in Fig.5. Solar PV system power can also be fed into the distribution or transmission grid after conditioning for grid integration [62].

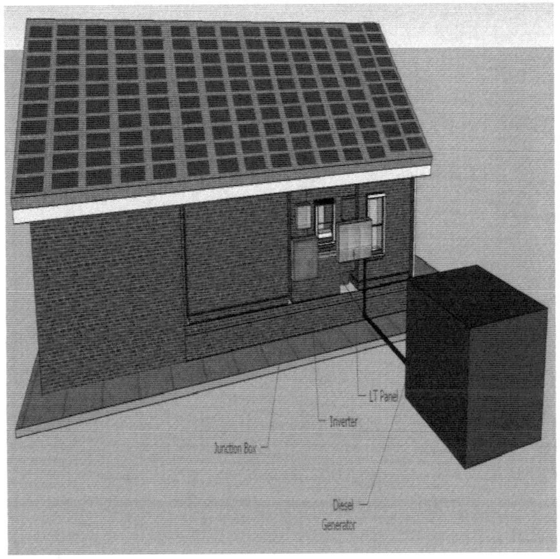

Figure 5: Schematic layout of roof top PV system [63]

The following are the benefits of a grid-connected rooftop solar PV system:

- Production of environmentally friendly energy.
- For his own electricity needs, the consumer becomes a generator.
- A reduction in grid electricity consumption.
- Exporting excess energy to the grid.

Table 4 shows the financial benefit estimates for grid-connected rooftop solar PV systems.

Table4: Financial benefits of the grid connected roof top solar PV systems [64]

System size	100 kWp
Systemcost	0.8 -0.9 crore
Subsidy	30%
Expected electricity generation	140000 - 160000 units
Payback period at Grid electricity cost	5-6 years
Payback period with Accelerated	4-5 years
Pay back at diesel power cost	3-4 years
Plant life	25 ears

4. Management of Solar Power

There is a huge potential of solar power generation in India as it passes through the line of equator, where there is availability of ample amount of solar radiant energy.[65] The Renewable energy potential and installed capacity as of November 2017 in India is elucidated in Fig.6.

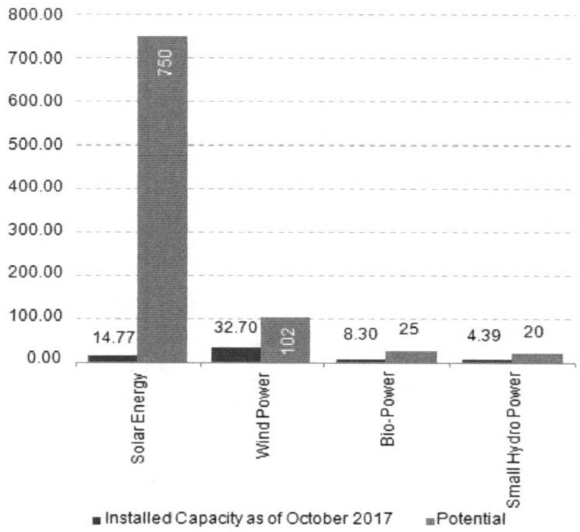

Figure 6: Renewable energy potential and installed capacity as of November 2017 (in GW) [66]

- The Indian Ministry of New and Renewable Energy (MNRE) began a large-scale solar resource monitoring and meteorological station network in 2011 as part of the Solar Radiation Resource Assessment (SRRA) project. The Solar Radiation Resource Assessment (SRRA) project includes the following components:
- SRRA field stations: 51 SRRA field sites in various parts of India have been installed with Automatic Solar Radiation and Weather Monitoring Stations (ASRWMS).
- A central receiving and processing station collects data, performs quality checks and controls, processes it, and disseminates it. [67].

5. Conclusion

Energy is inextricably linked to long-term growth and the resolution of the economic crisis. It encompasses almost all facets of development, such as livelihood, health, education, agriculture, and job creation. The production and consumption of energy are important components of India's long-term growth. Sustainable development, economic growth, and environmental protection all require significant amounts of energy. Energy generations' policies must be oriented toward the promotion of energy systems based on renewable energy, energy efficiency, and cleaner fossil fuel technologies, which can simultaneously address social, economic, and environmental issues. Increasing the utilization of the solar energy can also aid governments to achieve energy security and diversity.Solar energy is also under utilised, despite its vast untapped potential. On the other hand, the lower cost of solar power generation, advancements in solar energy technology, and governments' interest in supporting solar energy as a result of its positive environmental impact would enable solar energy to grow in India, resulting in long-term growth. The importance of solar energy as a source of energy in India's view is not only to increase power generation, but also to improve energy reliability by taking into account environmental, social, and financial considerations.

References

[1] SoniMS,GakkharNikhil.Techno-economicparametricassessmentofsolarpowerinIndia:asurvey. *Renew Sustain Energy Rev 2014*;40:326–34

[2] UNDP. World energy assessment 2000-Energy and the Challenge of Sustainability. *New York: UNDP; 2000* (ISBN9211261260).

[3] DincerI.Environmentalissues.II.Potentialsolutions.EnergySources,PartA: Recovery, Utilization, and Environmental Effects2001;23(1):83–92.

[4] Bilgen S, Kaygusuz K, Sari A. Renewable energy for a clean and sustain- able future. Energy Sources, Part A:Recovery,Utilization, and Environmental Effects 2004;26(12):1119–29.

[5] Rathore NS, Panwar NL. Renewable energy sources for sustainable develop- ment. New Delhi, India*: New IndiaPublishingAgency; 2007.*

[6] Zakhidov RA. Central Asian countries energy system and role of renewable energy sources. *Applied Solar Energy* 2008;44(3):218–23.

[7] Bergmann A, Colombo S, Hanley N. Rural versus urban preferences for renew- able energydevelopments. *Ecological Economics*2008;65:616–25.

[8] Reddy AKN, Subramanian DK. The design of rural energy centers. Indian Academy of Science, Bangalore 1980:109–30.

[9] Ravindranath NH, Hall DO. Biomass, energy, and environment: a developing country perspective from India. Oxford, United Kingdom: Oxford University Press;1995.

[10] Fridleifsson IB. Geothermal energy for the benefit of the people. Renewable and Sustainable Energy Reviews 2001;5:299–312.

[11] Dincer I. Environmental Issues. I. Energy Utilization. Energy Sources, Part A: Recovery, Utilization, and Environmental Effects2001;23(1):69–81.

[12] Sims REH. Bioenergy to mitigate for climate change and meet the needs of society, *the economy and the environment. Mitigation and Adaptation Strategies for Global Change*2003;8:349–70.

[13] Youm I, Sarr J, Sall M, Kane MM. Renewable energy activities in Senegal: a review. Renewable and Sustainable Energy Reviews2000;4(1):75–89.

[14] Demirbas A. Recent advances in biomass conversion technologies. Energy Educational Science and Technology2000;6:19–40.

[15] Hall DO, Mynick HE, Williams RH. Cooling the greenhouse with bioenergy. Nature1991;353:11–22.

[16] Nielsen JBH, Seadi TA, Popiel PO. The future of anaerobic digestion and biogas utilization. *Bioresource Technology*2009;100:5478–84.

[17] M.J. Montes, A. Abànades, J.M. Martınez-Val, Performance of a direct steam generation solar thermal power plant for electricity production as a function of the solar multiple, Sol. Energy 83 (5) (2009) 679–689.

[18] Zakhidov RA. Central Asian countries energy system and role of renewable energy sources. Applied Solar Energy 2008;44(3):218–23.

[19] Dincer I. Energy and environmental impacts: present and future perspectives. Energy Sources, Part A: Recovery, Utilization, and Environmental Effects 1998;20(4):427–53.

[20] Dincer I. Renewable energy and sustainable development: a crucial review. Renewable and Sustainable Energy Reviews2000;4(2):157–75.

[21] Central Electricity Authority (CEA), India,2017.

[22] Mills D. Advances in solar thermal electricity technology. Solar Energy 2004; 76:19–31. [22] Okoro OI, Madueme TC. Solar energy: a necessary investment in a developing economy. International Journal of Sustainable Energy2006;25(1):23–31.

[23] Jyotirmay M, Kumar NK, Wagner HJ. Energy and environmental correlation for renewable energy systems in India. Energy Sources, Part A: Recovery, Utilization, and Environmental Effects 2002;24(1):19–26.

[24] Xiaowu W, Ben H. Exergy analysis of domestic-scale solar water heaters. Renewable and Sustainable Energy Reviews 2005;9:638–45.

[25] Sharma A, Chen CR, Lan NV. Solar-energy drying systems: a review. Renewable and Sustainable Energy Reviews 2008;13(6–7):1185–210.

[26] Demirbas MF. Electricity production using solar energy. Energy Sources, Part A: Recovery, Utilization, and Environmental Effects 2007;29(6):563–9.

[27] Goswami DY, Vijayaraghavan S, Lu S, Tamm G. New and emerging developments in solar energy. Solar Energy 2004;76:33–43.

[28] Topcu YI, Ulengin F. Energy for the future: an integrated decision aid for the case of Turkey. Energy 2004; 29:137–54.

[29] Sakamoto Y, Zhou W. Energy analysis of a CO recycling system. International Journal of Energy Research 2000;24(6):549–59.

[30] Biermann E, Grupp M, Palmer R. Solar cooker acceptance in South Africa: results of a comparative field- test. Solar Energy 1999;66(6):401–7.

[31] D. Bishoyi, K. Sudhakar, Modelling and performance simulation of 100 MW LFR based solar thermal power plant in Udaipur, India, Resour. Effic. Technol. (2017), http://dx.doi.org/10.1016/j.reffit.2017.02.002.

[32] TaqiyeddineBoukelia, Mohamed-Salah Mecibah, Parabolic trough solar thermal power plant: potential, and projects development in Algeria, Renew. Sustain. Energy Rev. 21 (2013) 288–297.

[33] T.E. Boukelia, et al., Optimization, selection and feasibility study of solar parabolic trough power plants for Algerian conditions, Energy Convers. Manag. 101 (2015)450–459.

[34] L.I. García, J.L. Álvarez, D. Blanco, Performance model for parabolic trough solar thermal power plants with thermal storage: comparison to operating plant data, Sol. Energy 85 (10) (2011)2443–2460.

[35] M.J. Montes, A. Abànades, J.M. Martınez-Val, Performance of a direct steam generation solar thermal power plant for electricity production as a function of the solar multiple, Sol. Energy 83 (5) (2009) 679– 689.

[36] Dincer I. Energy and environmental impacts: present and future perspectives. Energy Sources 1998;20(4/5):427– 53.

[37] Rosen MA. The role of energy efficiency in sustainable development. Technol Soc 1996;15(4):21–6.

[38] International Energy Agency, "Methodology Guidelines on Life Cycle Assessment of Photovoltaic Electricity", IEA PVPS Task Subtask 20, LCA Report IEA-PVPS T12-01:2009 October2009.

[39] Dincer I, Rosen MA. A worldwide perspective on energy, environment and sustainable development. International J. Energy Res. 1998; 22(15):1305–21.

[39] www.worldwatch.org.

[40] Dincer I. Environmental impacts of energy. Energy Policy 1999;27(14):845–54.

[41] Colonbo U. Development and the global environment. In: Hollander JM, editor. The energy–environment connection.
Washington: Island Press; 1992. p. 3–14.

[42] Shuman, F., 1913. The most rational source of energy: tapping the sun's radiant energy directly. Sci. Am. 109, 350.

[43] Kalogirou, S. A. Solar thermal collectors and applications. Progress in Energy and Combustion Science, 2004a, 30,231–295.

[44] Samuel TD. Heat withdrawal from multi-layer thermal trap collectors. Solar Energy 1983;30(3):261–70.

[45]J.C.H. Phang, D.S.H. Chan and J.R. Phillips, "Accurate analytical method for the extraction of solar cell model parameters". Electronics Letters, Volume 20, Issue 10, 10 May 1984, p. 406 – 408, DOI: 10.1049/el: 19840281.

[46] Maximum photovoltaic power tracking for the PV array using the fractional-order incremental conductancemethod Original Research Article Applied Energy, Volume 88, Issue 12, December 2011, Pages 4840-4847.

[47] Chia-Hung Lin, Cong-Hui Huang, Yi-Chun Du, Jian-Liung Chen. "Maximum power tracking for photo voltaic power system: Development and experimental comparison of two algorithms". Renewable Energy, Volume 35, Issue 10, October2010,Pages2381-2387.

[48] Colonbo U. Development and the global environment. In: Hollander JM, editor. The energy environment connection. Washington: Island Press; 1992. p.3–14.

[49] B.S. Kumar, K. Sudhakar, Performance evaluation of 10 MW grid connected solar photovoltaic power plant inIndia, Energy Rep. 1 (2015)184–192.

[50] Deepak Bishoyi, K. Sudhakar. "Modeling and performance simulation of 100 MW PTC based solar thermal power plant in Udaipur India", Case Studies in Thermal Engineering 10 (2017)216–226.

[51] Jawaharlal Nehru National Solar Mission. MNRE. Website of Ministry of New & Renewable Energy, Government of India, http://mnre.gov.in/; 2010

[52] Garud S, Purohit I. Making solar thermal power generation in India a real- – overview of technologies, opportunities and challenges, The Energy and Resources Institute (TERI), Darbari Seth Block, IHC Complex, Lodhi Road, New Delhi 110003,India

[53] Indian SolarResources:National Renewable Energy Laboratory (NREL).http://www.nrel.gov/rredc/solar_data.html

[54] Ministry of New and Renewable Energy Source (MNERS) Policy for grid interactive renewablepower. http://www.mnes.nic.in

[55] State Wise Solar Parks: Andhra Pradesh, Solar Energy Corporation of India (SECI), A Government of India Enterprise,

[56] India Solar Resource Map: Solar Energy Centre, Ministry of New and Renew- able Energy, Government of India. Available: (http://mnre.gov.in/sec/solar-assmnt.htm).

[57] "State wise installed solar power capacity", Ministry of New and Renewable Energy, Govt. Of India. 13 July 2015. Retrieved 29 July 2015, Available: ⟨http://mnre.gov.in/file-manager/UserFiles/grid-connected- solar-power-projectinstalled-capacity.pdf⟩.

[57] A. Kumar, S. Gomathinayagam, G. Giridhar, V. Ramdhan, R. Meyer, I. Mitra, M. Schwandt, and K. Chhatbar, 'Solar Resource Assessment and Mapping of India', in SolarPACES, Marrakech, Morocco, 2012.

[58] NREL, 'Quality Assessment with SERI_QC',1993.

[59] C. N. Long and E. G. Dutton, 'BSRN Global Network recommended QC tests V2.0', Baseline Surface Radiation Network (BSRN),2002.

[60]C. Hoyer-Klick, D. Dumortier, A. Tsvetkov, J. Polo, J. L. Torres, C. Kurz, and P. Ineichen, 'MESOR Existing Ground Data Sets', D 1.1.2,2009.

[61] Ministry of Power .Available:http://www.powermin.nic.in

[62] Shukla, A.K., Sudhakar, K., Baredar, P., 2016b. Design, simulation and economic analysis of standalone roof top solar PV system in India. Sol. Energy 136, 437–449.

[63] http://seci.co.in/web-data

[64] https://mnre.gov.in/solar-rooftop-grid-connected

[65] Ministry of New and Renewable Energy Source, Policy support for grid interactive renewable power. (http://www.mnes.nic.in)

[66] Ministry of New and Renewable Energy (MNRE), Central Electricity Authority (CEA), IIT Chennai Study

[67] http://www.pib.nic.in/Pressreleaseshare.aspx?PRID=1555373

Effect of Die Design parameters on materials processed by Equal Channel Angular Pressing

Bhuwan Gupta[1*], Arshit Kapoor[1], Abhishek Singhal[1] and Krishna Mohan Agarwal[1]

[1]Mechanical Engineering Department, Amity University Uttar Pradesh, Noida, India
* Corresponding author. Tel: +91 7838780470, E-mail kbhuwan30@gmail.com

Abstract- Equal Channel Angular Pressing technique is used for grain refinement in bulk materials, where the material is passed through the die in which the two channels intersect at some angle, that result in finer grains. There are number of die design parameters which affects the production of ultra-fine grains. This paper presents the effect of die design parameters such as channel angle, corner angle, several passes and route followed for processing of the material for grain refinement. The results of different parameters are compared, and it was observed that the optimum values for producing maximum shear strain is for channel angle 90° and it reduces with the increase in channel angle. About 20° corner angles is found to suitable for filling the corner gap, around 4 to 6 passes, and route Bc leads to the ultrafine grained structure of homogeneous equiaxed grains separated by high angle grain boundaries in the materials. The sample of Al-TiO2 was prepared and its microstructure and mechanical properties were tested after equal channel angular pressing.

Keywords-SPD, ECAP, Grain refinement, Microstructure.

1 Introduction

In recent years, it has been felt by different industries that material should have high strength and at the same time the material must have low weight. To fulfill the combination of required properties, number of techniques have been utilized in recent years. The mechanical and physical properties of any material can be described by several factors, and one of them is average grain size. The average grain size of the material plays an important role and often the dominant role in defining the strength of the material. Hall – Petch equation given in (1) states that strength or grain boundary strengthening of the material can be increased by decreasing the average grain size of the material [1-5].

$$\sigma_y = \sigma_0 + kd^{-1/2} \tag{1}$$

In this equation, is known as friction stress, is a constant of yielding and is the mean grain size diameter. For changing a coarse-grained material into an ultra-fined grain material, it is important to produce a high shear strain to introduce a high level of dislocation, which causes re-arrangement of the grain boundaries. Ultra-fined grained material can be produced by using the techniques of SPD, where a very high strain is imposed on bulk solid without any change in the overall dimension of the work piece and which ultimately causes the grain refinement. The major SPD techniques which are already used to make UFG materials are accumulative roll-bonding (ARB), Multi-directional forging (MDF), High-Pressure Torsion (HPT), Twist Extrusion (TE), ECAP, etc. [6-12].

ECAP is one of the techniques of SPD in which a rod-shaped billet is passed through a die in which two channels intersect at some angle known as channel angle. The billet is pushed through the top channel of the die & taken out from the side channel and at intersection it must pass through the main deformation zone where high strain is imposed, which ultimately causes a grain refinement without any change in overall dimension. Since the dimensions of the billet remain unchanged, and that is the reason, the billet can be processed repeatedly for attaining exceptionally high strain [13, 15]. ECAP is known to be a homogenous process which simply means that workpiece has a uniform distribution of stress. There are factors which may affect its homogeneity. Broadly there are two factors that are internal and external factors. Internal factors are those which are dependent on type of material and its mechanical properties. External factors are those which are dependent on die design parameters like temperature, ram speed, friction, back pressure, rate of loading, geometry of die and passing routes. Internal factors majorly depend on material of workpiece and its properties.

Segal and his team introduced ECAP, also known as Equal Channel Angular Extrusion (ECAE), for the first time, in the early 1980s. Fig. 1 shows the processing of the materials through the ECAP die where the corner angle (a) $\Psi = 0$, (b) $\Psi > 0$.

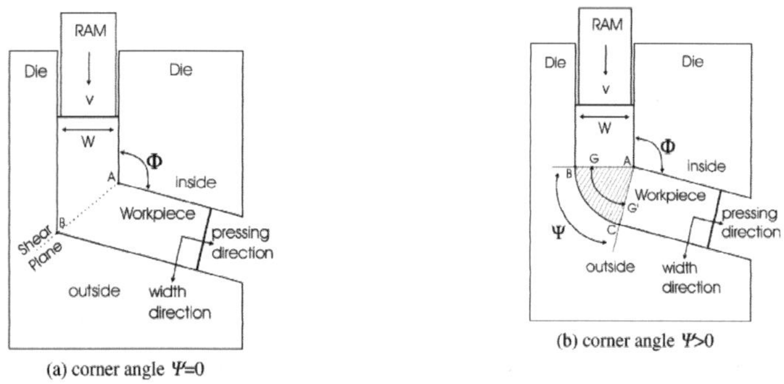

Fig.1. showing the processing of the material through the ECAP die

Composites are new advancement in getting light weight and high strength material in present days. They have the ability to fulfill the requirements of industry of getting the desired properties. Al-TiO2 is a composite which can be prepared by adding titanium in to the Aluminum by metallurgical process. The properties of these composites can be further enhanced with the help of ECAP. This paper studies the microstructure and mechanical properties of this sample after 0, 1, 2, 3 and 4 passes of ECAP.

2 Method of shear strain produced

This paper is mainly focused on the observation of the effect of different die design parameters such as channel angle, corner angle, number of passes & routes through which the material passes from the die [16,18,20]. The governing equation relating channel angle, corner angles and number of passes has been mentioned in Eq. 2 modified by Iwahashi, which was earlier given by Segal [17]. The equivalent strain after the N cycle is given by:

$$\epsilon_N = \frac{N}{\sqrt{3}}\left[2\cot\left(\frac{\emptyset}{2}+\frac{\Psi}{2}\right) + \Psi cosec\left(\frac{\emptyset}{2}+\frac{\Psi}{2}\right)\right] \tag{2}$$

Where ∈_N = Equivalent Shear Strain, where N=No. of passes, Ψ=Corner angle, ϕ =Channel angle. Hydraulic press helps to produce the strain at the point of intersection. Equation 3 can be used to calculate right amount of pressure required for pressing to avoid over pressing which may lead to several uncertainties. Where pressure required is denoted by P, m is friction coefficient, li and lo are the length of sample in entry and exit channel respectively, τo is the shear strength and other have their usual meanings.

$$P = \tau_o(1+m)\left[2cot\left(\frac{\varphi+\emptyset}{2}+\varphi\right)\right] + 4m\tau_o\left(\frac{l_i+l_o}{a}\right) \tag{3}$$

From the above equation, the conclusion can be made that friction also plays an important role apart from channel angle and corner angle in pressure requirement.

3 Results and discussions

The effect of different die design parameters such as channel angle, corner angle, number of passes and routes followed has been done by using computational techniques [19, 21-26]. Different channel angles varying from 90° to 135° were used to analyse the effect of channel angle on strain produced. Corner angle were varied from 0° to 40° and number of passes were 1 to 2. All routes were considered during the process.

3.1 Effect of die channel angle (ϕ)

Die channel angle is the angle where the two channels intersect, and this has great impact on the production of shear strain which ultimately leads to grain refinement. Table 1 and Table 2 shows the effect of channel angle on computational values of shear strain produced for different corner angles when n=1 and n=2 respectively. Fig. 2 and Fig. 3 illustrates the Effect of channel angle on shear strain produced when n = 1 and n = 2 respectively. The results are obtained by using equation 2 for equivalent shear strain produced for channel angle varying from 90° to 150° for corner angle 0° to 30°. The maximum value of shear strain is 1.154 for ϕ = 90° and Ψ = 0°. The graph is showing the decreasing trend of shear strain with the increase in channel angle for single pass. For double pass, the maximum value of shear strain is 2.308 for ϕ = 90° and Ψ = 0°. The graph is showing the similar decreasing trend of shear strain with the increase in channel angle for double pass also [27-31].

Table 1 Effect of channel angle on computational values of shear strain produced for different corner angles when n=1

S. No.	Channel angle	Equivalent Shear strain for Ψ = 0	Equivalent Shear strain for Ψ = 10	Equivalent Shear strain for Ψ = 20	Equivalent Shear strain for Ψ = 30

1	90	1.154	1.100	1.054	1.016
2	105	0.886	0.855	0.828	0.805
3	120	0.666	0.649	0.635	0.622
4	135	0.478	0.470	0.462	0.457
5	150	0.309	0.306	0.303	0.302

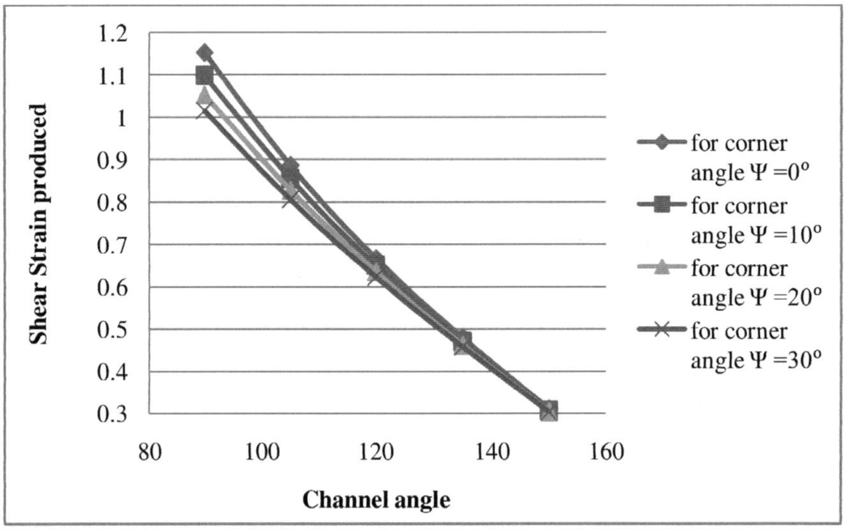

Fig. 2. Effect of channel angle on shear strain produced when n = 1

Table 2 Effect of channel angle on computational values of shear strain produced for different corner angles when n=2

S. No.	Channel angle	Equivalent Shear strain for Ψ = 0	Equivalent Shear strain for Ψ = 10	Equivalent Shear strain for Ψ = 20	Equivalent Shear strain for Ψ = 30
1	90	2.308	2.201	2.109	2.032
2	105	1.772	1.710	1.657	1.611
3	120	1.332	1.299	1.270	1.245
4	135	0.956	0.940	0.925	0.914
5	150	0.618	0.612	0.607	0.605

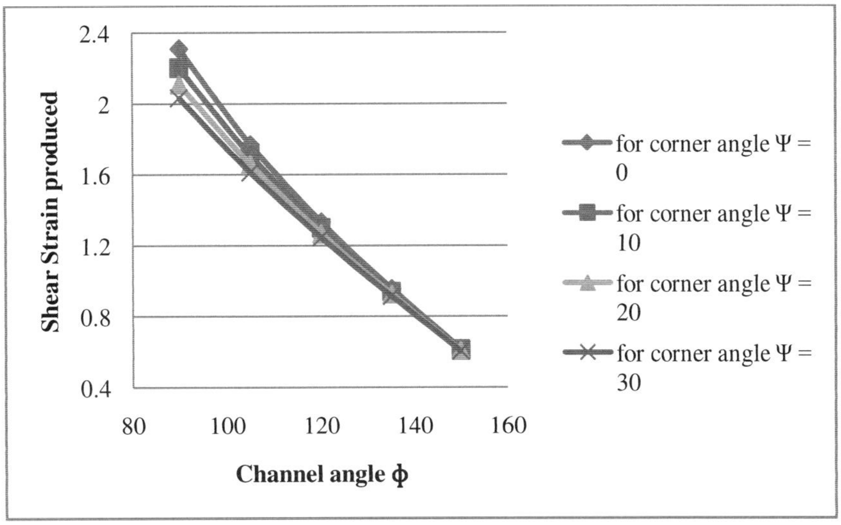

Fig. 3. Effect of channel angle on shear strain produced when n = 2

3.2 Effect of die corner angle (Ψ)

Die corner angle is the angle made by the curvature of two channels on the outer side. This has great impact on the smooth passage of material through the die. Table 3 and Table 4 shows the effect of corner angle on computational values of shear strain produced for different channel angles when n=1 and n=2 respectively. Fig. 4 and Fig. 5 illustrates the Effect of corner angle on shear strain produced when n = 1 and n = 2 respectively. The results are obtained by using equation 2 for equivalent shear strain produced for corner angle varying from 0° to 40° for channel angle 90° to 135°. The maximum value of shear strain is 1.154 for ϕ = 90° and Ψ = 00. The graph is showing the decreasing trend of shear strain with the increase in corner angle for single pass. For double pass, the maximum value of shear strain is 2.310 for ϕ = 90° and Ψ = 0°. The graph is showing the similar decreasing trend of shear strain with the increase in channel angle for double pass also. It can also be observed that corner angle has little effect on shear strain produced in the material.

Table 3 effect of corner angle on computational values of shear strain produced for different channel angles when n=1

Sr No	Corner angle	Equivalent Shear strain for ϕ = 90	Equivalent Shear strain for ϕ = 105	Equivalent Shear strain for ϕ = 120	Equivalent Shear strain for ϕ = 135
1	0	1.154	0.886	0.666	0.478
2	10	1.100	0.855	0.649	0.470
3	20	1.054	0.828	0.635	0.462
4	30	1.016	0.805	0.622	0.457

5	40	0.983	0.786	0.613	0.454

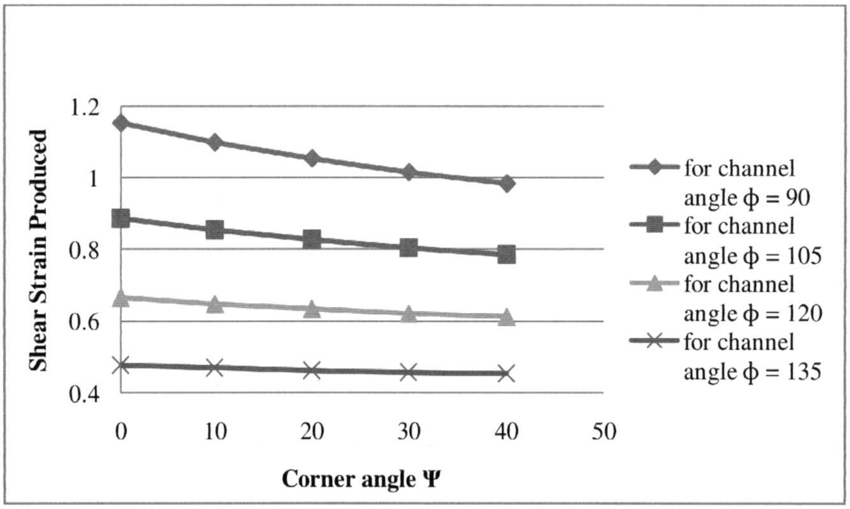

Fig. 4. Effect of corner angle on shear strain produced when n = 1

Table 4 effect of corner angle on computational values of shear strain produced for different channel angles when n=2

Sr. No	Corner angle	Equivalent Shear strain for ϕ = 90	Equivalent Shear strain for ϕ = 105	Equivalent Shear strain for ϕ = 120	Equivalent Shear strain for ϕ = 135
1	0	2.310	1.772	1.333	0.957
2	10	2.201	1.710	1.299	0.940
3	20	2.109	1.657	1.270	0.925
4	30	2.032	1.611	1.245	0.914
5	40	1.966	1.573	1.226	0.908

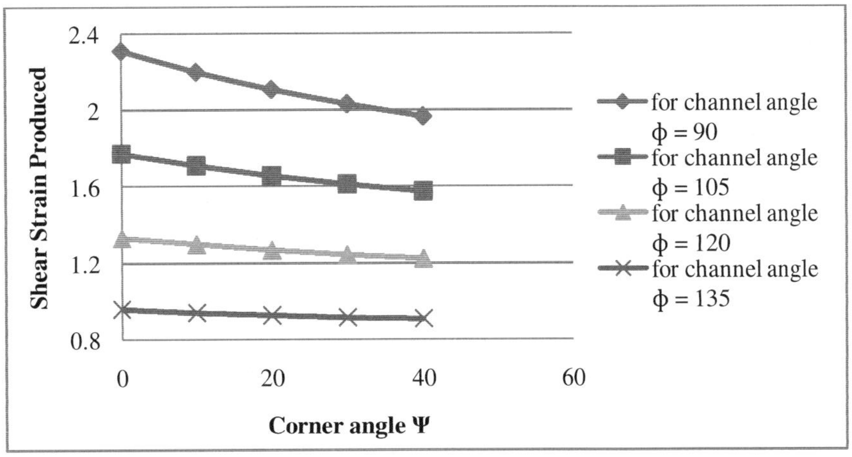

Fig. 5. Effect of corner angle on shear strain produced when n = 2

3.3 Effect of number of passes (n)

As per the equation 2, shear strain increases with the number of passes through the die but as per earlier study from other researchers, after 3 to 4 passes the further refinement of grains are not possible [27-30]. Sample of Al-TiO2 was prepared with TiO2 as 20% by weight and microstructure was analyzed under SEM after 0, 1, 2 and 3 passes. Figure 6(a-d) showed the optical microstructure of ECAPed experimental material up to 3 passes with magnification at 500X. The testing samples were prepared and micrographs were observed at room Temperature of 27°C. The etchant used was Keller's Reagent which is a mixture of nitric acid, hydrochloric acid and hydrofluoric acid, used to etch aluminum alloys to reveal their grain boundaries and orientations. Figure 6(a) showed the micrograph of as received material where dark spots were visible as alloying elements and Figure 6(b) explained that the grains were significantly refined after single pass of ECAP. Further Figure 6(c) illustrated that the grains were aligned in one direction with more reduction in grain size after second pass and Figure 6(d) showed the further refinement of grains with significantly homogeneous and equiaxed grain structure.

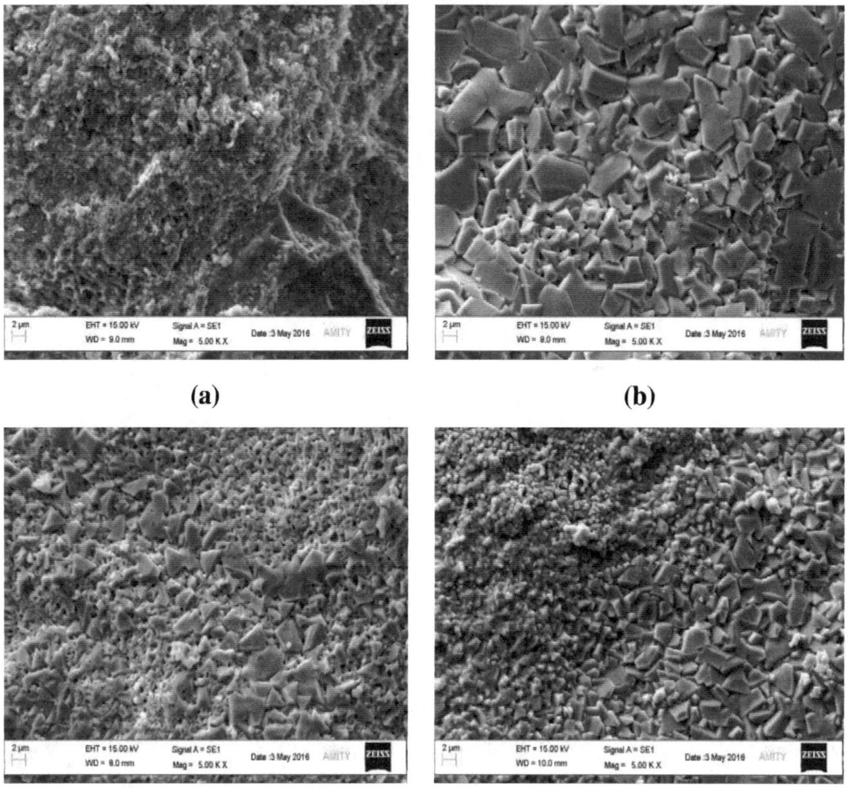

Fig. 6. micrographs of specimen after (a) 0, (b) 1, (c) 2and (d) 3 ECAP passes

3.4 Effect of Coefficient of Friction

Friction between sample and workpiece plays an important role in defining the homogeneity in microstructure and loading rate. F. Djavanroodi [32], states in his study that corner gap can be eliminated by high coefficient of friction which is basically formed at intersection of channels but the limitation is that it leads to increase of loading rate too. Atul Dayal [33], also reviewed the effect of friction and had made some conclusions that, there is almost no effect of friction on hard material's homogeneity but large effect can be seen in softer materials. More the friction in soft material, high the homogenous structure.

3.5 Enhancement in hardness of material

The sample of Al-TiO2 was tested for Vickers hardness test after several number of passes. There was 30% increment in VHN from as zero pass to 4 passes of ECAP. From 189 VHN to 245.7 was noted in the experiment. Figure 9 shows the graph of increment in VHN after each pass in ECAP.

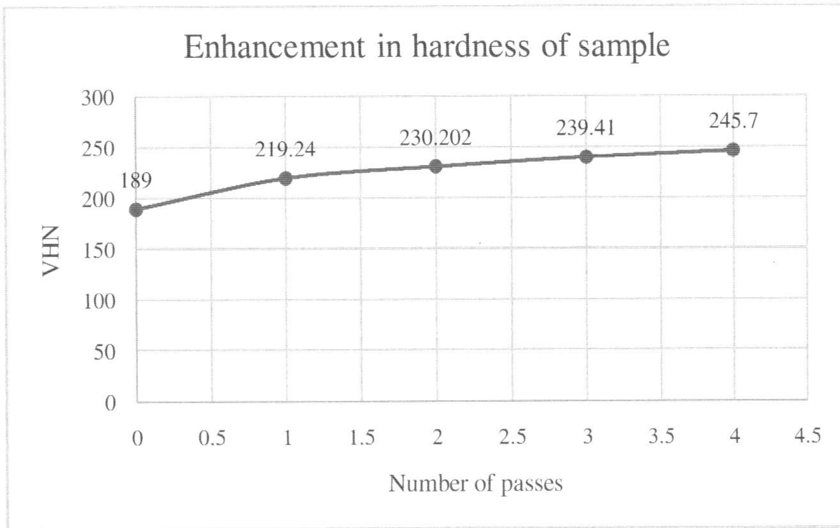

Fig. 7. Shows the enhancement of VHN with each pass of ECAP

3.6 Enhancement in tensile strength of material

The sample of Al-TiO2 was tested for tensile strength on UTM machine after several number of passes and it is been concluded that from 145MPa there was increment of 36% up to 197.11MPa. There is maximum enhancement after 1 pass i.e. 12% and decreasing gradually with number of passes. It is noted that there is 5% increment in 4th pass. Figure 10 shows the graph of increment if tensile strength with number of passes.

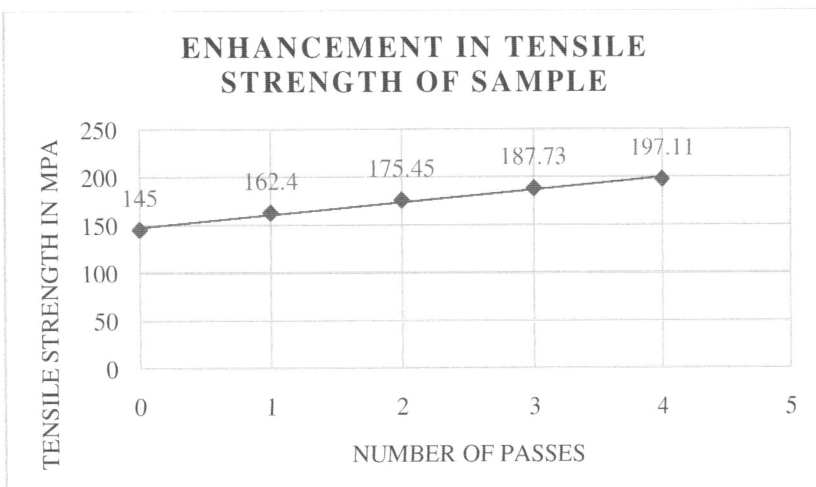

Fig. 8. Shows the enhancement of tensile strength with each pass of ECAP

4 Conclusion

The analysis was made to find out the effect of different die design parameters on grain refinement of materials. The following conclusion can be made regarding die design parameters: The channel angle, ϕ, is the most significant factor since it dictates the total strain imposed in each pass. A better strain homogeneity distribution has been obtained with less magnitude die channel angle close to 90°. The corner angle has less impact on strain distribution but it is used for smooth flow of material through the main deformation zone and filling the corner gap. The number of passes is important for producing the shear stain and results show that around 4 to 6 passes are required for optimum results for refining the grains. High coefficient of friction helps in minimizing the corner gap and better homogeneity in microstructure for softer material but high friction will increase pressing load requirement. Hardness of Al-TiO2 is increased 30% after 4th pass of ECAP upto 245.7 VHN. Tensile strength of Al-TiO2 is increased 36% after 4th pass of ECAP upto 197.11 MPa.

5 References

[1] HU H, ZHANG D, YANG M and DENG M 2011 Grain refinement in AZ31 magnesium alloy rod fabricated by extrusion-shearing severe plastic deformation process *Trans. Nonferrous Met. Soc. China - TRANS NONFERROUS Met. SOC CH*21 243–9

[2] Rosochowski A and Olejnik L 2007 Finite element simulation of severe plastic deformation processes *Proc. Inst. Mech. Eng. Part L-journal Mater. Appl. - PROC INST MECH ENG L-J MATER***221** 187–96

[3] Esmailzadeh M and Aghaie-Khafri M 2012 Finite element and artificial neural network analysis of ECAP *Comput. Mater. Sci.***63** 127–133[4] Szytula A and Leciejewicz J 1989 *Handbook on the Physics and Chemistry of Rare Earths* vol 12, ed K AGschneidner Jr and L Erwin (Amsterdam: Elsevier) p 133

[4] Pan Z, Xu F, Mathaudhu S, Kecskes L, Yin W, Zhang X, Hartwig K and Wei Q 2012 Microstructural evolution and mechanical properties of niobium processed by equal channel angular extrusion up to 24 passes *Acta Mater.***60** 2310–2323

[5] Kuhn T 1998 Density matrix theory of coherent ultrafast dynamics *Theory of Transport Properties of Semiconductor Nanostructures(Electronic Materials* vol 4*)* ed E Schöll (London: Chapman and Hall) chapter 6 pp 173–214

[6] Shaban M and Eghbali B 2014 Finite element simulation of cross equal channel angular pressing *Comput. Mater. Sci.***74** 124–128Ping

[7] Li P, Xue K-M, Wang X and Qian C 2014 Refinement and consolidation of pure Al particles by equal channel angular pressing and torsion *Trans. Nonferrous Met. Soc. China***24** 1289–1294

[8] Faraji G, Ebrahimi M and BushroaA R 2014 Ultrasonic assisted tubular channel angular pressing process *Mater. Sci. Eng. A***599** 10–15

[9] Kocich R, Macháčková A and Andreyachshenko V 2015 A study of plastic deformation behaviour of Ti alloy during equal channel angular pressing with partial back pressure *Comput. Mater. Sci.***101**

[10] Jafarlou D, Zalnezhad E, Ezazi A, Mardi N A and Hassan M A 2015 The application of equal channel angular pressing to join dissimilar metals, aluminum alloy and steel, using an Ag-Cu-Sn interlayer *Mater. Des.***87**

[11] Frint S, Hockauf M, Frint P and Wagner M 2016 Scaling up Segal's principle of Equal-Channel Angular Pressing *Mater. Des.***97**

[12] Agwa M, Ali M and Al-Shorbagy A 2016 Optimum processing parameters for equal channel angular pressing *Mech. Mater.***100**

[13] Agarwal K, Tyagi R, Chaubey V and Dixit A 2020 Comparison of different methods of Severe Plastic Deformation for grain refinement Comparison of different methods of Severe Plastic Deformation for grain refinement Severe plastic deformation SPD High Pressure Torsion HPT

Equal Channel Angular Pressing ECAP Cycling Extrusion and Compression CEC Accumulative Roll Bonding ARB Multidirectional Forging MF Twist Extrusion TE Repetitive Corrugation and Straightening RCS *IOP Conference Series: Materials Science and Engineering* vol 691

[14] Tohidlou E and Bertram A 2017 Effect of initial orientation on subgrain formation in nickel single crystals during equal channel angular pressing *Mech. Mater.* **114**

[15] Borodin E N and Bratov V 2018 Non-equilibrium approach to prediction of microstructure evolution for metals undergoing severe plastic deformation *Mater. Charact.* **141** 267–78

[16] Cao Y, Ni S, Liao X, Song M and Zhu Y 2018 Structural evolutions of metallic materials processed by severe plastic deformation *Mater. Sci. Eng. R Reports* **133** 1

[17] Segal V 2018 Review: Modes and Processes of Severe Plastic Deformation (SPD) *Mater. (Basel, Switzerland)* **11** 1175

[18] Nakashima K, Horita Z, Nemoto M and Langdon T 1998 Influence of channel angle on the development of ultrafine grains in equal-channel angular pressing *Acta Mater.* **46** 1589–99

[19] Agarwal K, Tyagi R K and Dixit A 2019 Theoretical analysis of equal channel angular pressing method for grain refinement of metals and alloys *Mater. Today Proc.* **25**

[20] Langdon T, Furukawa M, Nemoto M and Horita Z 2000 Using equal-channel angular pressing for refining grain size *JOM* **52** 30–3

[21] Kim H, Seo M and Hong S I 2001 Plastic deformation analysis of metals during equal channel angular pressing *J. Mater. Process. Technol.* **113** 622–6

[22] Kim H, Seo M and Hong S I 2002 Finite element analysis of equal channel angular pressing of strain rate sensitive metals *J. Mater. Process. Technol. - J MATER Process TECHNOL* **130** 497–503

[23] Alkorta J and Gil Sevillano J 2003 A comparison of FEM and upper-bound type analysis of equal-channel angular pressing (ECAP) *J. Mater. Process. Technol.* **141** 313–8

[24] Luis-Pérez C, Luri-Irigoyen R and Gastón-Ochoa D 2004 Finite element modelling of an Al–Mn alloy by equal channel angular extrusion (ECAE) *J. Mater. Process. Technol. - J MATER Process TECHNOL* **153** 846–52

[25] Semiatin S, Salem A and Saran M 2004 Models for severe plastic deformation by equal-channel angular extrusion *JOM* **56** 69–77

[26] Olejnik L and Rosochowski A 2005 Methods of fabricating metals for nano-technology *Bull. Polish Acad. Sci. Tech. Sci.* **53** 413–23

[27] Xu S, Zhao G, Ma X and Ren G 2007 Finite element analysis and optimization of equal channel angular pressing for producing ultra-fine grained materials *J. Mater. Process. Technol. - J MATER Process TECHNOL* **184** 209–16

[28] Agarwal K, Tyagi R K and Dixit A 2019 Theoretical analysis of equal channel angular pressing method for grain refinement of metals and alloys *Mater. Today Proc.* **25**

[29] Spirdione J and Ghonem H 2017 Dynamic Flow Stress of Fine Grain Material Processed Using Equal Channel Angular Pressing *Mater. Sci. Eng. A* **698**

[30] Muthiah S, Pillai R, Pai C and Brahmakumar M 2006 Equal channel angular pressing of pure aluminium - An analysis *Bull. Mater. Sci.*

[31] Anon 08 Book Chapter Springer Int Conf ICMET19, Galgotia Evolution of microstructure through various techniques of Severe Plastic Deformation.pdf

[32] Djavanroodi F and Ebrahimi M 2009 Effect of die channel angle, friction and back pressure in the equal channel angular pressing using 3D finite element simulation *Mater. Sci. Eng. A* **527**

[33] Dayal A, Hans Raj K and Sharma R S 2018 ECAP Die Design for Minimising Corner Gap *Mater. Today Proc.* **5** 1686–90

Prognostics and health management of induction motor by supervised learning classifiers

R Jigyasu[1], V Shrivastava[1], S Singh[1]

[1]Department of Electrical and Electronics Engineering, National Institute of Technology, Delhi-110040, India

Corresponding author's email- rajvardhan@nitdelhi.ac.in

Abstract. Current occasions are tied in with being modern and bringing down the expense of working the framework. Albeit the underlying interest in the framework can't be changed, which is fixed, the methodology is to bring down the expense of activity of the machine with the goal that the support cost is limited since the chance of disappointment will be lower. The machine might be in a distant area where physically checking the status or performing upkeep won't be effective. Accordingly, to improve execution and increment plant profitability, predictive support can be planned in such a way that numerous machines can be overhauled simultaneously. What's more, this will expand the general uptime of the plant all in all. For most machines, condition monitoring is a custom-fitted way to deal with predictive support. Also, late patterns show that AI is the quickest and most proficient approach to accomplish the ideal answer for this issue. In this work, predictive maintenance is done for two major occurring faults broker rotor bar and bearing faults. Time-domain analysis has been used to develop hundred feature vectors, which are utilized for training, testing, and validation. The best suitable classifier model to predict faults is developed after comparison of Artificial Neural Network, Ensemble, and Decision Tree smart classifiers based on model accuracy. Out of these, the maximum accuracy is obtained by Artificial Neural Network.

1. Introduction

With industrialization on one side, demand for human resources got reduced, but on the other side era of development had started. People started learning about the new technologies and their journey of skilled labor from just a worker started. With this advancement use of machines, especially 3 phase machines [1], increased because of their reliability, ruggedness, robustness, and less maintenance, etc. advantages. But as said by Heraclitus, "nothing is permanent in this world except change," machines also degrades, and little maintenance is required from time to time. Only repairing machines as possible during the initial days after loss, but with the advancement in technology, predictive maintenance [2] came into existence. With these preventive techniques, the loss of time, money, and lives got reduced to nearly no value.

There are different methods in this preventive maintenance, such as current signature analysis (MCSA), vibration analysis, acoustic analysis, thermal analysis, etc. [3]. These techniques are also known as monitoring techniques. In [4] MCSA technique,[5] vibration,[6] acoustic, [7] thermal analysis, and many more in [8-10] were used for the recognition of various kind of faults such as broken rotor bar (BRB), eccentricity, stator, bearing, etc. faults. According to [11], the percentages of occurrence of different faults are given in Figure 1.

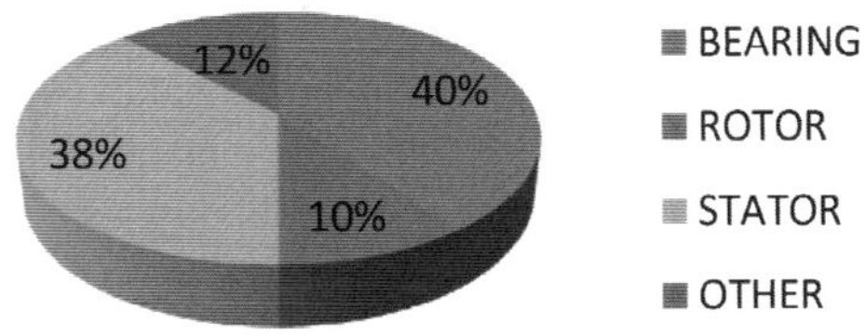

Figure 1. Percentage occurrence of faults.

There are several signal processing techniques available for preventive maintenance. Root mean square (RMS) Calculation, where RMS vibration velocity indicates vibration severity and RMS of stator current indicate motor loading. Time-Domain (TD) Analysis uses TD features such as RMS, Variance, Kurtosis, Peak value, Skewness, Median, etc. Frequency Analysis- FFT, which includes frequency domain (FD), features like average frequency (AF), median frequency (MF), Lower Band Power (LBP), Upper Band Power (UBP), etc. Time-Frequency Analysis, which is a combination of TD and FD analysis but better than them because of scaling and windowing techniques in it, and Stator Current Park's vector in which Stator faults diagnosed very effectively and works best in steady-state. This technique is a vector sum of 3 phase stator currents. For fault classification and detection, there are several classification methods such as an ANN, Support vector machine (SVM), Nearest Neighbour, Decision tree, etc. The above methods are used to identify different kinds of faults, such as bearing, rotor, eccentricity, soft foot, etc. For example, in [12], automatic recognition and classification of rotor cage defects in squirrel cage Induction motors (IM) took place, and 98.33% accuracy was achieved. In [13], 1-BRB, 2-BRB, and 3-BRB faults and 10% stator winding faults due to short circuits are diagnosed successfully using the SVM technique. In [14] Diagnosis of bearing failures has been done Using Weighted K-Nearest Neighbor, and an accuracy of 100% was achieved using Mahalanobis distance metric. In [15], the decision tree algorithm and vibration-based fault diagnosis model of a monoblock centrifugal pump with 100% accuracy were achieved.

This paper is divided into 6 sections starting from the beginning dedicated to condition monitoring and its applications. After that used techniques; Artificial Neural Network (ANN), Ensemble and Decision Tree, experimental unit, and results obtained in this paper are discussed, ended with a conclusion.

2. Technique Used

This paper focuses on the finding of the best classifier for induction motor current signals. And in this process, several techniques are combined to carry out the overall results. These techniques are validation scheme, ANN classifier, Ensemble classifier, and Decision Tree classifier which are explained below.

2.1. Artificial Neural Network

It is genuinely said that the working of ANN takes its foundations from the neural organization present in the human brain. ANN works on something referred to as Hidden State. The NN takes input information, train themselves to perceive patterns found in the information, and afterward predict the output for a new set of data. The ANN comprises of multilayer structure, where the signature goes through from between layers. These layers are made up of nodes called neurons. ANNs is capable to be trained and demonstrate non-straight and composite connections; it can sum up the model and anticipate on inconspicuous information. Numerous investigations have indicated that ANNs can more readily show heteroskedasticity. It is accepted that neural models having just two layers and sigmoid activation function can estimate any design boundary to subjective exactness [16].

The ANN has a vast range of applications such as in [17] ANN is used for translation of Sanskrit language to English and results are compared with Support Vector Machine and Naïve Bayes classifiers. In [18] ANN is used for predicting an alcohol user, this model achieved an accuracy of

98.7% which is quite good. In [19] model for the detection of the Brain, the tumor is developed with an accuracy of 99% using an advanced version of neural network i.e. Convolution Neural Network.

2.2. Decision Tree

A Decision tree is a classifier communicated as a recursive segment of the example space [20]. In a Decision tree, each inside node parts the case space into at least two sub-spaces as per a specific discrete capacity of the input attributes values. This method of classification deals with the impurity present in the sample of the data, here impurity means deviation from the desired value. And based on the level of impurity, each data sample from the dataset is divided into subcategories of a lesser number of impurities. Indecision trees classifier the method is slightly complex as each data sample has to be compared multiple times with the node available in a tree. The node of a tree is a base for the division of the data sample. For 2 class problem of the decision tree following equations are used:

$$\text{InformationGain(IG)} = -\left(\frac{J}{J+K} * \log_2\left(\frac{J}{J+K}\right) + \frac{K}{J+K} * \log_2\left(\frac{K}{J+K}\right)\right) \quad (1)$$

$$\text{Entropy(E(A))} = \sum_{i=1}^{v} \frac{J_i + K_i}{J+K} * I(J_i K_i) \quad (2)$$

$$\text{Gain} = \text{IG-E(A)} \quad (3)$$

In the above conditions, J and K are the quantities of potential results in entire objective examples; v is the number of tests in a single section. This gain chooses the root node in the grouping tree.

2.3. Ensemble

This technique works on the voting principle. There are two types of Ensemble Classifiers first homogeneous type and other heterogeneous types [21]. In homogenous type, the output is decided using the same type of classifier but the algorithm followed will be different. And heterogeneous type the output is decided using different classifiers such as decision tree, nearest neighbor, random forest, etc. The training and testing are done using the bagging and boosting techniques. Basic architecture is shown in Figure 2.

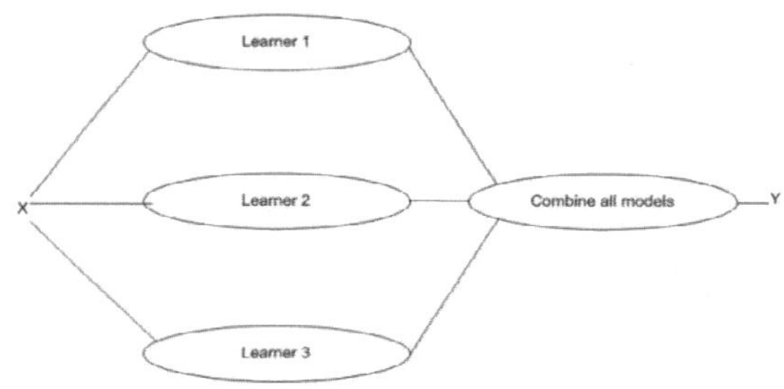

Figure 2. The basic architecture of the Ensemble Classifier

3. Experimental Setup

The test machine is a 3 phase squirrel cage induction motor rated at 1 hp (Table 1 shows the motor specifications). The instrumentation system is mounted close to the drive end bearing shown in figure 3. The various conditions shown in figure 4 introduced in the induction motor are described as follows: 1) healthy; 2) with outer race fault in bearing; 3) with inner race fault in bearing (Table 2 shows the bearing details); 4) 1 Broken Rotor Bars; 5) 3 Broken rotor bars. The current transducer LA 55P is used to alter the current signal into a scope of the MYDAQ data acquisition device. The fault is delivered on the bearing by cutting a 2mm distance across the two sides of the bearing. Broken rotor bar fault has been made by exhausting openings of 6mm distance over the rotor to eliminate the rotor bar.

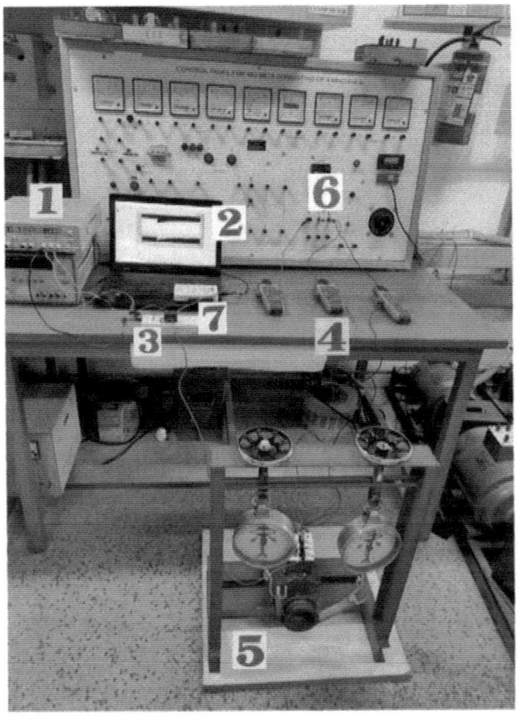

(1)DC Supply (2)Laptop (3)Current Sensor (4)Clamp Meters (5)3-phase IM loading arrangement (6)3-phase supply (7)Data Acquisition System(MYDAQ)

Figure 3. Experimental Setup

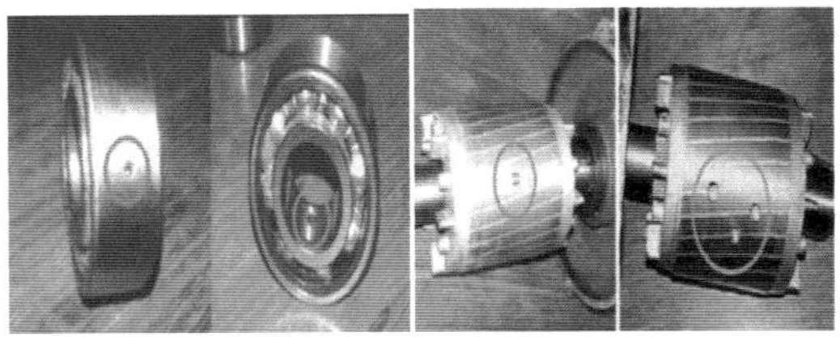

(a) (b)

Figure 4. (a) Bearing Defects; (b) Rotor defects

Table 1. Motor Specifications

Sl. No.	Parameters	Description
1	Company Name	Bharat Bijlee
2	Type	IE2 High Efficiency-2H (3 ph. SCIM)
3	Serial Number	U1S54767|2H08453
4	Duty Type	S1 (Continuous)
5	Efficiency	79.6 %
6	Bearing Number	Drive End: 6004 2Z Non-drive End: 6004 2Z
7	Motor Power	0.75 kW/1HP
8	Frequency	50 Hz ±5%
9	Supply Voltage	415 V±10%
10	Power Factor	0.77
11	Speed	1410 RPM
12	Insulation Class	F
13	Temperature	50 °C (temperature rise: B)
14	Protection Type	IP55

Table 2. Bearing Specifications

Sl. No.	Parameters	Description
1	Bearing Number	6004 2Z
2	Type	Shielded Deep Groove Ball Bearing
3	Contact Angle	0°
4	Inner Race Diameter	20 mm
5	Outer Race Diameter	42 mm
6	Width	12 mm
7	Material	Steel
8	Cage material	Steel
9	Race type	Plain

IOP Conf. Series: Materials Science and Engineering 1168 (2021) 012006 doi:10.1088/1757-899X/1168/1/012006

4. Signal Processing and Feature Extraction

Features are the qualities of a signal. This paper has utilized time-domain specifications for feature extraction. The technique of this experiment has appeared in Figure 5. Features are the qualities of a specific signal. The Time-domain features are mean, variance, kurtosis, skewness, and so on are utilized.

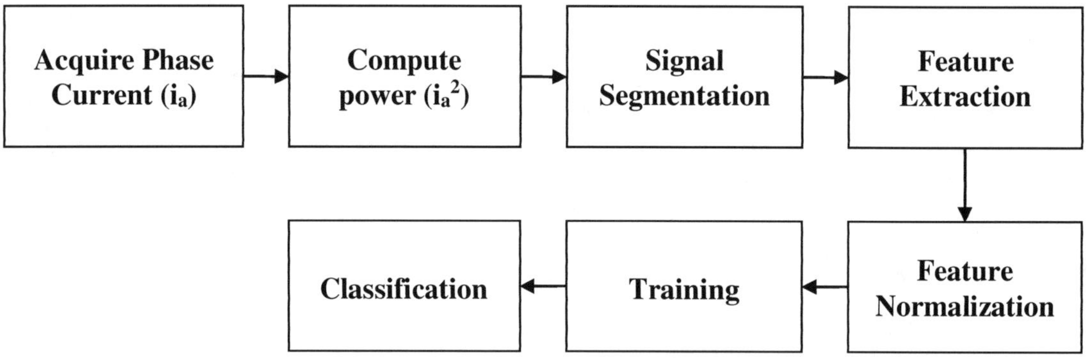

Figure 5. Methodology block diagram.

Before feature extraction signal division happens. In this progression, the got signal was confined into twenty overlapping pieces. All of them contain 8000 samples. After feature extraction normalization is done before utilizing them for training. The normalization was in the degree of [0, 1] except for skewness, which was to the extent of - 1 to 1.

5. Result and Discussion

The smooth running of motor saves from loss of lives and economic disturbances. To classify the faults Artificial Intelligence-based classifiers were used. The output results are shown below in terms of accuracy attained by the classifier with different validation schemes. Three classifiers are compared concerning accuracies attained under different schemes. The classifiers are trained and tested using experimentally acquired data. There are 4 types of faults classified in this database based on the different time-domain features.

5.1. Results obtained using ANN

The basic architecture of ANN is based on the human brain. It is a supervised learning method follower classifier. The power signal was acquired from the healthy and faulty motor and training and testing were performed. Table 3 and Table 4 show the obtained results for different activation functions.

From Table 3. it can be seen that the training functions trainlm trainscg, traincgp, and traincgf based classifiers performed best for no validation case. The Traingdm based classifier achieved the worst percentage accuracy i.e. 22%.

Table 3. Percentage accuracy for different transfer functions without validation.

S.No.	Transfer Function	% Accuracy
1	trainscg	98.7
2	trainbfg	95.6
3	traincgb	97.8
4	traincgf	98.4
5	traincgp	98.5

6	traingda	95.8
7	traingdx	96
8	trainlm	98.9
9	trainoss	97.9
10	trainrp	97.7

The Overfitting problem of classifiers is solved using the validation schemes. From Table 4., the trainlm based classifier performed with a percentage accuracy of 99.1875% with a different number of hidden neurons.

Table 4. Percentage accuracy for a different transfer function with validation.

S.No.	Transfer Function	% Accuracy
1	trainscg	89
2	trainbfg	66.7
3	traincgb	89.6
4	traincgf	79.7
5	traincgp	91.7
6	traingda	89.2
7	traingdx	94.22
8	trainlm	99.18
9	trainoss	82.4
10	trainrp	90.2

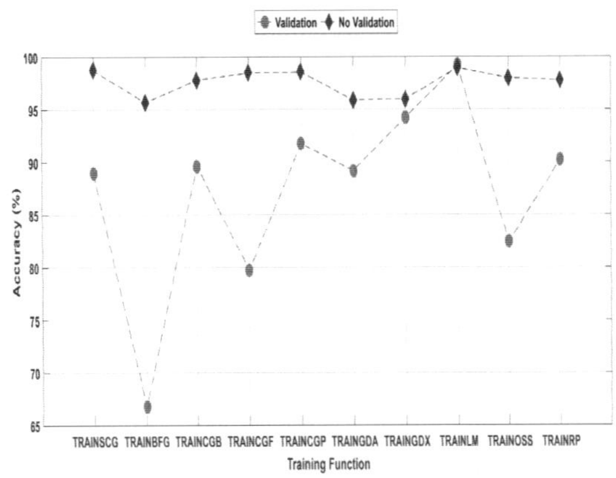

Figure 6. Average accuracies for different training functions.

From Figure 6. the mean percentage accuracies attain by different training functions based classifiers, the trainlm based classifier performed best in both cases. The training and testing of NN were done for a different number of neurons in the hidden layer and Validations. For the validation case, 10% data was used for validation, 70% for training, and 20% for testing. For no validation case, 75% data for training and 25% for testing from total samples. The classification was done using 'patternnet' NN having a different number (2, 4, 6, 8, 10, 12, 14, 16) of hidden layer neurons.

5.2. Results obtained using ensemble classifier
This category of classifiers is a special case, where the classification approach of two classifiers is combined and categorized as an ensemble classifier. Table 5. furthermore, Table 6. shows distinctive accuracies acquired from the Ensemble Classifier on different validation schemes. From the outcomes, it tends to be seen that the precision acquired without validation achieved maximum accuracy of 74.66% with the Subspace Discriminant algorithm while validation achieved 98.9% with the Bagged type algorithm.

Table 5. Percentage accuracies using ensemble classifier without validation.

S.No	Type	% Accuracy
1	Boosted	43.66
2	Bagged	69.66
3	Subspace Discriminant	74.66
4	Subspace KNN	61.33
5	RUS Boosted Trees	44.66

Table 6. Percentage accuracies using ensemble classifier with validation.

S.No	Type	% Accuracy
1	Boosted	73.33
2	Bagged	98.9
3	Subspace Discriminant	79
4	Subspace KNN	98.8
5	RUS Boosted Trees	64.33

5.3. Results obtained using Decision Tree
It is the way toward picking a game-plan from among choices to accomplish the ideal objective. A Decision tree is a graphical portrayal of potential answers for a choice dependent on specific conditions. It's known as a Decision tree since it begins with a solitary box (or root), which at that point diverges into a few arrangements, actually like a tree. A Decision tree has just blasted nodes (parting ways) however no sink nodes (combining ways). The results of with and without validation schemes have appeared in Table 7., and Table 8. Table 7 shows the percentage accuracy achieved by various Decision trees without validation. The most accomplished percentage accuracy was 62.33% with a complex algorithm. It can be seen that the accuracy obtained with validation is better than no validation, which is 83.66% with complex and medium both.

Table 7. Percentage accuracies using decision tree classifier without validation.

S.No	Type	% Accuracy
1	Complex	62.33
2	Medium	62.33
3	Simple	51.66

Table 8. Percentage accuracies using decision tree classifier with validation.

S.No	Type	% Accuracy
1	Complex	83.66
2	Medium	83.66
3	Simple	61.66

6. Conclusion
A methodology of fused techniques has been presented in this work, to aid in predictive maintenance of induction motor by developing an instrument that performs continuous condition monitoring using the current signature of the machine. There are several factors on which the classifier's accuracy

depends for example data size, features, data quality means low data redundancy, etc. This paper finds out the way of detection as well as diagnosis of two faults BRB and bearing. Out of the three compared techniques, the ANN performed best with an accuracy of 99.18%. For the Ensemble classifier maximum attained accuracy was 98.9% and for the decision tree its 83.66%. In the case of ANN, for trainlm, the accuracy was nearly invariable for a different number of processing elements. However, the average accuracy was improved for different training functions with increasing network size with processing elements. For future scope, more techniques can be compared along with optimization in feature selection.

7. References

[1] Lei, Y., Yang, B., Jiang, X., Jia, F., Li, N. and Nandi, A.K., 2020. Applications of machine learning to machine fault diagnosis: A review and roadmap. *Mechanical Systems and Signal Processing*, *138*, p.106587.

[2] West, J.G., 1994. DC, induction, reluctance, and PM motors for electric vehicles. *Power Engineering Journal*, *8*(2), pp.77-88.

[3] Lu, B., Durocher, D.B. and Stemper, P., 2009. Predictive maintenance techniques. *IEEE Industry Applications Magazine*, *15*(6), pp.52-60.

[4] Jigyasu, R., Sharma, A., Mathew, L. and Chatterji, S., 2018, June. A review of condition monitoring and fault diagnosis methods for induction motor. In *2018 Second International Conference on Intelligent Computing and Control Systems (ICICCS)* (pp. 1713-1721). IEEE.

[5] Zarei, J., Tajeddini, M.A. and Karimi, H.R., 2014. Vibration analysis for bearing fault detection and classification using an intelligent filter. *Mechatronics*, *24*(2), pp.151-157.

[6] Glowacz, A., Glowacz, W., Glowacz, Z. and Kozik, J., 2018. Early fault diagnosis of bearing and stator faults of the single-phase induction motor using acoustic signals. *Measurement*, *113*, pp.1-9.

[7] Glowacz, A. and Glowacz, Z., 2017. Diagnosis of the three-phase induction motor using thermal imaging. *Infrared Physics & technology*

[8] Kumar, B., Chauhan, Y.K. and Shrivastava, V., 2014, May. Performance analysis of induction motor drive with optimal rotor flux for energy-efficient operation. In *2014 IEEE International Conference on Advanced Communications, Control and Computing Technologies* (pp. 319-322). IEEE.

[9] Kumar, B., Chauhan, Y.K. and Shrivastava, V., 2014. Assessment of a fuzzy logic-based MRAS observer used in a photovoltaic array supplied AC drive. *Frontiers in Energy*, *8*(1), pp.81-89.

[10] Kumar, B., Chauhan, Y.K. and Shrivastava, V., 2012. Efficacy of different rule-based fuzzy logic controllers for induction motor drive. *International Journal of Machine Learning and Computing*, *2*(2), p.131.

[11] Morsalin, S., Mahmud, K., Mohiuddin, H., Halim, M.R. and Saha, P., 2014, May. Induction motor inter-turn fault detection using the heuristic noninvasive approach by an artificial neural network with Levenberg Marquardt algorithm. In *2014 International Conference on Informatics, Electronics & Vision (ICIEV)* (pp. 1-6). IEEE.

[12] Soualhi, A., Clerc, G. and Razik, H., 2012. Detection and diagnosis of faults in induction motor using an improved artificial ant clustering technique. *IEEE Transactions on Industrial Electronics*, *60*(9), pp.4053-4062.

[13] Keskes, H., Braham, A. and Lachiri, Z., 2013. Broken rotor bar diagnosis in induction machines through stationary wavelet packet transform and multiclass wavelet SVM. *Electric Power Systems Research*, *97*, pp.151-157.

[14] Sharma, A., Jigyasu, R., Mathew, L., and Chatterji, S., 2018, May. Bearing fault diagnosis using a weighted k-nearest neighbor. In *2018 2nd International Conference on Trends in Electronics and Informatics (ICOEI)* (pp. 1132-1137). IEEE.

[15] Sakthivel, N.R., Sugumaran, V. and Babudevasenapati, S., 2010. Vibration-based fault diagnosis of monoblock centrifugal pump using a decision tree. *Expert Systems with Applications*, *37*(6), pp.4040-4049.

[16] Bishop, C.M., 1995. *Neural networks for pattern recognition*. Oxford university press.

IOP Conf. Series: Materials Science and Engineering 1168 (2021) 012006 doi:10.1088/1757-899X/1168/1/012006

[17] Koul, N. and Manvi, S.S., 2019. A proposed model for neural machine translation of Sanskrit into English. *International Journal of Information Technology*, pp.1-7.
[18] Kumari, D., Kilam, S., Nath, P. and Swetapadma, A., 2018. Prediction of alcohol abused individuals using artificial neural network. *International Journal of Information Technology*, *10*(2), pp.233-237.
[19] Kalaiselvi, T., Padmapriya, S.T., Sriramakrishnan, P. and Somasundaram, K., 2020. Deriving tumor detection models using convolutional neural networks from MRI of human brain scans. *International Journal of Information Technology*, pp.1-6.
[20] Quiroz, J.C., Mariun, N., Mehrjou, M.R., Izadi, M., Misron, N. and Radzi, M.A.M., 2018. Fault detection of broken rotor bar in LS-PMSM using random forests. *Measurement*, *116*, pp.273-280.
[21] He, W., Zi, Y., Chen, B., Wu, F. and He, Z., 2015. Automatic fault feature extraction of a mechanical anomaly on induction motor bearing using ensemble super-wavelet transform. *Mechanical Systems and Signal Processing*, *54*, pp.457-480.

Simulation of centrally loaded GrFRP and analysis of fixed (both ends) supported stress concentrated JFRP beam

Peeyush Maheshwari[1*], A Misra[1], T S Samant[1], P K Vishnoi[1] and P S Gautam[1]
[1]Department of Mechanical Engineering, College of Technology,
G.B. Pant University of Agriculture and Technology Pantnagar, India
*peeyush.maheshwari@outlook.com

Abstract— Natural fiber polymer composites are generally used in the furniture, home appliances, construction works. Present work is based on the simulation of the Graphite Fiber Reinforced Polymer (GrFRP) composite laminates through numerical model. In simulation we get maximum 7% error with the available experimental results. Analyzed a beam which is made from natural fiber reinforced polymer composites. It consists of eight stacks of natural fiber lamina, a central hole, both the ends supported as fixed and loaded by a uniformly distributed load (UDL). A numerical tool have been developed to analyze the Jute Fiber Reinforced Polymer (JFRP) composites. In this analysis, four different sequencing pattern of lamina has been used. The numerical failure index and failure strength for the different tacking sequences of lamina is determined. The maximum failure strength for the (60°/-60°/60°/-60°)s laminate is 22.51MPa. We found that in the simulation the results of the model are very close to experimental results. Its method of analysis is more convenient and economical over the experimental work. Therefore, the method was found very suitable for modelling and analysis in future.

Keywords—Simulation, Analysis, ANSYS, Composites, JFRP

1. Introduction

From the last several years polymer composites are used in the furniture, home appliances, construction. The composite are fabricated with the help of natural waste or fibers also known as natural fiber polymer composites [1]. These composites are cheap in cost, light in weight and corrosion resistant. Now a days, the analysis of reliability of the composites are expensive because they required specific environment and loading conditions. Therefore the analysis of the natural fiber composite based on the loading conditions. It may be done through the help of numerical tool such as ANSYS, ABAQUS, and FORTRAN [2]. The failure behaviour of the simply supported rectangular plate has been determine through Maximum stress and Tai-Wu failure criteria, when it's loaded in uniaxial direction [3]. Failure load of conoidal shell graphite laminated composite has been investigated through Maximum stress, Maximum strain, Hoffman, Tsai-Hill, Tai-Wu failure criteria, when it's loaded in uniformly distributed load (UDL) in transverse direction [4]. Tensile stress and modulus has been investigated experimentally and calculated with the help of Rule of Mixture for composite [5]. This paper has two part i.e. simulation of model and analysis of jute fiber reinforced polymer (JFRP) composite beam. The JFRP beam has fixed at both the ends with central hole and loaded by uniformly distributed load.

2. Materials and Methods

2.1. Material properties

The Graphite/epoxy and Jute/epoxy lamina elastic and strength properties was taken from Kam *et al.* 1999 and Maheshwari *et al.* 2018. The lamina have isotropic properties. The lamina properties as given in Table1.

IOP Conf. Series: Materials Science and Engineering 1168 (2021) 012007 doi:10.1088/1757-899X/1168/1/012007

2.2. Method for Simulation and Analysis

The method use to simulation and analysis is known as computational method in ANSYS. Finite element analysis is containing three basic steps i.e. pre-processing, processing and post-processing. In first step, pre-processing is require to define the elastic properties, strength properties, boundary and loading condition. In second step, processing ANSYS analyse the condition and solve the problem. In third step, post-processing, we seen different types of results i.e. stress distribution, strain distribution and failure strength or load etc. All the steps are shown in below figure1. Stacking sequences of lamina in the graphite fiber reinforced polymer composites for (0°/90°/0°/90°)s laminate in the figure 2 [8-10].

Table1: Properties of Graphite/epoxy and Jute/epoxy

	Modulus	Graphite/epoxy lamina [6]	Jute/epoxy lamina[7]
Elastic Modulus	E_1 (GPa)	142.50	11.73
	$E_2 = E_3$ (GPa)	9.79	7
Modulus of Rigidity	$G_{12}=G_{13}$ (GPa)	4.72	2.7154
	G_{23} (GPa)	1.19	2.6069
Poision ratio	$\nu_{12} = \nu_{13}$	0.27	0.3250
	ν_{23}	0.25	0.3910
Strength	X_t (MPa)	2193.50	310
	X_c (MPa)	2457.00	310
	$Y_t=Z_t$ (MPa)	41.30	34
	$Y_c=Z_c$ (MPa)	206.80	34
	R (MPa)	61.28	35
	S = T (MPa)	78.78	35

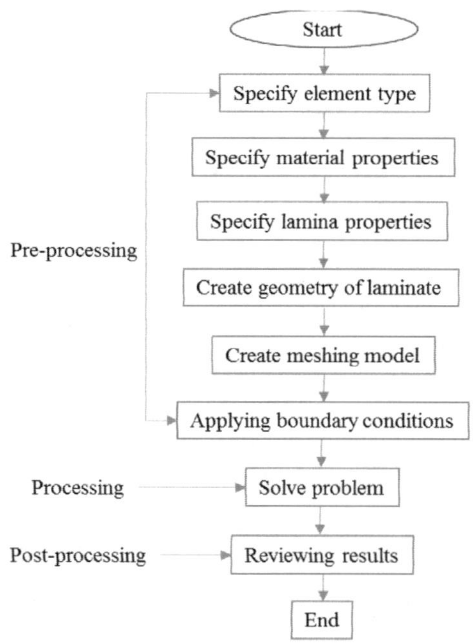

Fig.1. Analysis processing chart [7]

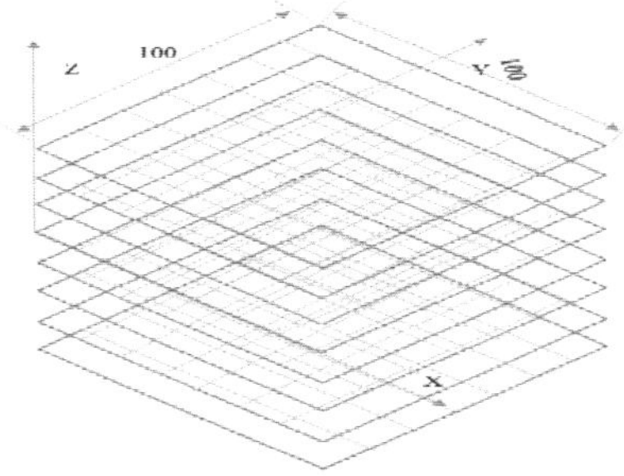

Fig. 2. GrFRC plates stacking sequence (0°/90°/0°/90°)s

3. Results and Discussion

3.1. Simulation

For the simulation of the Graphite Fiber Reinforced Polymer (GrFRP) we design a numerical model through ANSYS. We take the experimental results and theoretical results from Kam et el. for sequence which shown in table3.

Table2. Comparison of ANSYS results with experimental data for (0°/0°/90°/90°)s laminate

Failure Criteria	First Ply Failure Load (N)			Error (%)
	Theoretical[6]	**ANSYS**	**Experimental [6]**	
Max. Stress	229.11	235.84	253.6	7.00
Tsai-Wu	238.6	241.25		4.87

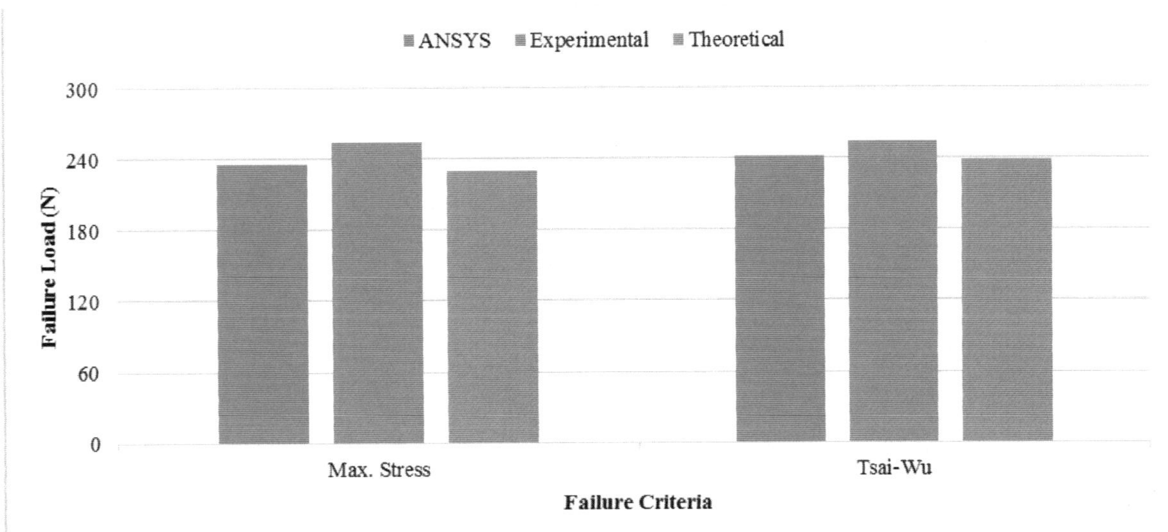

Fig.3. Compression of failure loads for $(0^{\circ}/0^{\circ}/90^{\circ}/90^{\circ})_s$ laminate

From the table2 and figure3, we obtain first ply failure load by ANSYS for Maximum stress and Tsai-Wu failure theories. We find the maximum error through Max. Stress failure criteria and minimum error through Tsai-Wu failure criteria. The value of maximum and minimum error is 7.00% and 4.87% respectively.

In figure 4(a) and (b) shows the stress distribution of laminates for sequence (0°/0°/90°/90°)s and failure index. Maximum failure index for Max. Stress failure criteria and Tsai-Wu failure criteria are 0.00424 and 0.00416 respectively.

Table 3 Comparison of ANSYS results with experimental data for (0°/90°/0°/90°)s laminate

Failure Criteria	First Ply Failure Load (N)			Error (%)
	Theoretical[6]	ANSYS	Experimental[6]	
Max. Stress	290.12	313.67	317.74	1.28
Tsai-Wu	304.13	321.33		1.13

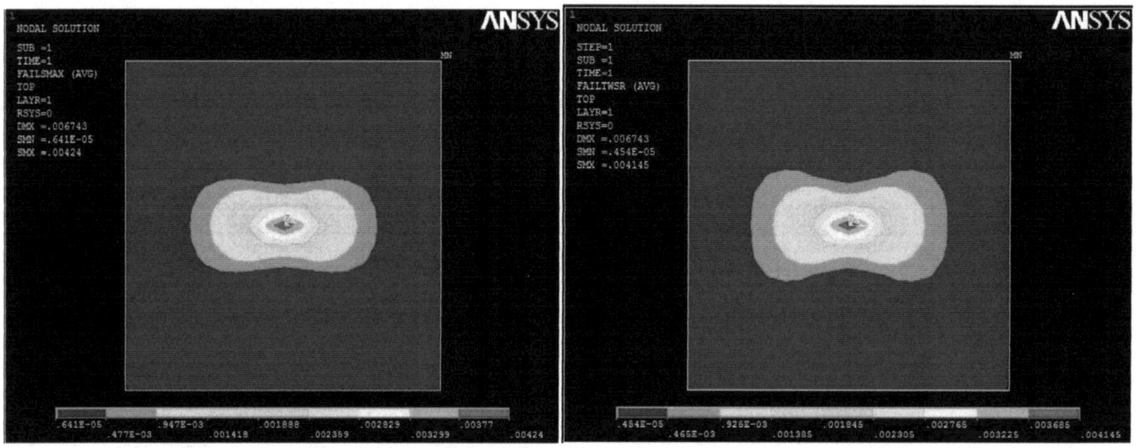

Fig. 4. Failure index for (0°/0°/90°/90°)s laminates (a) Max. Stress failure criteria and (b) Tsai-Wu failure criteria

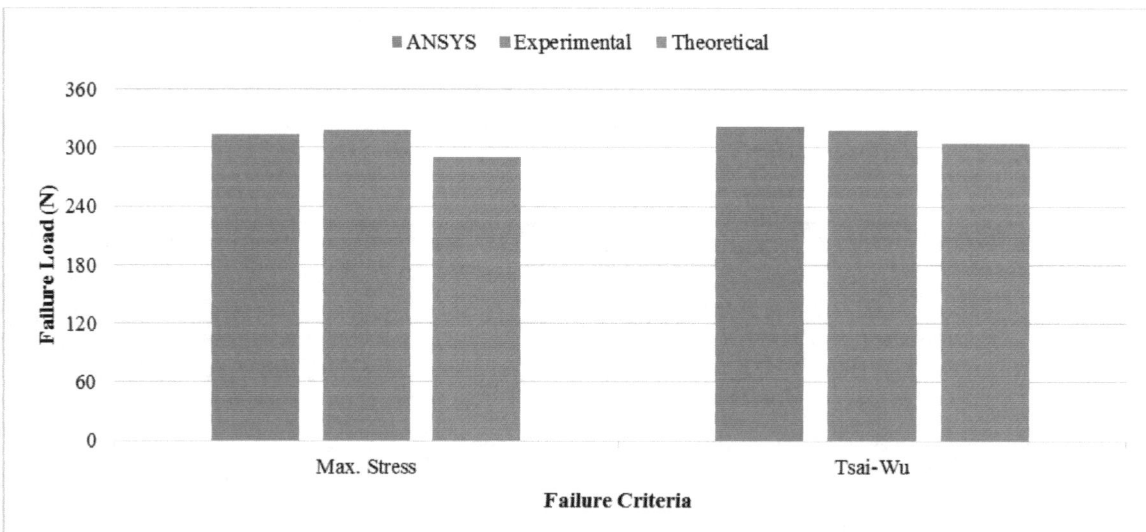

Fig5. compression of failure loads for (0°/90°/0°/90°)$_s$ laminate

From the table3 and figure5, we take the experimental results and theoretical results from Kam et al for GrFRP. We obtain first ply failure load by ANSYS for Maximum stress and Tsai-Wu failure criteria. We find maximum error through Max. Stress failure criteria and minimum error through Tsai-Wu failure criteria. The value of maximum and minimum error is 1.28% and 1.13% respectively.

IOP Conf. Series: Materials Science and Engineering 1168 (2021) 012007 doi:10.1088/1757-899X/1168/1/012007

In fgure 6 (a) and (b) shows the stress distribution of laminates for sequence (0°/90°/0°/90°)s and failure index. Maximum failure index for Max. Stress failure criteria and Tsai-Wu failure criteria are 0.003188 and 0.003188 respectively.

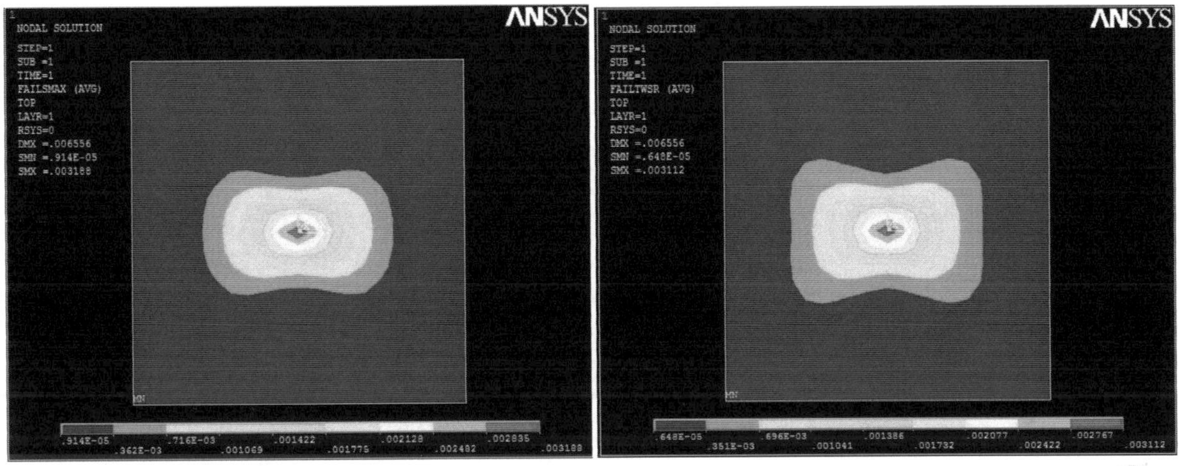

Fig. 6. Failure index for (0°/90°/0°/90°)s laminates (a) Max. Stress failure criteria and (b) Tsai-Wu failure criteria

3.2. Analysis of Jute fiber reinforced polymer laminated composite beam (both ends fixed) with central hole

The model have been design for specific boundary and loading condition as shown in figure7. The dimensions are length- 100mm, height- 50mm, thickness- 1mm while the diameter of central hole is 20mm.

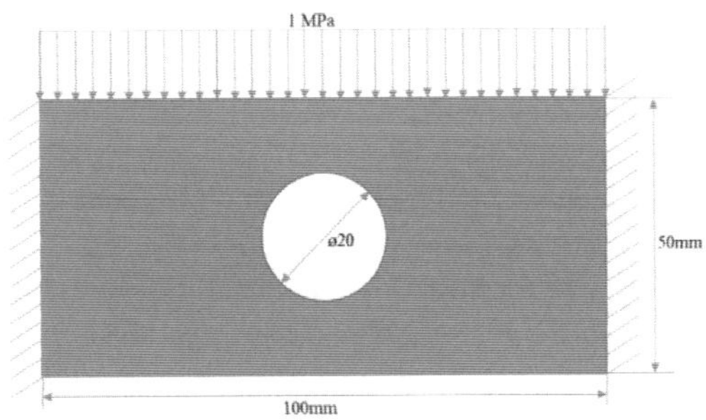

Fig. 7 JFRP laminated composite beam with central hole subjected to uniformly distributed load

In Figure 8 and figure 9 shows comparison of failure index for different stacking sequences of laminates on the basis of Maximum stress and Tsai-Wu failure theories respectively. From figures it have been observed that for (θ°/-θ°/ θ°/-θ°)s, (θ°/90-θ°/ θ°/90-θ°)s and (θ°/θ°/ θ°/θ°)s laminate failure index decreases when the value of θ° increases upto 60°, 60° and 75° respectively. In figures it have been observed that approximate no change in failure index due to increase in ply angle upto 90° for (0°/θ°/ 0°/θ°)s laminate.

In Figure 10 and figure 11 shows comparison of failure strength for different stacking sequences of laminates on the basis of Maximum stress and Tsai-Wu failure theories respectively. From figures it have been observed that for (θ°/-θ°/ θ°/-θ°)s, (θ°/90-θ°/ θ°/90-θ°)s and (θ°/θ°/ θ°/θ°)s laminate failure strength increase when the value of θ° increases upto 60°, 60° and 75° respectively. In figures it have been observed that approximate no change in failure strength due to increase in ply angle upto 90° for (0°/θ°/ 0°/θ°)s laminate.

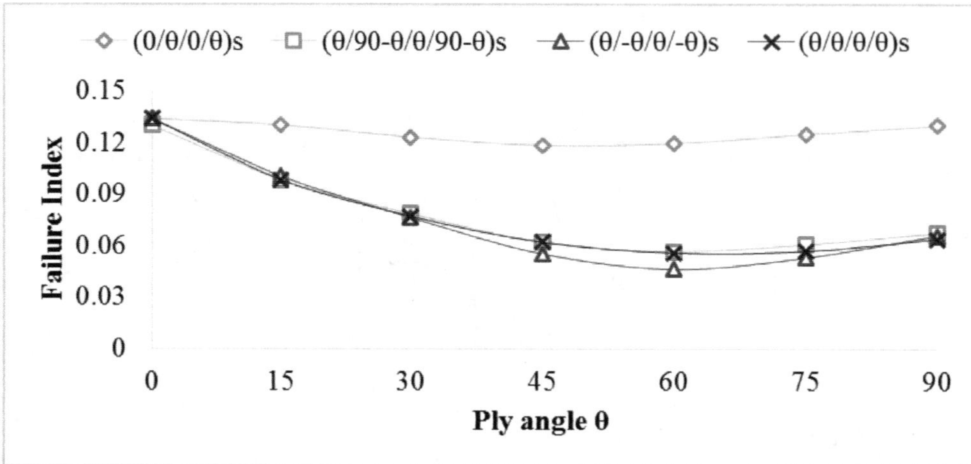

Fig. 8. Failure index (Max. Stress failure criteria) for different sequences of laminates

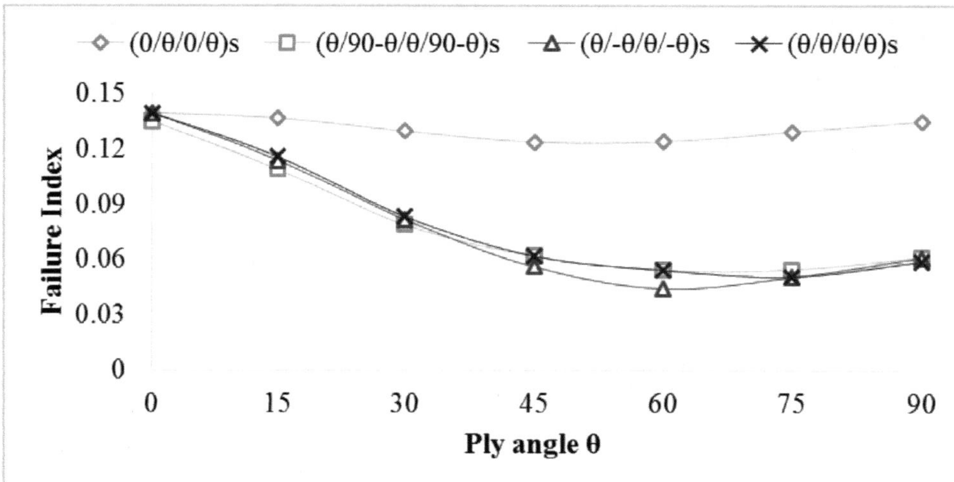

Fig. 9. Failure index (Tsai-Wu failure criteria) for different sequences of laminates

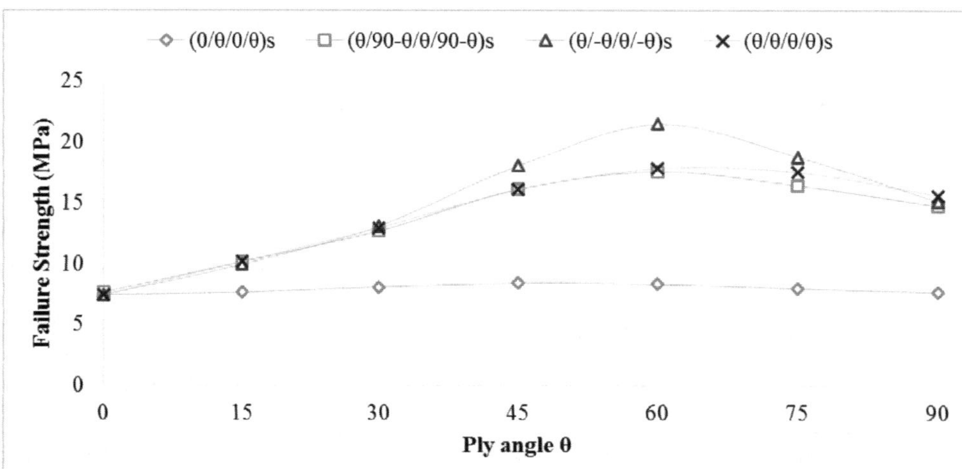

Fig. 10. Failure Strength (Max. Stress failure criteria) for different sequences of laminates

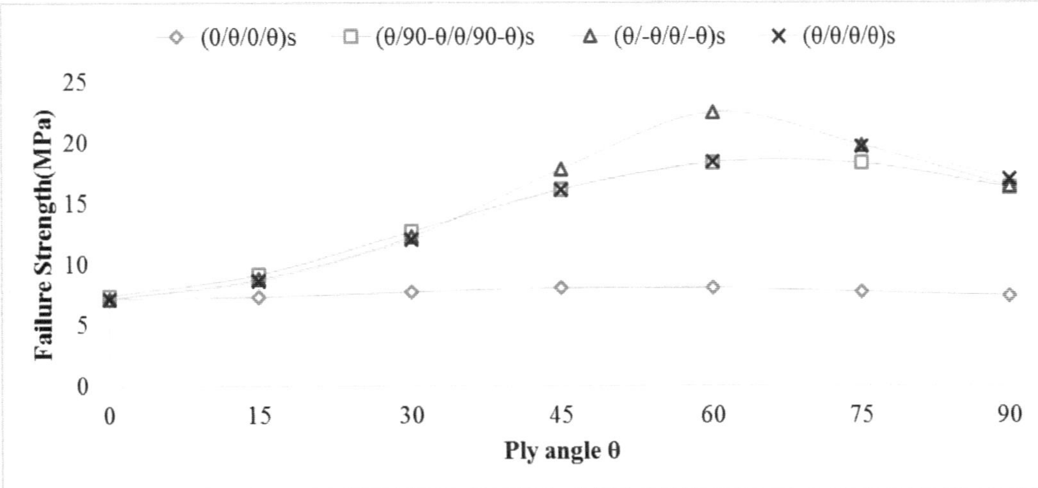

Fig. 11. Failure Strength (Tsai-Wu failure criteria) for different sequences of laminates

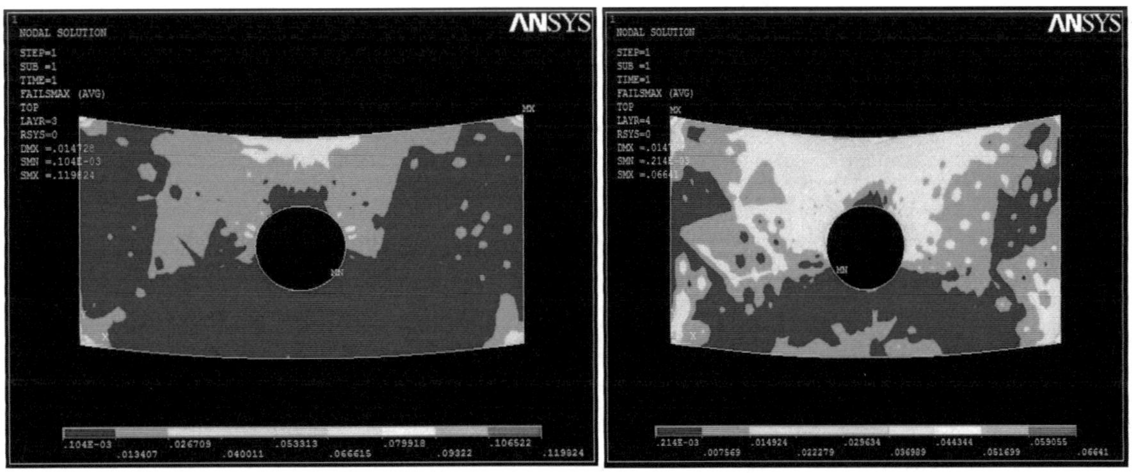

Fig. 12. Stress Distribution of (0°/60°/0°/60°)s laminate in (a) 0° layer (b) 60° layer

In figure 12(a) and 12(b) shows the stress distribution of (0°/60°/0°/60°)s laminate in 0° layer and 60° layer respectively.

Form figure 8 to figure 11 it has be concluded that cross ply laminates or laminates having 90° layers can take higher load comparison to laminates having other layers. Laminates in which layers are oriented at 45° are strongest among all other configurations due to higher values of shear stresses. Under given loading condition it is recommended to use layers with orientation of 45°. Results obtained using maximum stress theory and Tsai-Wu failure theories are very close.

4. Conclusion

The computational technique has been validated for first ply failure load of graphite-epoxy laminate, with the experimental results published in literature [6]. Tsai-Wu failure theory predicts better results than Maximum stress failure theory. Maximum error of 7.00 % has been observed between Max. Stress failure criteria and experimental results for (0°/0°/90°/90°)s laminate. Maximum error of 1.28 % has been observed between Maximus Stress failure criteria and experimental results for (0°/90°/0°/90°)s laminate. A fixed beam containing centre hole, made up of Jute-epoxy composite system has been taken under consideration. A uniformly distributed load of 1MPa acts on the upper face of a beam. Failure index for

different lamination schemes has been calculated. First ply failure index for (0°/45°/0°/45°)s, (60°/30°/60°/30°)s , (60°/-60°/60°/-60°)s and (75°/75°/75°/75°)s laminates exhibit minimum failure index. First ply failure strength for (0°/45°/0°/45°)s, (60°/30°/60°/30°)s , (60°/-60°/60°/-60°)s and (75°/75°/75°/75°)s laminates exhibit maximum failure strength. The maximum failure strength for the (60°/-60°/60°/-60°)s laminate is 22.51MPa.
We found in the simulation the results of model are very close to experimental results. Its method of analysis is more convenient and economical over the experimental work. Therefore, its method is very suitable for modelling and analysis in future.

References

[1] Maheshwari P., Misra A., Singh A. K., Rawat K. S., Verma A. and Singhal K., 2020, Simulation of CoFRP analytical and computational model for various fiber volume fraction and Investigation of the local stresses for different fiber orientation angle. IOP Conf. Series: Material Science and Engineering. 802: 012006.

[2] Nirbhay M., Dixit A., Misra R. K. and Mali H. S., 2014, Tensile test simulation of CFRP test specimen using finite elements. International Conference of Advances in Manufacturing and Materials Engineering, AMME2014 5:267-273.

[3] Rahmani H., Najafi S. H. M. and Ashori A., 2015, Elastic Properties of Carbon Fibre-Reinforced Epoxy Composites. Polymers and Polymer Composites. 23: 475-481.

[4] Bakshi K. and Chkravorty D., 2014, First ply failure study of thin composite conoidal shells subjected to uniformly distributed load. Thin Walled Structures 76:1-7

[5] Wallace C. A., Saha G. C., Afzal M. T. and Lloyd A., 2019, Experimental and computational modeling of effective flexural/tensile properties of microwave pyrolysis biochar reinforced GFRP biocomposites. Composites Part B. 175: 107180

[6] Kam T. Y., Sher H. F. and Chao T. N., 1996, Predictions of deflection and first-ply failure load of thin laminated composite plates via the finite element approach. International Journal of Solids Structure. 33(3):375–98

[7] Maheshwari P., Misra A. and Kumar A., 2018, Investigate the Variation in Elastic Constants and Failure Strength of Lamina with FVF and FOA for JFRP Lamina. International Journal of Applied Engineering Research. 13 157-161.

[8] Barbero E. J., 2007, Finite Element Analysis of Composite Materials, CRC Press, Boca Raton, 225p.

[9] Maheshwari P. and Misra A., 2017, Failure Test Simulation of Graphite Fiber Reinforced Polymer using Finite Element Analysis International Journal of Innovative Trends in Engineering 31:29-33.

[10] Daniel I. M. and Ishai O., 2006, Engineering Mechanics of Composite Materials, Oxford University Press, New York, 463p.

An investigation into Parametric Modelling and FDM based Rapid Prototyping of Lumbar Spine

Manikandan H[1], S Anjani Kumar Ojha[2]

[1] Assistant Professor, Department of Mechanical Engineering,
CMR Institute of Technology, Bengaluru 560037, Karnataka, India.
2 Senior Software Engineer PLM/CAD, GE Power Systems India Pvt Ltd,
Sanand, Ahmedabad 382170.

Abstract. Low back pain is the acute problem faced by workforce in the world today. Four out of five adults experience significant low back pain sometimes during their life. Mostly, low back pain problem can only be solved by operations. Each time doctors carrying out operations need to study the patient spine. The objective of the present work is to generate parametric model of lumbar spine using event driven programming approach of Visual Basic 6.0 and CATIA V5 R10. In parametric modelling, the features are parameterized by giving them meaningful labels and relationships between parameters in the entire model rather than giving only fixed numeric values. The output of the developed model is compared with the specimen samples of lumbar spine. Additive manufacturing using FDM (Fused Deposition Modelling) is used for 3D printing the spine model. Thus, the present work will give an insight into additive manufacturing of patient specific model and analyze the outcome before the clinical procedures.

Keywords: Lumbar spine, parametric modelling, Rapid prototyping, FDM.

1. Introduction

The adult spine typically consists of 33 vertebrae most of which are united by a series of intervertebral articulations to form a firm but flexible shaft. The spinal column is divided into five sections. In cephalad-caudal direction (transverse direction), these are cervical (C1-C7), thoracic (T1-T12), lumbar (L1-L5), five sacral vertebrae forming the sacrum and the four coccygeal vertebrae that fuse together as coccyx in adults. In the sagittal plane the spine has four physiological curves. These curves are convex anteriorly in the cervical and lumbar regions. In the thoracic and sacral regions these curves are convex posteriorly. Curvatures of the cervical and lumbar regions are largely due to the wedge-shaped intervertebral discs. These curves increase the flexibility of the spinal column and augment its shock absorbing capacity. The vertebrae in each region of the spinal column have distinguishing characteristics that reflect the differences in the physiological and biomechanical properties of each region. Intervertebral disc degeneration is widely believed to be a main cause of low back pain. It involves gross structural disruption as well as cell-mediated changes in matrix composition of the disc. Treatment of back pain due to disc degeneration is done either by non-surgical or by surgical means. Lumbar fusion or arthrodesis is one of the most commonly performed operative procedures to get relief from back pain. It is often performed when there is significant degeneration or instability which causes significant low back pain or discomfort. Various methods like a simple fusion with a bone graft, fusion using a bone graft along with instruments like pedicle screws and platting systems, or fusion using interbody cages are being employed to fuse the diseased lumbar segments. Lumbar Interbody Fusion is a common treatment alternative for low back pain. To assist clinical assessment of these procedures and devices, and to study their biomechanical effects on the spinal column, various research techniques are employed. Traditionally, research on spine biomechanics has involved two types of approaches viz.

experimental and analytical. It is difficult to conduct experimental studies to determine the effect of pathological conditions like disc degeneration, osteoporosis and surgical interventions like interbody fusion. The experimental difficulties include collecting the specimens and the question of control with regard to the repeated testing of the same spine under various loads. Moreover experimentations are unwieldy and expensive. The analytical approach, on the other hand, bypasses many of these problems. It also permits to simulate the spine behavior under different physiological situations (sound, pathological, instrumented) using sets of mathematical equations. Finite element method is one such approach and is most suitable for complex structures like spine. According to Shih-Hao Chena et al [2], used FEA to study spine discs and explored bio-mechanical aspects. KuSurgical [3] suggested treatments for lower back pain based on fusion (arthrodesis) and disc replacement (arthroplasty). A 3D model of a lumbar segment was created and simulated through static analyses using FEA software. Dong Suk Shina et al. [4] in their study, developed a finite element model of the human lumbar spine for the parametric study and thereby range of motion was investigated for a set of spine discs. Peter J. Evans et al. [5] found out that the degeneration of spine discs serves as a source of pain in elderly patients and in their study, developed a disc model possessing scaffold like geometry to study disc degeneration. According to Horak Z et al [6], degenerative disc disease constitutes one significant reason for low lumbar back pain. In their work, range of motion was studied for the model developed based on the input data received from a class of patients having the disease. Ayturk, U. M., & Puttlitz, C. M[7], in their work generated a FE model of human lumbar spine and performed an extensive validation using kinematic and stress-strain data.

2. Methodology

The present work has undergone the following stages of which the data collection forms the very tedious stage, as it involves various stake holders from the common man till the medical professionals. The stages that are followed are as follows:

1. Anatomy of Lumbar spine has been studied and specimen (original lumbar spine) models were collected for understanding of lumbar spine morphology.
2. Based on the study and understanding of morphology, parameters governing bone structure were decided and measured using vernier caliper.
3. Model of vertebra is then created using CATIA and Visual Basic.
4. After creating the model, STL file of the model is imported to prototyping machine and a prototype has been created.

The lumbar spine surfaces are not a regular geometry surfaces, it contains lot of curves and projections and sizes of parts of lumbar spine differs from body to body. There are medical imaging software available in the market to develop a model, but each and every time doctor needs a healthy bone to create patient specific model. To overcome the problem, the present study can be very beneficial as models can be generated quite easily only by assigning parameters value. Patient specific model can be easily generated

2.1. Modelling approach

To create the parametric model of this part it is necessary to know the parameters on which it is depended. For this, bones of different human beings had been collected and studied, on the basis of the study parameters were collected. Each bone is then measured with the vernier caliper to get the parameters value. The modeling of lumbar vertebra is then carried out in CATIA. The solid model of lumbar vertebra including all the and parts is modeled in CATIA based on the lumbar anatomical studies. The process and parts were defined based on the parameters considered in the present work and are listed in following Table 1.

Table 1. Parameters defining lumbar vertebra.

Body	Part
Vertebral Body	Major and Minor axis length at superior plane
	Major and Minor axis length at mid plane
	Major and Minor axis length at inferior plane
	Vertebral height
Pedicle	Pedicle width
	Pedicle height
	Pedicle length
	Distance between outer layer of pedicles
Spinous process	Spinous process length
	Width at the top of spinous process
Transverse process	Distance between transverse process
FACET	Facet height
	Facet width

In vertebra, the body forms the predominant part possessing the shape of a cylinder like. It is shaped like an hourglass, thinner in the center with thicker ends. So, for the ease of the modelling it is assumed that the body contains three major planes namely superior plane, mid plane and inferior plane as shown in Figure 1a. The major and minor axis length at each of the above mentioned plane is measured with the vernier caliper. The pedicles change in the morphology from upper lumbar to the lower lumbar. The parameters considered are pedicle width, height, length from the center of body and the distance between outer layers of the pedicles. All the parameters are measured by vernier caliper. The pedicles are shown in figure 1b. The spinous process of a vertebra is directed backward and downward from the junction of the laminae (in humans) and serves for the attachment of muscles and ligaments. The parameters are length of spinous process and width at the top of spinous process, each parameter are measured with vernier caliper. The spinous processes are shown in figure 2

The transverse or costal processes of a vertebra is as shown in figure 3a which serves as a medium for the attachment of muscles and ligaments. The only parameter considered is distance between the transverse process. An articular facet (or articular surface) is a surface where two anatomical structures (usually bones) meet. The parameter are facet height and facet width. Again the parameters are measured with the help of vernier caliper. The articulate facet are shown in figure 3b.

IOP Conf. Series: Materials Science and Engineering 1168 (2021) 012008 doi:10.1088/1757-899X/1168/1/012008

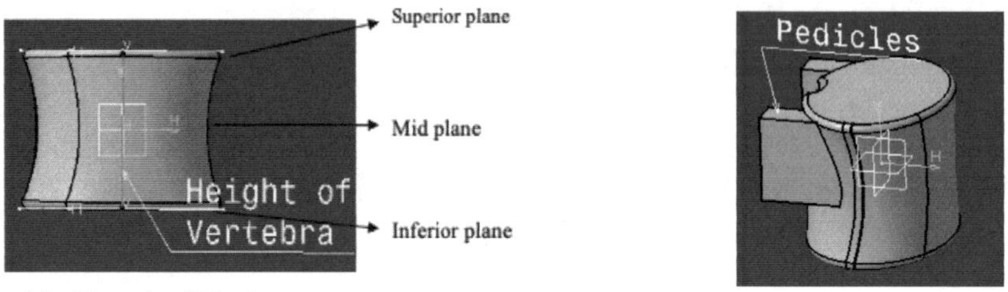

(a) Vertebral Body

(b) Pedicle

Figure 1 Vertebral body and Pedicle

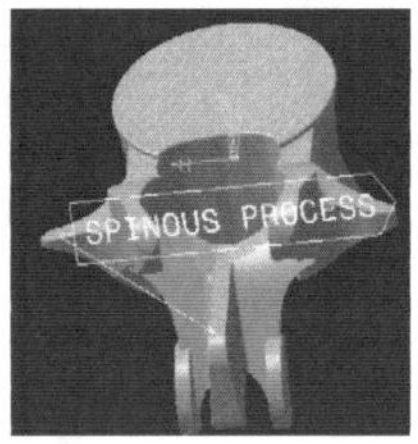

Figure 2 Spinous process

(a) Transverse Process

(b) Facet

Figure 3 Transverse Process and Facet

The user interface had been developed using the above-mentioned parameters and the different views of model (L1 vertebra). The parameter values has been mentioned in table 2. The fifth vertebra L5 has not been modelled because of complicated structure due to its relation with sacrum. The table 2 consists of parameters of L1 vertebra.

Table 2. L1 Vertebra - Parameters and values

Parameter	Value (in mm)
Height of vertebra	24.68
Superior plane major axis length	19.84
Superior plane minor axis length	16.25
Mid plane major axis length	16.95
Mid plane minor axis length	15.65
Inferior plane major axis length	20.24
Inferior plane minor axis length	16.05
Pedicle length	19.34

Pedicle height	17.86
Pedicle width	6.65
Distance between outer layer of pedicle	36.29
Width of spinous process at top	3.04
Length of spinous process	21.56
Height of facet (Inferior)	5.05
Width of articulate facet (Inferior)	3.00
Articulate facet (Inferior) Major Axis	7.05
Articulate facet (Inferior) Minor Axis	6.25
Distance between tips of transverse process	66.64

3. Rapid prototyping of spline

The prototype of lumbar spine has been developed to compare the model with the original human lumbar spine specimen. The machine used was STRATASYS-FORTUS 360 mc. In the preprocessing stage the 3D CAD model data in the form of STL file is feed into the software. Software automatically slices the model and calculates support structures, creates toolpaths and calculates the time required. In the second stage parts are build layer by layer in an additive process and extrusion heads lay down thermoplastics to create each layer. After completion of second stage support structures are dissolved in a water-based solution or snapped off by hand.

4. Results and discussion

Each lumbar vertebra model was compared with the specimen (original human bone). The deviation of the 3D printed model is compared with the original specimen bone and the deviations are listed in tables 3 and 4 respectively. The deviation is found to be very minimal or negligible as evident from the tables.

Table 3 Comparison between modelled bone and dry vertebra

PARAMETER VALUES OF L1 VERTEBRA(in mm)			
Parameter	Dry vertebra	Modeled vertebra	Deviation
Height of vertebra	24.68	24.68	0
Superior plane major axis length	19.84	19.573	-0.267
Superior plane minor axis length	16.25	13.202	-3.048
Mid plane major axis length	16.95	16.79	-0.16
Mid plane minor axis length	15.65	14.23	-1.42
Inferior plane major axis length	20.24	20.03	-0.21
Inferior plane minor axis length	16.05	15.95	-0.1
Pedicle length	19.34	19.34	0
Pedicle height	17.86	17.86	0
Pedicle width	6.65	6.63	-0.02

Distance between outer layer of pedicle	36.29	36.29	0
Width of spinous process at top	3.04	3.15	0.11
Length of spinous process	21.56	22.53	0.97
Height of facet (inferior)	5.05	5.05	0
Width of articulate facet (inferior)	3.02	3.01	-0.01
Articulate facet (inferior) Major axis	7.05	7.05	0
Articulate facet (inferior) MINOR AXIS	6.25	6.25	0
Distance between tips of transverse process	66.64	66.64	0

The spine model that has obtained after the present study can be more useful for the doctors and the spine surgeons in carrying out their studies on the spine. The present study will reduce the dependency on the animals and the dead bodies. The study can be also be used as computer based tutorial for the students. Patient specific model can be generated with giving the parameters values that can be obtained through scan. Patient specific model of spine can be used to determine the type and size of implants required for a particular patient for operations, before hand. The modelled spine can be useful for generation of the finite element model which is useful for different kind of analysis. The present study can help in better case information to reduce operating time, enhance patient and physician communication and improve patient outcomes.

TABLE 4 Comparison between modelled bone and dry vertebra

L1 VERTEBRA			
Parameter	ORIGINAL BONE (in mm)	PROTOTYPE	DEVIATION (in mm)
Height of vertebra	24.68	24.88	0.2
Superior plane major axis length	19.84	19.46	-0.38
Superior plane minor axis length	16	15.21	-0.79
Mid plane major axis length	16.95	17.04	0.09
Mid plane minor axis length	15.65	14.92	-0.73
Inferior plane major axis length	20.24	19.91	-0.33
Inferior plane minor axis length	16.02	15.23	-0.79
Pedicle length	19	19	0
Pedicle height	17.86	16.17	-1.69
Pedicle width	6.65	5.88	-0.77

IOP Conf. Series: Materials Science and Engineering 1168 (2021) 012008 doi:10.1088/1757-899X/1168/1/012008

Distance between outer layer of pedicle	36.29	34.08	-2.21
Width of spinous process at top	3.04	3.02	-0.02
Length of spinous process	21.56	22.28	0.72
HEIGHT OF FACET (inferior)	5.05	5.96	0.91
Width of articulate facet (inferior)	3.20	3.1	-0.1
Articulate facet (inferior) Major axis	7.03	7.33	0.3
Articulate facet (inferior) Minor axis	6.25	6.04	-0.21
Distance between tips of transverse process	66.64	66.31	-0.33

References

1. Pokras, R., and Kubishke, K.: Diagnosis related groups using data from the national hospital discharge survey: Hyattsville, MD, National center for health statistics,82(1982),pp 265-570
2. Shih-Hao Chena, Zheng-Cheng Zhongb, Chen-Sheng Chenc: Biomechanical comparison between lumbar disc arthroplasty and fusion: Medical Engineering & Physics 31(2009),pp 244-253
3. Guilhem Denozie` re, David N. KuSurgical: Biomechanical comparison between fusion of two vertebra and the implantation of an artificial intervertibral disc:Journal of Biomechanics 39(2006),pp 766-775
4. Dong Suk Shina, Kunwoo Leea,, Daniel Kimb: Biomechanical study of lumbar spine with dynamic stabilization device using finite element method: Computer Aided Design 39(2007),pp 559-567
5. Peter J. Evans, Binil Starly and Wei Sun:Computer Aided Tissue engineering for the design and evaluation of Lumbar spine arthroplasty:Computer-Aided Design & Application 3(2006),pp 771-778
6. Horak Z., Jirkova L., Tichy P., Sedlacek R:Experimental measurement of Lumbar spine stiffness:Clinical biomechanics 15(2005),pp 549-558
7. Ayturk, U. M., & Puttlitz, C. M. (2011). Parametric convergence sensitivity and validation of a finite element model of the human lumbar spine. Computer methods in biomechanics and biomedical engineering, 14(8), 695-705.

Stress and Failure Analysis of Composite Prosthetic Leg

Prashant S. Hatti[1], Anupama B. Somanakatti[2], Rakshith M[3]

[1]Assistant Professor, Department of Mechanical Engineering,
CMR Institute of Technology, Bengaluru 560037, Karnataka, India.
[2]Assistant Professor, Department of Mechanical Engineering
APS College of Engineering, Bengaluru – 560082, Karnataka, India.
[3]Assistant Professor, Department of Mechanical Engineering,
Cambridge Institute of Technology - North Campus, Bengaluru – 562110, Karnataka, India.

Abstract. Prosthesis is an artificial substitute for body parts such as limbs, eye, tooth, etc. Generally, the prosthesis is referred to as a fabricated substitute used to assist the damaged part of the body. The devices are made up of light materials like plastics, aluminium, and composites materials. The prosthetic technology is aimed to replicate the body part which is damaged irreversibly due to unfortunate incidents. The present work deals with creation of an optimized design for the prosthetic leg proposed by Saito et al. [2]. The material of prosthetic leg proposed here is made up of Carbon Fiber Reinforced Polymer (CFRP) composite laminate, having unidirectional and cross-ply stacking sequence, which is analyzed for static stress and failure loads. From the results, it is evident that, while a uni-directional laminate can withstand a failure load of 4910 N, a cross-ply laminate withstands 4282 N. Also, the individual ply versus stress is studied in this work.

Keywords: Prosthesis legs, static stress, failure load, ply.

1. Introduction

Several people all over the globe require lower leg prosthetic legs. But due to the high cost of the prosthesis component, people are not able to offer the high-cost prosthesis leg. Thus, people were forced to resort to crude limitations constructed from wood or bamboo without regard to comfort or safety standards. The reasons for this high-cost prosthesis limbs are due to the very small production of the prosthesis and custom-made designs. Hence the mass production of these prosthesis and optimization of design that incorporates low manufacturing costs, high strength, great durability, and smooth walking will be the key in reducing the manufacturing cost of the prosthetics.

Saito et al.[2] had developed a low-cost transtibial prosthesis made of fiber-reinforced plastic (FRP) which is as shown in figure 1.

IOP Conf. Series: Materials Science and Engineering 1168 (2021) 012009 doi:10.1088/1757-899X/1168/1/012009

Figure 1 Original prosthesis as designed by Saito et al. (1).

The design consisted of a foot component, an axial structural member, and a cosmetic cover. Later the prosthesis' cosmetic cover is attached to the foot component to give the regular shape of the foot. The prosthetic leg is so designed that, it can reproduce the motion of the intact leg adequately and also ankle of the prosthetic leg should follow the same path of the biological joint during walking. The axial member can be adjusted for the optimum fit. The two aluminium stiffeners were incorporated into the foot as shown in figure1 which are used to increase the strength and load-bearing capacity of the prosthetic leg.

Jill Hahl et.al [3] performed a failure test on the fabricated prototype; it showed that it is capable of withstanding 6,600 N load before failure. It surpasses the failure load of 5,500 N of the original prosthesis and also satisfies the static test outlines in the ISO standard 10328. They found that the optimized design provides sufficient strength and also reduced the cost of manufacturing.

Jill Hahl et.al [4] studied the design developed by Saito et al. (1). They modified it and proposed an optimized design by performing FE analysis on various modified prosthesis models and it was tested by using FEA software. The model has the maximum von- mises stress and maximum shear stress of 56.3MPa and 3.6MPa respectively. Also, the failure analysis was performed and the results obtained were congruent with the ISO standard 10328 of the prosthesis.

Glen K et.al [5] studied the influence of multiple factors viz., abnormal loading of the residual-limb soft tissues, abnormal or excessive musculoskeletal loading of the residual-limb proximal structures, and/or abnormal or excessive loading of the intact limb musculoskeletal structures, on biomechanical performance of the prosthesis and resultant loading impact on to the residual limbs. Also, mechanical properties of the vertical shock-absorbing pylons and prosthetic foot ankle systems and the impact on amputee gait were discussed.

Priya Sharma et. al [8] performed numerical analysis of Jaipur (JP) foot and Polyurethane (PU) foot to test the stress bearing capacity, and to compare the displacement, crack propagation and durability of both the foot and concluded that JP Foot has a better stress bearing capacity and material strength than PU Foot.

2. Stress and failure analysis

The main objective of this study is stress and failure analysis of the low-cost transtibial prosthesis leg. The numerical analysis is performed to determine the stress concentration on the FRP foot and determine the onset of failure load at which the failure takes place. Finite element Modelling is done and analysis is performed using commercial FEA software.

Stress analysis can be performed by adopting any one of the following techniques, such as classical mathematical techniques, analytic mathematical modelling or computational simulation, or experimental methods. The technique adopted here is based on classical mathematical technique i.e classical laminated plate theory. The design of composite structures requires the prediction of the maximum load that a laminate can withstand before it fails. For this purpose, usually, first ply, the second ply, and last ply failure criteria are mostly used for failure analysis of composite structures. The entire process of failure prediction helps in finding the maximum stresses, maximum strains, maximum load, or the failure load that a composite structure can withstand. To determine the safe life of a composite structure, designing below the failure load values is important. Thus failure analysis is the preliminary investigation to be carried out before proceeding to any kind of analysis. Thus, the present study deals with linear static stress analysis for composite laminate made of carbon epoxy material for stacking sequences of {0}20s,{0/90/0/90/0/90/0/90/0/90}s on the basis of last ply failure adopting Tsai-Wu composite failure criteria [3] and failure loads for different lamina are obtained.

2.1 FE model development

The analysis of the prosthesis leg is done using ABAQUS/CAE software, a finite element program. The number of elements in a 3D FE model is around 3300 elements. The element type used was a 3D shell element. This was a 4 node, doubly curved general-purpose shell, reduced integration, hourglass control, finite membrane strains (S4R) elements.

2.2 Geometry

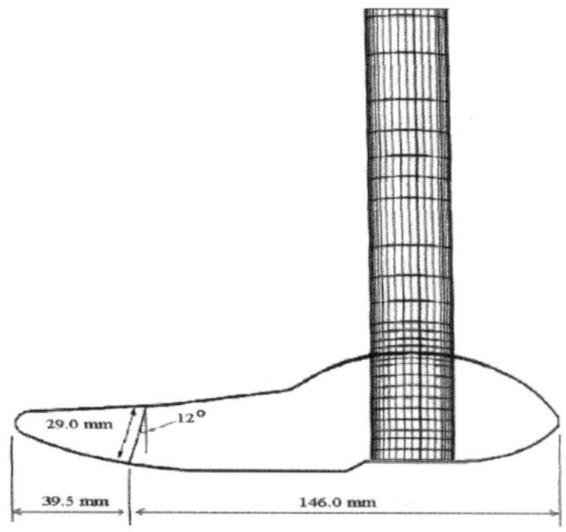

Figure 2 Proposed optimum geometry of the prosthetic leg.

The model mainly consists of a composite foot, aluminium pipe, and a rigid part (ground surface). The model was used to investigate a series of loadings and boundary conditions, intended to simulate the stages of loading during the stance phase of gait. The composite foot posses a single integrated stiffener. The optimum stiffener placement was positioned at 39.5 mm from the toe with a clockwise base rotation of 12^0 as shown in figure 2. The foot component is made of a laminated composite structure. The total number of plies in the composite foot is around 20. Each ply is of thickness 0.25 mm/ply, having a total thickness 5 mm. The foot has a width of 60 mm. A 6061-T6 aluminium pipe of diameter 28 mm and a height of 300 mm was inserted and fitted into the hole cut at the top of the prosthetic composite foot.

2.3Material Properties

The prosthetic model shown in figure 2 is having mainly 3 parts, composite foot, aluminium pipe and a rigid part. The composite foot is made of Carbon Fibre Reinforced Polymer (CFRP) laminate with 20 plies, each ply having thickness of 0.25 mm, total of 5mm thick. The axial member is made of 6061-T6-aluminium pipe of dia 28 mm. The material properties are shown in table 1 and table 2.

Table 1: Material properties for CFRP laminate

Material	Engineering constants for laminate								
	E_1 (GPa)	E_2 (GPa)	E_3 (GPa)	G_{12} (GPa)	G_{23} (GPa)	G_{13} (GPa)	υ_{12}	υ_{23}	υ_{13}
CFRP	170	9	9	4.8	4.5	4.8	0.34	0.5	0.34

Table 2: Material properties for Aluminium 6061 T6

Material	Engineering constants for isotropic material	
	E (GPa)	υ
Aluminium 6061 – T6	68.9	0.35

2.4 Loading and Boundary Condition

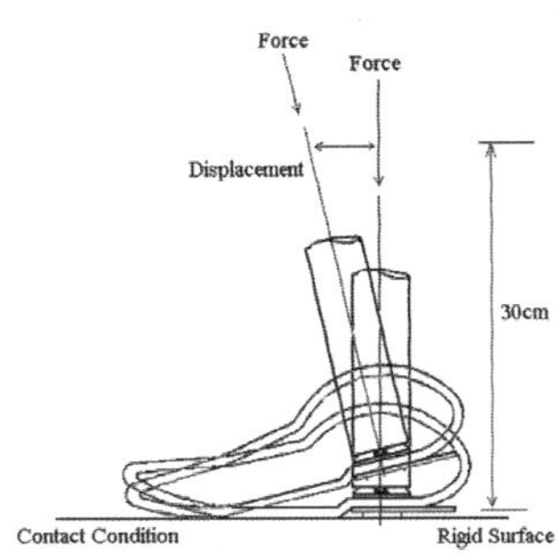

Figure 3 Loading and boundary conditions used in the finite element model.

A rigid surface is used as a ground surface and a rigid surface contact condition was used to simulate the ground surface. The height of the prosthetic model is fixed at 300 mm from the ground surface. The maximum weight of the patient is assumed to be 100kg. Hence a force of 1000N is applied on the top of prosthesis. Horizontal displacement ranging from 0 to 110 mm with 5mm thickness is used to simulate walking

2.5 Interaction Property

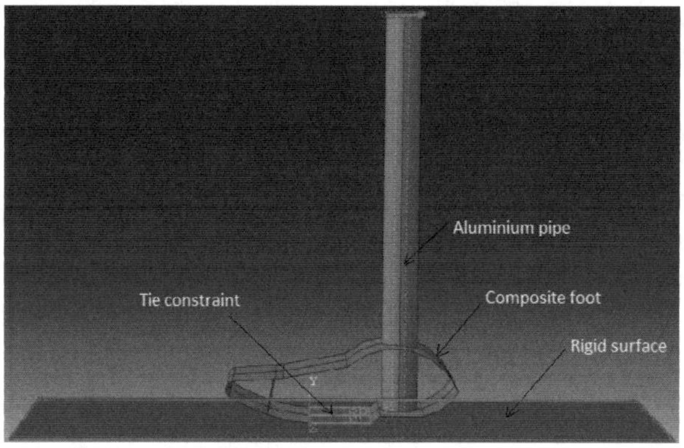

Figure 4 Interaction properties in the FE model

Figure 4 shows the various interaction properties in the FE model. A rigid contact was fixed at a height of 300 mm from the ground surface. A tie constraint is used to give contact between the ground surface and the foot surface.

2.6Meshing

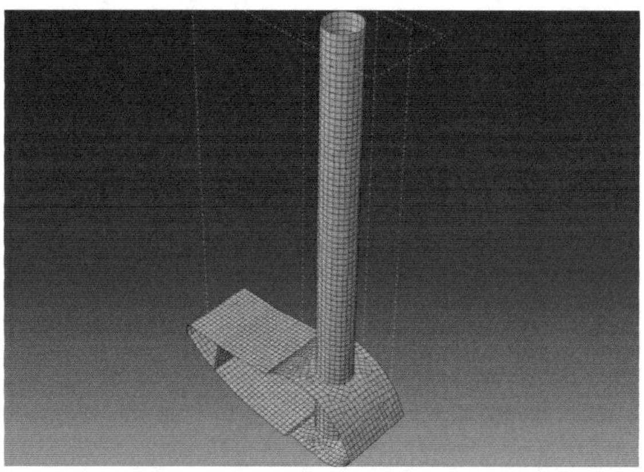

Figure 5 Meshed FE model

The element type used is a 4 node doubly curve general shell element, with reduced integration, hourglass control, and finite membrane strains. Figure 6 shows S4R element and location of integration point. The total number of elements and nodes in the FE mode are 3363 and 3432 respectively.

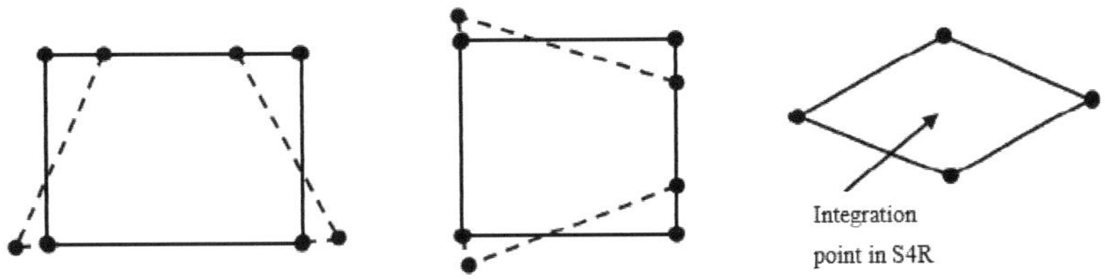

Figure 6 S4R element and location of integration point

3. Results and Discussion

The FE model was developed and subjected to stress and failure analysis. The static load [3] is applied on each ply (for both unidirectional and cross ply orientations) and the stresses developed on each ply is found and recorded. Graphs are plotted for on the recorded values and ply number as shown in figures 7 and 8 respectively. It is found that, individual ply is able to with stand the static load as evident from table 3 or in other words, the composite prosthesis model is able to with stand the failure load and hence the design is safe for the specified purpose. The material used in static analysis are CFRP laminate with unidirectional and cross-ply orientations.

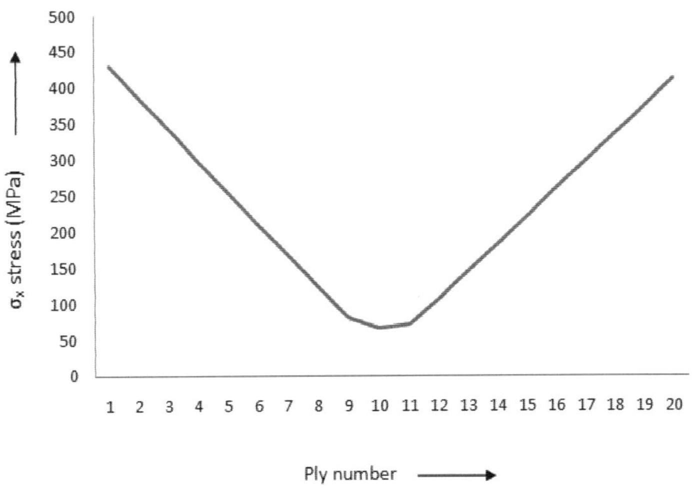

Figure 7 Stresses σ_x at each ply for a unidirectional laminate

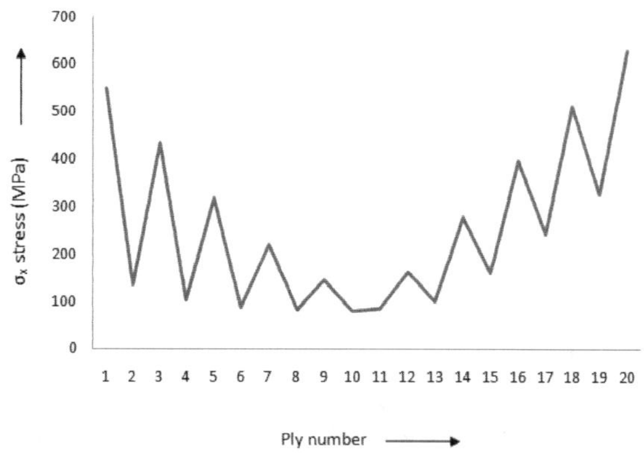

Figure 8 Stresses σ_x at each ply for a cross ply laminate

All the values obtained in the analysis for both unidirectional and cross-ply orientation of CFRP laminate are less than the values of lamina strength shown in table 3.

Table 3 Values of lamina strengths

	σ_x (MPa)	σ_y (MPa)	σ_{xy} (MPa)
Tension	2050	62	81
Compression	1200	190	81

Where σ_x, σ_y, σ_{xy} are the ultimate strengths in the longitudinal, transverse, and shear directions, respectively.

Figures 9 and 10 shows static failure analysis for CFRP composite laminate having 20 plies with the unidirectional sequence was performed using Tsai-wu failure criteria [3]. The maximum load that unidirectional CFRP laminate can withstand is around 4910N. Similarly, maximum load that the cross-ply CFRP laminate can withstand is around 4282N.

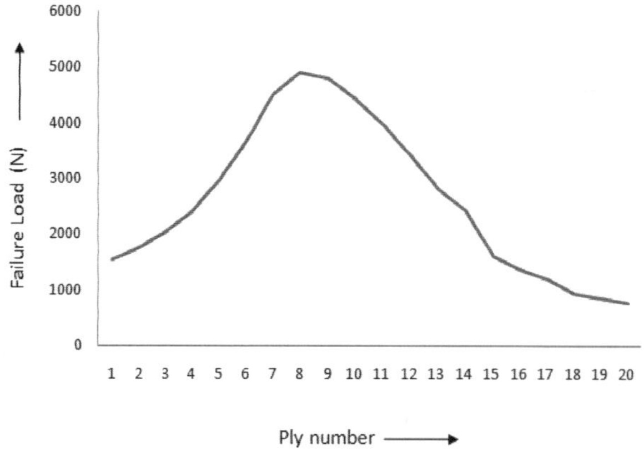

Figure 9 Ply number Vs failure load per ply for CFRP laminate with unidirectional orientation

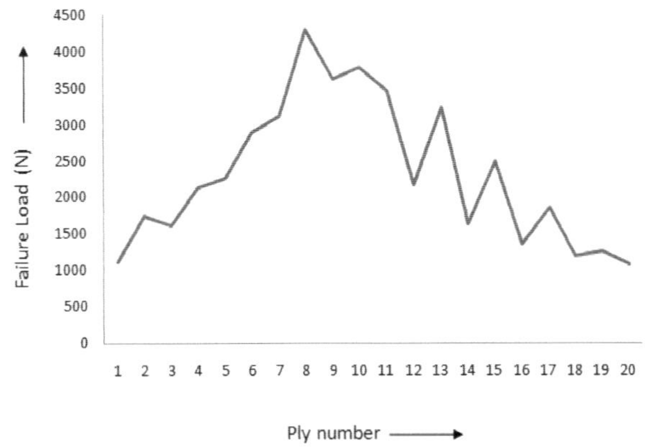

Figure 10 Ply number Vs failure load per ply for CFRP laminate with Cross-ply orientation

4 Conclusion

Static stress analysis and static failure loads were performed for a prosthetic leg made of CFRP laminate having unidirectional and cross-ply stacking sequences. As per the failure results, the maximum load that unidirectional CFRP laminate can withstand is around 4910N and the maximum load that cross-ply CFRP laminate can withstand is around 4282N respectively, which is congruent with ISO standard 10328 for the prosthesisrespectively, which is congruent with ISO standard 10328 for the prosthesis.

References

1. Gibson RF. Principles of composite materials mechanics. New York: McGraw-Hill Inc.; 1994.
2. Saito M, Sawamura S, Carroll B, "Mass-produced prosthesis uses pultruded fiber reinforced plastic", Modern Plastics (1997), No. 74 :175-7.
3. Jill Hahl, Makoto Saito and Minoru Taya, "Experimental and numerical predictions of theultimate strength of a low-cost composite transtibial prosthesis", University of Washington (2000), Vol. 22, No. 5: 91-98.
4. Jill Hahl, Makoto Saito and Minoru Taya, "Optimization of mass produced transtibial prosthesis made of pultruded fiber reinforced plastic", Journal of Rehabilitation Research and Development (2000), Vol. 37, No. 4: 405–413.
5. Glenn K. Klute, Carol F, Kallfelz, MEngr and Joseph M. Czerniecki, "Mechanical properties of prosthetic limbs: Adapting to the patient", Journal of Rehabilitation Research and Development (2001), Vol. 38, No. 3: 299–307.
6. Cochrane H, Orsi K, Reilly P "Lower limb amputation Part 3: Prosthetics - a 10 year literature review". Prosthetics and Orthotics International. 2001; 25(1):21-28.
7. Victoria S. Richardson and Erin J. Vozzola, "Analysis of a Lower Limb Prosthesis", Biomechanics concentration, 2008.
8. Priya Sharma, Shubhdha Sharma, S Vidhy, M K Mathru, "Comparative finite element analysis of Jaipur foot and poly urethane foot", Journal of Engineering, 2013, Volume 5, 518-521.
9. Angelica Ramirez, "FE Analysis on transtibial prosthetic running leg to determine critical stress and strain" (2014) ME, 404-71.

IOP Conf. Series: Materials Science and Engineering 1168 (2021) 012010 doi:10.1088/1757-899X/1168/1/012010

An Application of Analytical Hierarchy Process in Selection of Coating Material Composition in Lost Foam Casting Process

Gagan Varshney[1], Syed Qaisar Husain[2], Avinash Ravi Raja[3], Girendra Bhati[4]
[1,2,3,4]Greater Noida Institute of Technology, Greater Noida, U.P. 203207, India
E-mail: gagan.me@gnit.net.; meesam.amu123@gmail.com; avinash.me@gniot.net.in; girendrabhatime@gmail.com.

Abstract. In modern industrial era, the dimensional accuracy and surface finish are two major criteria in the selection of casting process. To achieve this paradise, lost foam casting is one of the casting processes. In this casting process, the shell is made by removal of expandable polystyrene or foam pattern with the application of heat. During the foam decomposition process, many problems have been generated such as bending, expansion, distending and crack of the shell. These problems may be eliminated by selecting the optimum ratio of coating materials. In this research work, zircon and aluminium silicate refractory coating materials with sodium silicate binder have been used with different composition and prepared the four test samples from them. For selection of best shell material composition, analytical hierarchy process is used.

Keyword: *Foam Pattern, Zircon, Alminium Silicate, Sodium silicate, Analytical Hierarchy Process*

1. Introduction:

In automobile sectors, castings made of Aluminium and its alloys are extremely applicable because these are replacing the heavy forged steel and cast. It became very important to produce defect free aluminium castings without using secondary machining. Investment casting is one of the oldest casting processes. It is used for producing better dimensional accuracy of complex shape products. Another important process is Lost Foam Casting (LFC). In this process lost foam pattern is used. It is a cavity less and binder less process. With the help of this process, we can fabricate the automobile and IC engine components such as cylinder head, piston, connecting rod, brake shoes etc. as well as agriculture equipment [1-2].
The lost foam selected for LFC process have 92% C and 8% H. C6H5 benzene ring in lost foam is comparatively stable and –CH= CH2- chain tends to decompose first. The benzene ring remains in liquid phase and reacts with melt causing casting defects. Polymethylmethacrylate (PMMA) decomposed mostly into gas phase, like 80% at 700° C while polystyrene only 40% at 700 °C [3-6]. Coating materials such as siliminite, quartz, aluminum silicate, zircon flour etc have been used by various researchers and the binder for cost effectiveness. Zircon and aluminum silicate have high density and viscosity, low dielectric constant, and pH value nearer to neutral refractory [7-10].

The selection of binding agent depends on the shape and size of particles of refractory materials so that the bonding and adhesive strength of particles and binders has been increased on the surface of foam pattern. Sodium silicate and potassium silicate with

water have been used as a solvent. The optimal density of ceramic slurry was 2000 kg/m3 [11-12]. Pattnaik et al. (2012) reviewed the different primary coating materials in investment casting process with wax pattern. They analysed the layer thickness of Ti−1100 casting and ZrO2 primary coating (38 μm) was higher than Y2O3 (Yttrium Oxide) primary coating (18 μm) [13]. Lu et al. (2018) analyzed the influences the firing time and firing temperature on the firing bonding strength of fibered reinforced ceramic shells in investment casting process. They have taken aluminium silicate and glass fibres as reinforced material. The maximum bending strength of reinforced shells has been achieved at 950 °C temperature and firing to 120 minutes [14]. Venkat et al. (2020) fabricated the ceramic shell for the blisk of low-pressure turbine rotor by using wax pattern-based investment casting process. They used aluminium silicate and zircon as coating material with cobalt aluminate [15].
The Analytical Hierarchy Process (AHP) is an important mathematical tool which is used to find the priorities of various decision options through pair-wise comparisons of decision elements with respect to common criteria [16-17].

Table 1. The Saaty Rating Scale

Intensity	Definition	Description
1.	Equal importance	Two factors contribute equally to the objective
3.	Moderate importance of one over another	Experience and judgment favour one factor over another
5.	Essential or strong Importance	Experience and judgment strongly favour one factor over another
7.	Very strong importance	An factor is strongly favoured and its dominance demonstrated in practice
9.	Extreme importance	The evidence of favouring one factor over another is of the highest possible order of affirmation
2, 4, 6, 8	Intermediate values,	When compromise is needed

Usually, the patterns are made up of wax in investment casting. But wax has low melting point and also, it is not easy to handle the big and complex shapes of wax pattern. These problems have been eliminated in lost foam casting process [18]. The different combinations of coating materials (zircon and aluminium silicate with sodium silicate or potassium silicate binder) have been optimized by using Analytical Hierarchy Process (AHP).

I. **Selection of Materials**

a) Coating materials: -

In this work, zircon and aluminium silicate sand are used as coating materials, in different ratio and composition. Sodium silicate is used as binder. The following compositions of refractory powder have been selected for primary coating.

Sample: 1 – Zircon (0 %) + Aluminium silicate (50 %) + Sodium Silicate (50%) Sample: 2 – Zircon (20 %) + Aluminium silicate (30 %) + Sodium Silicate (50%) Sample: 3 – Zircon (30 %) + Aluminium silicate (20 %) + Sodium Silicate (50%) Sample: 4 – Zircon (50 %) + Aluminium silicate (0 %) + Sodium Silicate (50%)

For secondary coating compositions, aluminium silicate (50 %) with Sodium Silicate (50%) is selected.

b) Pattern Material:-

In lost foam casting process, lost foam patterns are used. The density of lost foam pattern is 20 kg/m3. For this analysis, step shaped pattern has been shown in Figure 1.

II. **Methods**

c) Slurry Preparation: Slurry of different composition of refractory powder materials has been uniformly mixed and made through mechanical stir. Compositions of coating materials have stirred at 1000 rpm for 15 minutes.

d) Coating on lost foam pattern: After the preparation of primary and secondary slurries, lost foam pattern has been dipped in primary slurry. The process is repeated three times. After it, secondary coating has been placed on the foam pattern. The thickness of coating (primary and secondary) has found 5-7 mm for all samples.

e)

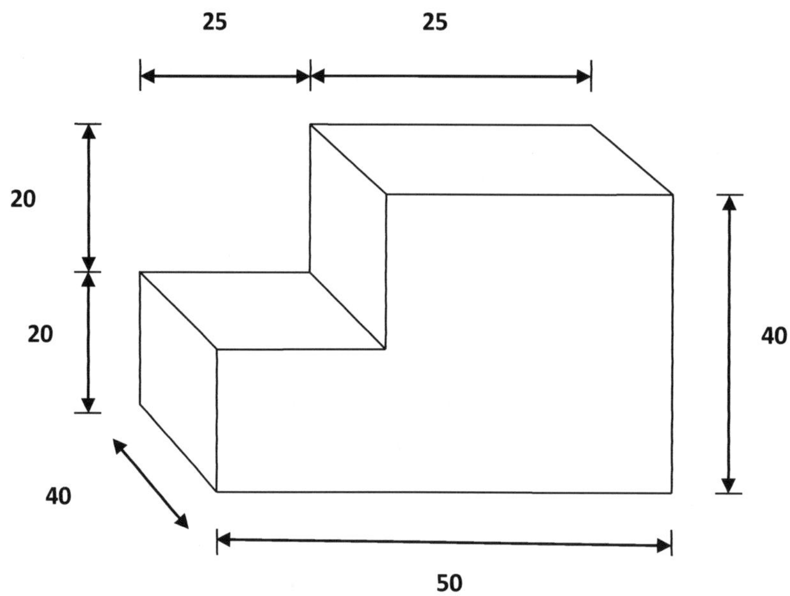

All Dimensions are in mm

Figure 1: Lost Foam Pattern

f) **Decomposition of coated expandable pattern:** Shell has been generated by removal of lost foam pattern by applying the heat in furnace. When coated foam pattern is placed in the furnace, the foam starts to melt. At high temperature, foam starts to evaporate. After fully decomposition of foam, cavity has been created. This process is called foam pattern removing. In this method, first, shell is preheated at 80°C for 30 minutes. After preheating, the temperature raises up to 120°C at a rate of 4°C/ minute. Shells are heated at constant temperature (120°C) for 20 minutes. Again the temperature rises up to 160°C at a rate of 4°C/ minute for 10 minutes and then heated at constant temperature for 20 minutes. This process is repeated till foam evaporates completely.

2. Result and Discussion:

a) Selection of Best Shell by using Analytical Hierarchy Process (AHP):

There are many criteria (properties of shell) on which the shell quality depends, as following.
Dimensional accuracy criteria: Bending and Distending Resistance.
Surface Finish criteria: Internal and External Surface Finish of shell.
Cost: Cost of coating powder materials and Binder.
Strength of Shell: Crack Resistance.
For these alternatives and criteria, Saaty's Hierarchy Tree has been shown in figure 2. In this investigation, the most important properties (bending and distending resistance, internal surface finish, crack resistance of shell) are considered. Following steps are used for finding the best alternative as per AHP.
STEP 1: (Consideration)
Select most important properties of shell (bending resistance as criteria A, distending resistance as criteria B, Internal surface finish as criteria C, and crack resistance as criteria D) and four shells (S1, S2, S3 and S4).

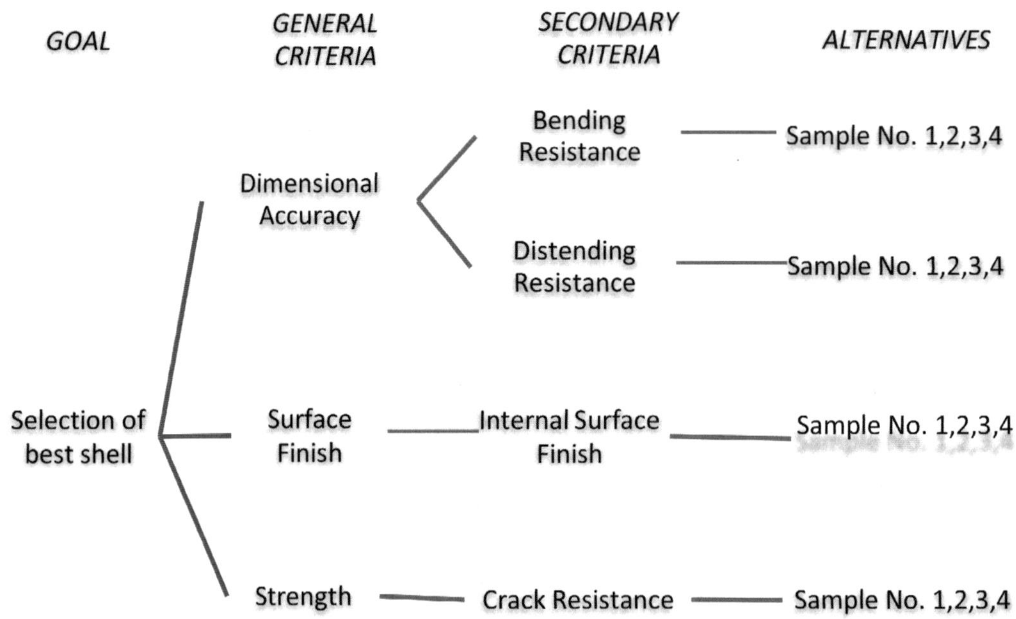

Figure 2. Saaty's Hierarchy Tree

STEP 2: (Pair wise comparision)
In this step, the pair wise comparision of individual criteria with respect to all criteria has been takes place. Similarly, the pair wise comparision of individual alternative with respect to all alternatives for each criteria, has been analyzed.

Pair wise comparision of criteria:
- Criteria A is 3 times important than criteria B.
- Criteria B is 3 times important than criteria C.
- Criteria D is 1.5 times important than criteria C.
- Criteria A is 6 times important than criteria C.
- Criteria A is 5 times important than criteria D.
- Criteria B is 2 times important than criteria D.

Pair wise comparision of alternatives:

- **For criteria A**
- Alternative S2 is 2 times important than alternative S1.
- Alternative S3 is equal important than alternative S4.
- Alternative S3 is 1.5 times important than alternative S1.

- Alternative S2 is 1.5 times important than alternative S3.
- **For criteria B**
  - Alternative S1 is 1.5 times important than alternative S3.
  - Alternative S2 is equal important than alternative S4.
  - Alternative S3 is 1.5 times important than alternative S1.
  - Alternative S2 is 2 times important than alternative S1.

For criteria C

- Alternative S2 is equal important than alternative S3, S4.
- Alternative S2 is 2 times important than alternative S1.
- **For criteria D**
  - Alternative S2 is equal important than alternative S1.
  - Alternative S3 is equal important than alternative S4.
  - Alternatives S2 and S1 is 2 times important than alternative S3 and S4.

STEP 3: (Generation of matrix as per pair wise comparision)

In this step, the matrix form has been generated as per pairwise comparision.

- **For all criteria:**

	A	B	C	D
A	1.000	3.000	6.000	5.000
B	0.333	1.000	3.000	2.000
C	0.167	0.333	1.000	0.667
D	0.200	0.500	1.500	1.000

- **For all alternatives:**

For Criteria A:

	S_1	S_2	S_3	S_4
S_1	1.000	0.500	0.667	0.667
S_2	2.000	1.000	1.500	1.500
S_3	1.500	0.667	1.000	1.000
S_4	1.500	0.667	1.000	1.000

For Criteria B:

	S_1	S_2	S_3	S_4
S_1	1.000	0.500	0.667	0.500
S_2	2.000	1.000	1.500	1.000
S_3	1.500	0.667	1.000	0.667
S_4	2.000	1.000	1.500	1.000

For Criteria C:

	S_1	S_2	S_3	S_4
S_1	1.000	0.500	0.500	0.500
S_2	2.000	1.000	1.000	1.000
S_3	2.000	1.000	1.000	1.000
S_4	2.000	1.000	1.000	1.000

For Criteria D:

	S_1	S_2	S_3	S_4
S_1	1.000	1.000	0.500	0.500
S_2	1.000	1.000	0.500	0.500
S_3	2.000	2.000	1.000	1.000
S_4	2.000	2.000	1.000	1.000

STEP 4: (Finding the eigen vector for each matrix)

In this step, the eigen vectors are calculated by squaring the matrix and then adding each row as describe below.

	A	B	C	D
A	1.000	3.000	6.000	5.000
B	0.333	1.000	3.000	2.000
C	0.167	0.333	1.000	0.667
D	0.200	0.500	1.500	1.000

Squaring the matrix

$$\begin{bmatrix} 1.000 & 3.000 & 6.000 & 5.000 \\ 0.333 & 1.000 & 3.000 & 2.000 \\ 0.167 & 0.333 & 1.000 & 0.667 \\ 0.200 & 0.500 & 1.500 & 1.000 \end{bmatrix} * \begin{bmatrix} 1.000 & 3.000 & 6.000 & 5.000 \\ 0.333 & 1.000 & 3.000 & 2.000 \\ 0.167 & 0.333 & 1.000 & 0.667 \\ 0.200 & 0.500 & 1.500 & 1.000 \end{bmatrix}$$

SQ is square matrix

SQ =					Total of each row	Eigen Vector
	4.000	10.498	28.500	20.002	= 63.001	63.001/ 108.762=0.579
	1.567	4.000	10.998	7.666	= 24.229	24.229/ 108.762=0.223
	0.578	1.501	4.000	2.835	= 8.915	8.915/ 108.762=0.082
	0.817	2.100	5.700	4.000	= 12.517	12.517/ 108.762=0.116
					Total = 108.762	Total = 1.000

A	0.579
B	0.223
C	0.082
D	0.116

Again finding eigen vectors of the square matrix we get the following matrix

A	0.579	→ Most important criteria
B	0.223	→ Second most important criteria
C	0.082	→ Least most important criteria
D	0.116	→ Third most important criteria

Similarly **For criteria A**

0.166	→ Least important alternative
0.352	→ Most important alternative
0.241	→ Second most important alternative
0.241	→ Second most important alternative

For criteria B

0.153	→ Least important alternative
0.315	→ Most important alternative
0.217	→ Second most important
0.315	→ Most important alternative

For criteria C

0.142	→ Least important alternative
0.286	→ Most important alternative
0.286	→ Most important alternative
0.286	→ Most important alternative

For criteria D

0.167	→ Least important alternative
0.167	→ Least important alternative
0.333	→ Most important alternative
0.333	→ Most important alternative

STEP 5: (Finding the best alternative)

In this step, the best alternative is found by multiplying the criteria alternative matrix andcriteria rank.

	A	B	C	D	*	Criteria Rank	=		
S1	0.166	0.153	0.142	0.167		0.578		0.161	
S2	0.352	0.315	0.286	0.167		0.223		**0.317**	→ **BEST**
S3	0.241	0.217	0.286	0.333		0.082		0.250	
S4	0.241	0.315	0.286	0.333		0.116		0.272	

The result of AHP shows that alternate sample no. 2 is selected as the best shell. This isbecause the important factors are considered as priority.

3. Conclusions:

By incorporating the all-important properties like bending resistance, distending of shell, internal surface finish and crack resistance, sample no. 2 has found as the most optimized. So, optimum composition of coating material (sample no. 2) is aluminium silicate 30%, zircon 20% with binder sodium silicate 50 % for primary coating.

References:

1) R. C. Reed (2006) "The superalloys fundamentals and applications". Cambridge University Press;

2) S. M. Afazov, A. A. Becker, T.H. Hyde (2011) "FE prediction of residual stresses of investment casting in a Bottom Core Vane under equiaxed cooling" Journal of Manufacturing Processes 13. 30–40.

3) S. U. Behm, K. L. Gunter, J. W. Sutherland (2003) "An Investigation into the Effect of Process Parameter Settings on Air Emission Characteristics in the Lost Foam Casting Process" Transactions of the American Foundryman Society ,1-10

4) T.V. Moilbog, H. Littleton, (2001) "Experimental simulation of pattern degradation in lost foam". Am. Foundryman Soc. Trans. 109, 1523–1555.

5) Dong Kyu Kim and Waltraud M. Kriven (2004) "Mullite (3AI203 2SiO2) AluminumPhosphate (AIPO4), Oxide, Fibrous Monolithic Composites. J. Am. Ceram. Soc., 87, 151794-803.

6) T.V. Molibog and H. Littleton (2001) "Proceedings of Lost Foam Casting", AFS, Dallas, Texas, 2001, 71.

7) H. Tanaka, S. Sawai, K. Morimoto, K. Hisano (2001) "Measurement of spectral emissivity and thermal conductivity of zirconia by thermal radiation calorimeter". Journal of Thermal Analysis and Calorimetry; 64, 867–72.

8) Lj. Trumbulovic, Z. Acimovic, Z. Gulisija, Lj. Andric, (2004) "Correlation of technological parameters and quality of castings obtained by the EPC method", Mater. Lett. 58, 1726–1731

9) Wei Zhang and Ming Leu, (2000). Investment Casting with Ice Patterns Made by Rapid Freeze Prototyping" 11th Annual Solid., 2000 - utwired.engr.utexas.edu, 66-72.

10) Zhao, Er-tuan, Kong, Fan-tao, Chen, Yu-yong, Li, Bao-hui, 2011. Interfacial reactions between Ti−1100 alloy and ceramic mould during investment casting" Trans. NonferrousMet. Soc. Chine, 21, 348-352.

11) S. Kumar, P. Kumar, H.S. Shan, (2004) "Effect of filler material in zircon flour coating used in evaporative pattern casting process", Indian Foundry J. 50, 34–40.

12) Zilin Xu, Jiangwei Zhong, Xianglin Su, Qingyan Xu and Baicheng Liu, (2019), "Microstructure evolution and mechanical behaviors of alumina-based ceramic shell for directional solidification of turbine blades", Journal of Materials Research andTechnology, Vol. 8, Issue 1, pp 876-86.

13) S. Pattnaik, D.B. Karunakar, P.K. Jha, (2012) Developments in investment casting process - a review. Journal on Material Processes and Technology, 212, 2332–2348 (2012)

14) Kai Lu, Xiangdong Liu, and Zehai Duan (2018) Effect of firing temperature and time on hybrid fiber-reinforced shell for investment casting, International Journal of Metal Casting 13(29), 1-8.

15) Y. Venkat, K.R. Choudary, D.K. Das, A.K. Pandey, Sarabjit Singh (2020), "Ceramic shell moulds for investment casting of low-pressure turbine rotor blisk" Ceramics International, 272-8842(20) 31-34.

16) T.L. Saaty (1980), "The Analytic Hierarchy Process", McGraw Hill, New York, NY.

17) T. L. Saaty (1994), "Fundamentals of Decision Making and Priority Theory with the Analytic

Hierarchy Process", RWS Publications, Pittsburgh, PA.

18) Girendra Bhati, Sudhir Kumar, Ajay Kumar, Sanjeev Kumar (2020) "Optimization of process parameters of A-359 aluminium alloy in EPS assisted-investment casting processusing Taguchi method", IOP Conference Series: Materials Science and Engineering, 2020.

CFD simulation of NACA airfoilsat various angles of attack

Y Seetha Rama Rao[1],M V N SrujanManohar[2]and S V V SivaPraveen[3]

[1]Associate Professor, Dept of Mechanical Engineering, Gayatri Vidya Parishad College of Engineering (A), Visakhapatnam, Andhra Pradesh, India

[2]Research scholar, Pondicherry Engineering College, Pondicherry, India

[3]Bachelor of Technology in Mechanical, Gayatri Vidya Parishad College of Engineering(A), Visakhapatnam, Andhra Pradesh, India

Abstract. The need for more efficient design has become a main concern to work with a good range for angle of attack (AOA) to reduce the stall formation. In this project, two types of NACA series airfoils i.e., symmetrical and asymmetrical profiles depending on their usage in various sectors are compared. The profile of the airfoils is imported into ANSYS Fluent from the data files. The 2D geometry of the airfoils is simulated under two different wind velocitiesi.e., at 3m/s and 15m/s respectively. The graphs are plotted for comparing coefficients of lift (C_L) and drag (C_D)withrespect to the angle of attack(α) for symmetrical and asymmetrical airfoils. The stall formation with change of geometry and angle of attack is also observed.

Keywords: *NACA airfoil, Angle of Attack,Stall, Ansys Fluent, CFD Simulation*

1. INTRODUCTION

An airfoil is the cross-sectional shape of a wing, blade (of a propeller, rotor, or turbine) or sail.It is the most fundamentaland prime element in building a wind turbine blade. An aerodynamic force is produced when an airfoil-shaped body is moving through a fluid. The component of this force perpendicular to the direction of motion is called Lift. The component parallel to the direction of motion is called Drag. The lift developed on an airfoil is basically the outcome of its angle of attack.[1]. The onrushing airgets deflected by the airfoil, when aligned at a suitable angle. This results in creating a force on the airfoil that is in the direction opposite to the deflection. The key factors altering both of these coefficients are the angle of attack (α) between the incoming wind direction and the airfoil chord, the geometry of the airfoil and the Reynold's number. Behaviour of the airfoil due to movement of air is divided into three phases: (1) the attached flow phase, (2) the high lift/stall development phase and (3) the flat plate/fully stalled phase.

The NACA airfoils are the shapes of different airfoils designed for aircraft wings developed by the National Advisory Committee for Aeronautics (NACA). The shape of the these airfoils is designated using a series of numbers followedby "NACA". With the help of equations by entering these digits, the properties of airfoil can be calculated. The cross-section of the airfoil can also be generated easily.

The NACA series4-digited airfoil section definesits profile as follows:

- First digit indicates maximum camber as percentage of the chord.
- Second digit provides the distance of maximum camber from theleading edge of the airfoil in tenths of the chord.
- Last two digits indicates maximum thickness of the airfoil as a percentage of the chord.

Airfoils of this series have max. thickness at 30 percent of chord from the leading edge.

A. Meana-Fernandezet.al [2] proposed an airfoil shape usedin the applications of vertical-axis wind turbine. Various airfoil geometries have been examined using Java Foil, a panel method software. The performance of the airfoil generated is optimized using the results obtained from the analysis. At different angles of attack, the CFD simulations were ran in order to give detailed perception regarding field of fluid and other mechanisms useful to enhance the performance of the presented airfoil.The two designs related to VAWT are compared with the primary and presented airfoil to verify the improvement in performance.Ultimately, the feasibility of Java Foil had been tested by comparing geometries of various airfoils because it is convenient and has the ability to obtain results for a large range of flow environments in a relativelyless amount of time. However, in case of higher angles of attack i.e., beyond stall, the results always deviate from the original outcome.

Anagha S Gowda [3] performed aerodynamic performance characteristics comparison between two airfoilsNACA 2412 and NACA 4412 under similar flow conditions at a Reynolds Number of 2 million. The creation of geometry and meshing were done using ANSYS software and FLUENT fulfilled the purpose as solver. The performance of every airfoil with aerodynamic parameters was compared with respect to angle of attack and from that set, a particular airfoil which is optimal for precise aerodynamic characteristics were selected.

Castineira-Martinez Set.al [4] performed a large number of simulations to arrive at for a typical profile for a wind turbine. The solutions of these simulations were calculated over various sixes of mesh by a 2-dimensional Reynolds-Average Navier-Stokes (2D-RANS) approach. Three other turbulence models like k-ε and k-Ω were also used in the simulations. For each and every model and size of mesh, over four angles between -2.5 and 12.5,coefficients of lift, drag and momentum were calculated and compared them with the experimental values. As a result of all these attempts, a workable model that holds good standards (i.e.,accord with both numerical & experimental data) and can be used for further investigation as a basic model was obtained.

2. TURBULENCE MODEL &GEOMETRY

Turbulence modelling is the construction and use of a mathematical model to predict the effects of turbulence.K-epsilon (k-ε) turbulence model is the most common model used in computational fluid dynamics (CFD) to simulate mean flow characteristics for turbulent flow conditions. It is a two-equation model which gives a general description of turbulence by means of two transport equations (PDEs).

The original momentum for the K-epsilon model was to upgrade the mixing-length model, and to find an auxiliary to algebraically prescribe turbulent length scales in moderate to high complexity flows.

Realizable k- ε Model: The benefit of the realizable k-ε model is that it provides advanced predictions for the diffusion rate of both planar as well as round jets. It also shows superior performance for flows involving boundary layers and rotation under strong adverse pressure gradients,recirculation and separation. In every measure of comparison, Realizable k-ε signifies a superior potential to express the mean flow of the complex structures.

Geometry of NACA airfoils were acquired as co-ordinate vertices i.e., texts file and imported into the CATIA. Few modifications were made to those files to adjust the geometry and prepared avalid surface model for further steps.

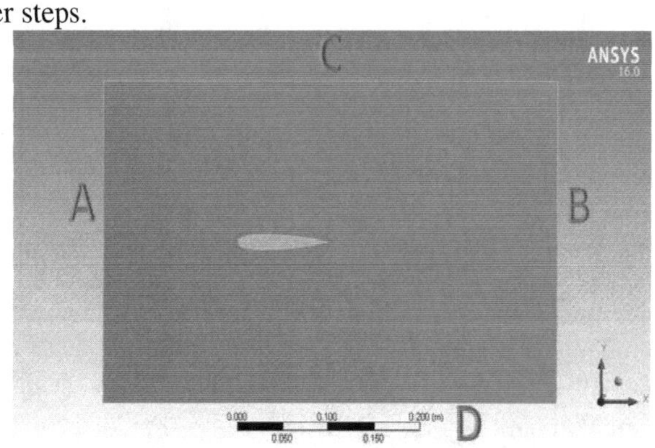

Figure 1. Geometry of airfoil

FLUENT is essential in performing of CFD analysis as it creates the working environment where the object is simulated. Thecrucial part of this is to create the mesh throughout the object. This has to be extended in every direction possible to apply the physical properties of the surrounding fluid(air). In order to apply the boundary conditions that are required to carry out analysis, mesh and edges must be grouped together for the effective output.At first, import the coordinates of the airfoil and create the curve using 2D analysisand launch the design model. Now, draw the required size of the domain for the airfoil section. The coordinate system can be created either at the leading edge or at the trailing edge of the airfoil which would be helpful in creating the rectangular mesh domain geometry using dimension tool and sketcher toolbox.

3. MESHING &FLUENT SETUP

A domain consisting of a rectangular profile of adequate dimensions is to be designedwith airfoil as an interior surface. The mesh that is generated has to be very fine at regions close to the cross section of the airfoil. For this airfoil, a structured rectangular mesh was used.But depending upon the accuracy and limitations in the software, the mesh at certain areas of the domain can be varied accordingly.

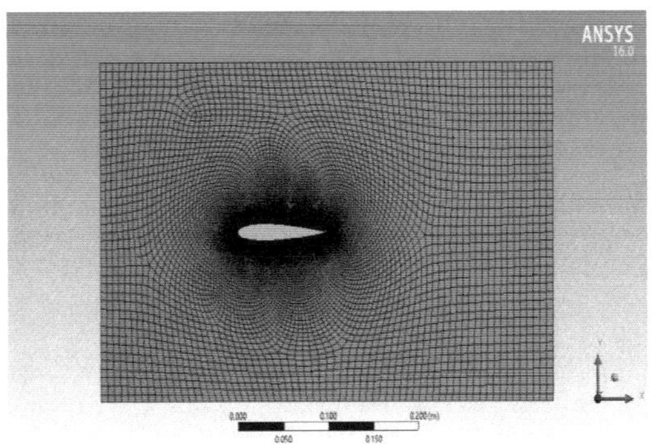

Figure 2. Meshed airfoil domain

A fine mesh implies a higher number of calculations which in turn makes the simulation use longer time to finish. For the NACA airfoils, starting from the leading edge,the distance between the nodes increases. An even number of points are distributed and analysed from the point of maximum thickness on the airfoil to the trailing edge. The meshed rectangular domain shown in Fig 2. consists of nodes and elements. Each of the airfoil sections used are meshed and contains 82,833 nodes and 81,017 elements on an average.

Assigning properties to the respective geometries is a crucialpart of the simulation to make it work. In this case, the airfoil itself was treated as interior surface. The boundaries conditions to the mesh were set as velocity as inlet at the entry and as the end boundary oroutlet, pressure at zero-gauge had been simulated. All the other surfaces are treated as walls that are stationary and having no slip conditions. The boundary conditions that are provided for this simulation is tabulated referring to Figure 1 as shown below.

Table 1. Boundary conditions of fluid domain.

EDGE	NAME	CONDITION APPLIED
A	INLET	VELOCITY
B	OUTLET	PRESSURE
C	TOP WALL	STATIONARY&NO SLIP
D	BOTTOM WALL	STATIONARY&NO SLIP

Figure 3. Flow of the air in the domain

The meshed component was imported into FLUENT, and the parameters of the system were setto Standard and Double precisionto make certain that it provides ample accuracy.Generally,as default it possesses single precision but as the solution has to be meticulous, double precision is advisable. The residuals for the various turbulence modelvariables were set to 10e and the maximum iteration count to 100.If the values of the coefficients are stabilized then also the process can be terminated.

An important part is selection of suitable viscous model for the simulation toproceed and in case of FLUENT, many choices are available.From the available viscous models, Realizable k-epsilon model is selected which uses two set of equation to arrive at the solution. Further in the selection of solution methods, the pressure-velocity coupling is SIMPLE.Then the initialisation of the solution is done in standard method from the inlet and the calculation is initiated.

Table 2. Working Parameters in Ansys Fluent.

Working Fluid	Air
Density of Air	1.225 kg/m^3
Viscosity of Air	$1.7894*10^{-5}$ kg/m-s
Solver Type	Pressure -Based
Viscous Model	Realizable k-epsilon
Pressure -Velocity Coupling	SIMPLE
Momentum	Second Order Upwind

4. GRAPHS AND OBSERVATION

The coefficients of lift and drag are used to plot graphs with respect to angle of attack at two different velocities namely at 3 m/s and 15 m/s i.e., at Reynold's number of 20000 and 100000 respectively to compare the effect of variation in flow over each of the airfoil profiles.

4.1. Comparison of Symmetrical airfoils

The above-mentioned procedure is applied to a set of symmetrical airfoils namely NACA 0012,0015,0018 and 0021. These four profiles are simulated under two velocities and the respective graphs are plotted with the results obtained for comparison with one another.

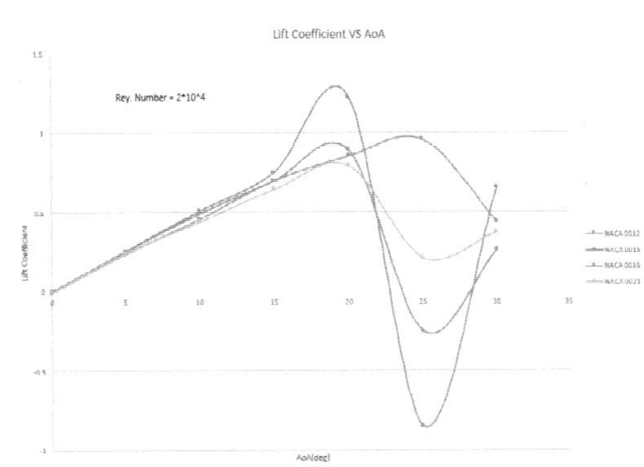

Figure 4 C_D Vs α at 3 m/s

From the graph shown in Figure 4, at a velocity of 3 m/s, the lift coefficient for all the four symmetric airfoils varies linearly till an angle of attack of 15° and then attains a peak angle (stall) and gradually decreases thereafter.

Similarly, for the plot between drag coefficient and angle of attack in Figure 5, the variation of drag is almost same for a certain angle and then changes for each of the profiles.

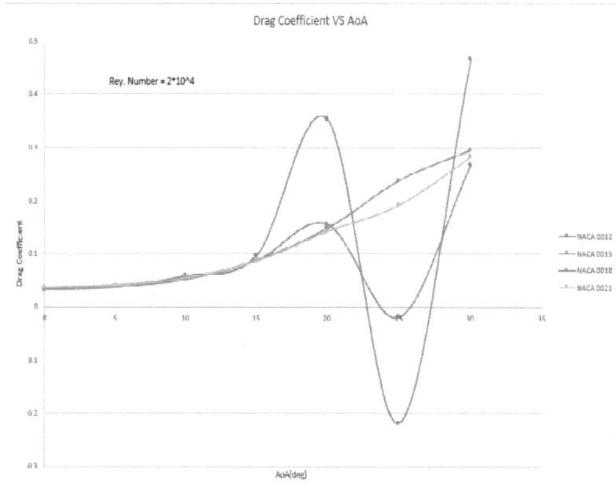

Figure 5. C_D Vs α at 3 m/s

However due to low velocity acting on the airfoil, the variation of the forces may not be as accurate as required. So, for proper outputs the same set of airfoils are analysed for a velocity of 15 m/s i.e., at a Reynold's number of100000 and the respective graphs are plotted as shown below.

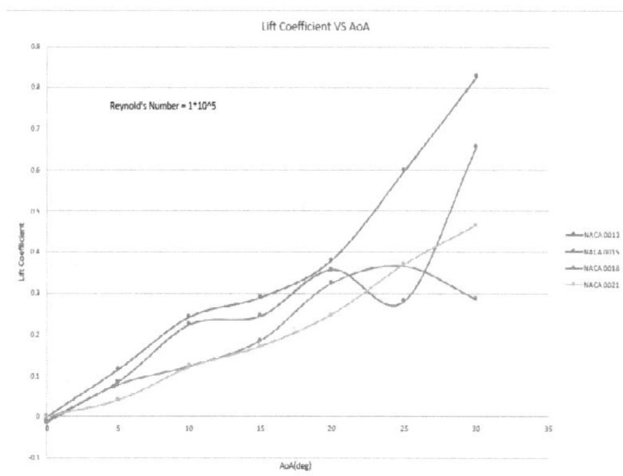

Figure 6. C_L Vs α at 15 m/s

From the Figure 6, it is observed that the curves obtained at this velocity of 15 m/s are more linear when compared with 3 m/s. The variation of lift coefficient with respect to angle of attack is minimum but airfoil NACA 0021 shows the more linear behaviour compared to the other profiles.

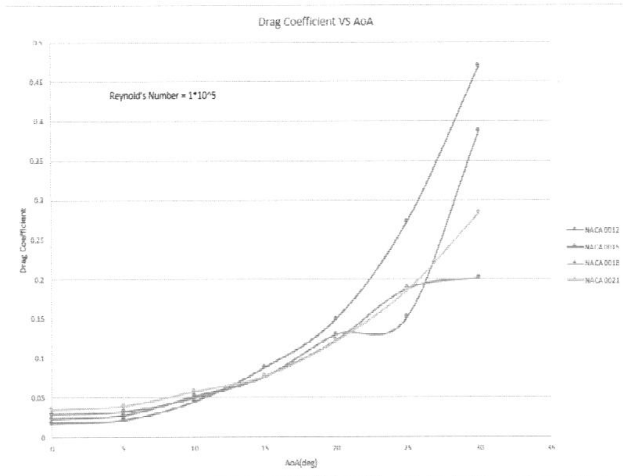

Figure 7. C_D Vs α at 15 m/s

This graph in Figure 7. shows the drag coefficient variation, also depicts that the airfoil profile NACA 0021 gives the lower drag with increase in angle of attack compared to the other three profiles in the set.

4.2. Comparison of non-symmetrical airfoils

Similar to the symmetrical airfoils, a set of non-symmetrical airfoils are taken with same thickness and chord length. The airfoils considered are NACA 0012,2412,3412 and

4412 and using lift and drag coefficients graphs are plotted against angle of attack.

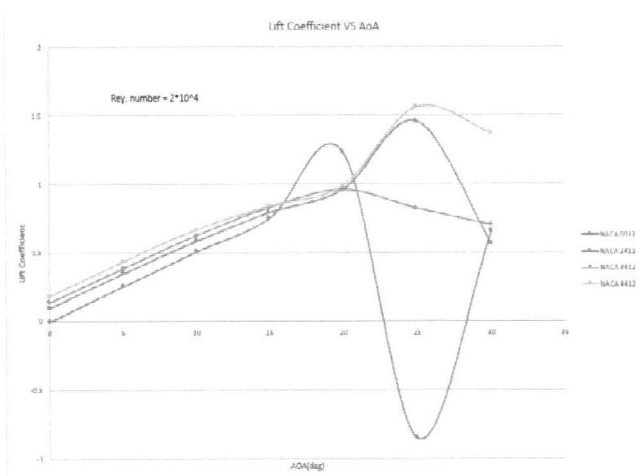

Figure 8. C_L Vs α at 3 m/s

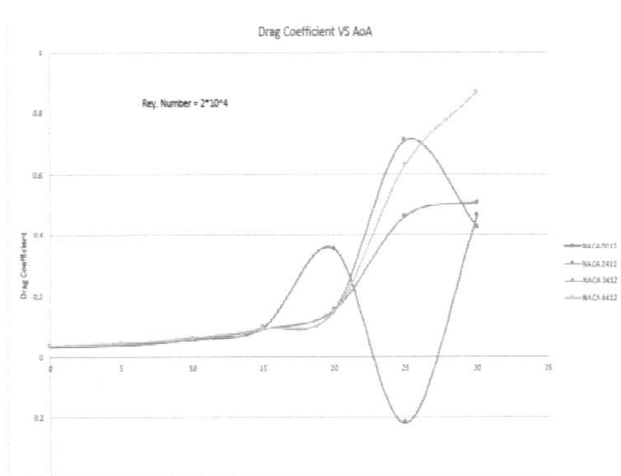

Figure 9. C_D Vs α at 3 m/s

Figures 8 & 9. shows the graphs plotted between lift and drag coefficients of non-symmetrical airfoils with respect to angle of attack. Compared to symmetrical airfoils, these profiles provide less drag forces.

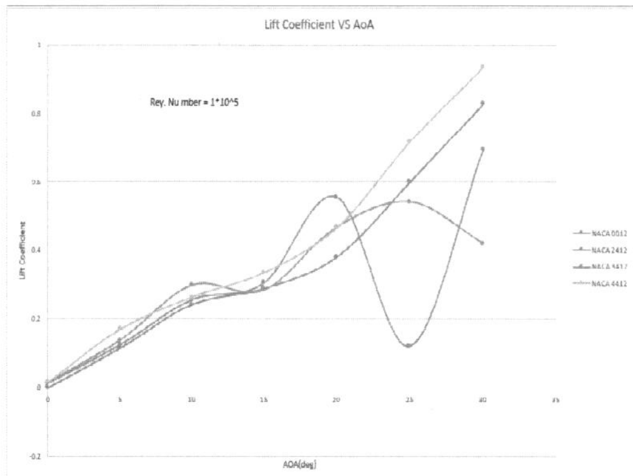

Figure 10. C_L Vs α at 15 m/s

At a velocity of 15 m/s, the variation of lift with the angle of attacks seems to be stabilised as it gives curves with less fluctuations. From the graph in figure 7, NACA 2412 shows the highest value at an angle of 20° approximately and falls thereafter. For the profiles NACA 3412 and 4412 it keeps on increasing.

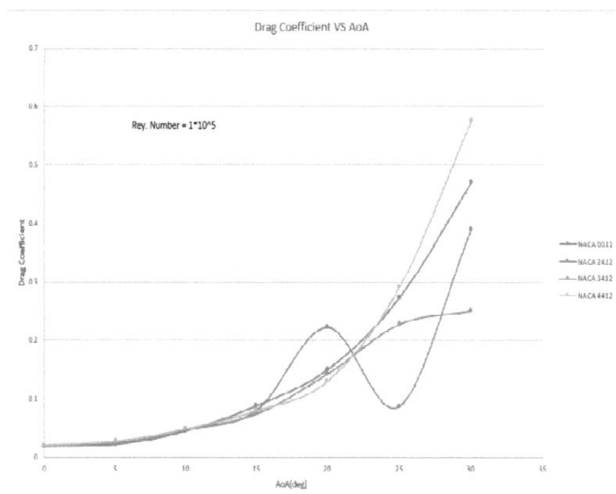

Figure 11. C_D Vs α at 15 m/s

The following are the pressure and velocity contours of two airfoils namely NACA 0021 and NACA 2412.With the help of these contours, the variation of pressure and velocity over the cross-section of the airfoils can be clearly observed and understood. The difference caused by the application of camber is distinctly visible from these contours of 4-digit airfoils. This also helps in making changes to airfoil geometry if necessary, to alter the pressure or velocity on its cross-section.

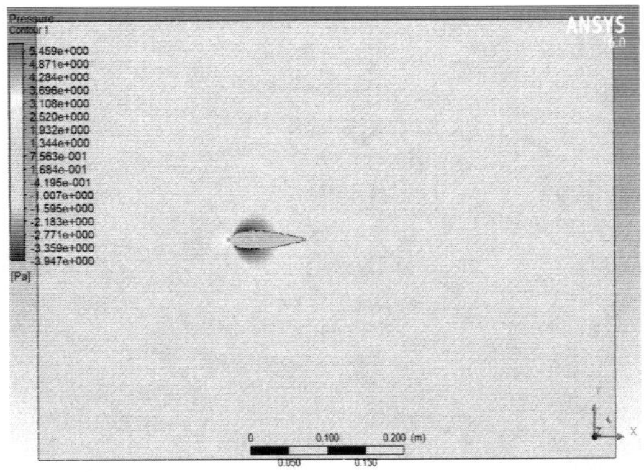

Figure 12. Pressure Contour of NACA 0021 airfoil

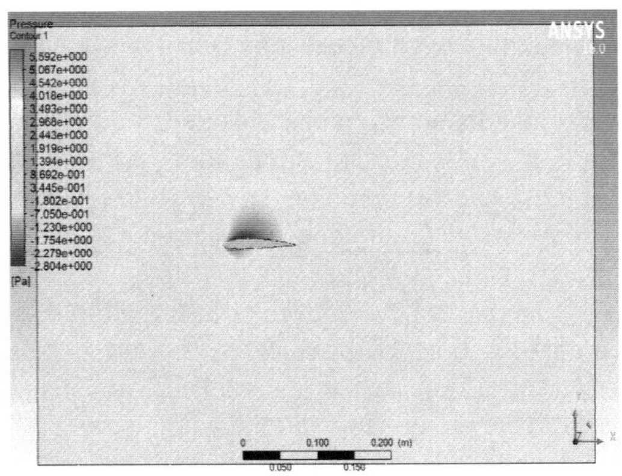

Figure 13. Pressure Contour of NACA 2412 airfoil

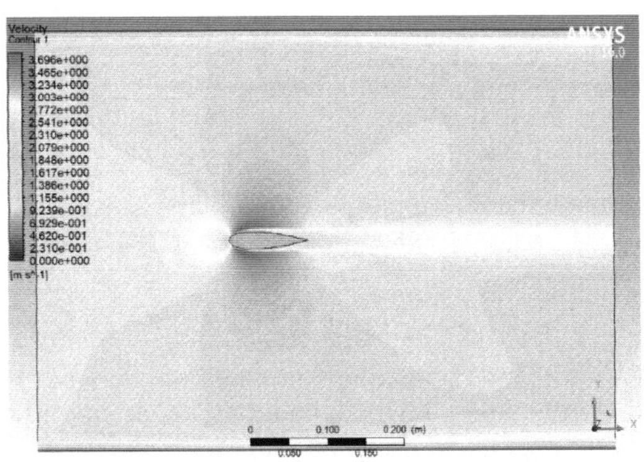

Figure 14. Velocity Contour of NACA 0021 airfoil

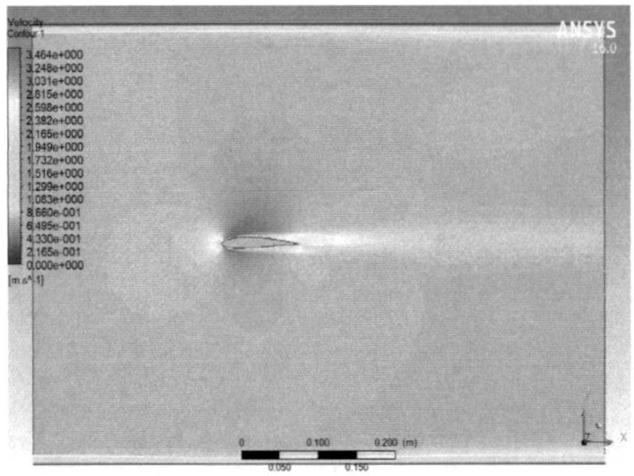

Figure 15. Velocity Contour of NACA 2412 airfoil

5. CONCLUSION

From the CFD analysis conducted on the symmetrical and non-symmetrical airfoils having same chord length at Reynold's number of20000 and 100000 respectivelywithangles of attack of0° to 30°, the following are the observations drawn from the results obtained,

1. In symmetrical airfoils, the only variant being the thickness of airfoil as it increases, the coefficient of lift as well as stall formation angle also widens. Additionally, with the increase in Reynolds number, increase and decrease in coefficients of lift and drag is observed respectively. This also in turn improves the performance of the airfoil.Also, due to increase in stall angle, it provides a wide range of angles to be used for the airfoil alignment.
2. In case of cambered airfoils, the performance of the airfoil gets improved as the value of Reynolds number increases. Compared to symmetrical airfoils, the lift coefficient values of cambered airfoils are a cut above for same thickness and also gives almost same or slightly low values of stall angle.
3. With arise in thickness,the increase in coefficient of lift and angle of stall formation is about 10%.
4. Increasing camber value to 2%in airfoils results in increaseof lift and decrease of drag comparedto the symmetrical airfoil. A slight reduction in the stall angle can be observed but still cambered airfoils perform better compared to symmetric airfoils. Eventually, there is an increment in stall angle and C_L whereas decrement in $C_D$as Reynolds number increases.
5. In order to enhance lift and delay stall formation, thickness of airfoil can be increased but doing thiscould result in some limitations like increase in cost and weight of the airfoil and also put excess load on the turbine.

REFERENCES

[1] Hurt, H. H., Jr, "Aerodynamics for Naval Aviators". *U.S. Navy Aviation Training Division*. January 1965. NAVWEPS 00-80T-80. pp.21-22.

[2] A. Meana-Fernandez, L. Diaz-Artos, J.M. Fernandez Oro & S. Velarde-Suarez., "An optimized airfoil geometry for vertical-axis wind turbine applications", *International Journal of Green Energy,*2020.

[3]. Anagha S Gowda, Comparison of Aerodynamic Performance of NACA 4412 and 2412 using Computational Approach, *International Journal of Engineering Trends and Technology (IJETT)* - Volume **67** Issue 4 - April 2019.

[4] Castineira-Martinez, E., I. Solis-Gallego, J. Gonzalez, J. Fernandez Oro, K. Arguelles Diaz, and S. Velarde-Suarez., "Application of computational fluid dynamics models to aerodynamic design and optimization of wind turbine airfoils". *Renewable Energy and Power Quality Journal*, 2014. Volume **1**,pp 370–75.

[5].Vad Mathiesen, Brian., "Smart Energy Systems for coherent 100% renewable energy and transport solutions". *Applied Energy*. 2015.Volume **145**,pp 139–154.

[6].MH Mohamed. "Performance investigation of Darrieus turbine with new airfoil shapes". *Renewable Energy Lab. of Mechanical Power Engineering*. Energy **47**(2012),pp 522-530.

[7] Mohammad Jafari, Alireza Razavi, Mujtaba Mir Hosseini,. "Effect of airfoil profile on aerodynamic performance and economic assessment of H-rotor vertical axis wind turbines", *ELSEVIER Journel*,15 December 2018, Volume **165** Part A, pp. 792-810.

[8].Bertin, John J.; Cummings, Russel M., Pearson Prentice Hall (ed.). *Aerodynamics for Engineers* (5th ed.).2009. p. 199.

[9] Jianlong Ma, Yafan Duan, Ming Zhao, WenchunLv, Jianwen Wang, Qilao Meng Ke and Yongfeng Ren., "Effect of Airfoil Concavity on Wind Turbine Blade Performances", *Hindawi Shock and Vibration*,Volume 2019, pp 1-11.

[10] Houghton, E. L.; Carpenter, P.W. Butterworth Heinmann (ed.). *Aerodynamics for Engineering Students* (5th ed.).2003. p. 17.

[11] Sorensen. N and J. Michelsen.,"Drag prediction for blades at high angle of attack using CFD". *Journal of Solar Energy Engineering*,2004.**126**,pp 1011–16.

[12]. Argyropoulos, C., and N. Markatos., "Recent advances on the numerical modelling of turbulent flows". *Applied Mathematical Modelling*,2015. Volume **39**,pp 693–732.

[13].Hidesada Kanda,KenshuuShimomukai., "Numerical study of pressure distribution in entrance pipe flow". *ELSEVIER Journal of Complexity*, June 2009, Volume **25**, Issue 3, pp 253-267.

IOP Conf. Series: Materials Science and Engineering 1168 (2021) 012012 doi:10.1088/1757-899X/1168/1/012012

Heat Transfer Analysis And Optimisation Of 2-Wheeler Engine Cyclinder Head Fins Using FEA

Anubhav Tiwari, Ankit Kumar, Himanshu Kumar, Dhananjay, Harendra patel, Manoj Sharma and Avinash Ravi Raja

Department Of Mechanical, Greater Noida Institute Of Technology, Knowledge Park-II Greater Noida, U.P 201306, India

Abstract. This research describes about the thermal analysis of fins by FEA method. Fins are extended surfaces which help to increase the heat transfer rate[1]. In this research we observe that when heat is produced in IC engine when fuel is burned, will have to be at higher level so that thermal efficiency increases, but to fend off from damages like thermal damage, useless or excess heat have to be removed from the engine. In air cooled engine, fins are placed at upper layer or on boundary of engine cylinder to increase heat transfer rate, because of this analysis of fins is very important and help to enhance the performance of engine and also increase the life span of the engine parts. The main aim of this research is to analyse the thermal properties of fins by varying certain conditions i.e. doing some modifications in conventional model, in our research work we have done two modifications in the fins geometry. In modified model-1 we have increased the number of fins, to achieve this we decreased the gap between the fins of conventional model. And in modified model-2 we have done circular perforation of 4mm diameter on the faces of all the fins of the conventional model. The cylinder head with fins are modelled on Soildworks software 2019 version and all model's analysis is performed on Ansys software. We have used Steady State Thermal of Ansys mechanical to perform our analysis.And as per the analysis the results showed that modified model-1 has the maximum heat transfer rate as compared to other models.

KEYWORDS: Thermal analysis, Fins, FEA, cylinder block, steady state thermal analysis

1. Introduction

Fins are extended surfaces which are used to increase the heat transfer from the surface and then cool various surfaces through convection process[4]. In general heat transfer by Fins having limitations because of system's design though this problem is to be rectified by doing some modifications in design framework of fins. In engine, cooling part is very important also cooling mechanism is depend upon fin design of cylinder head. In this paper we tried to increase the heat flow rate by changing the different properties in existing model such as by increasing no of fins by decreasing gaps between the fins and by doing circular holes in fin. Through 3 D modelling software solid works we design a cylinder block by considering the data of existing model and then the thermal analysis is done on ANSYS.

1.1 Technical Specifications

In this research we have bought a cylinder head of a two-wheeler from the market and measured all its dimensions with the help of measuring instruments, and all the information about the material and boundary conditions of this cylinder block is taken from a published research paper[5], the technical specifications are given below in Table 1.1

Table 1.1 Technical specifications.

S.no	Model Name	Hero Honda Splendor
1	**Swept Volume**	97.2 mm^3
2	**Stroke**	49.5 mm
3	**Bore**	50 mm
4	**Fin Material**	Cast Iron
5	**No. of fins**	11-8-11

Figure 1.1 Photo of existing model.

1.2 Equations And Assumptions used in analysis using Steady State Thermal of ANSYS Mechanical:

For a steady-state (static) thermal analysis in Simulation, the temperatures {T} are solved for in the matrix below:

$$[k(T)]\{T\}=\{Q(T)\} \quad (1)$$

This results in the following assumptions:
-No transient effects are considered in a steady-state analysis
-[K] can be constant or a function of temperature
-Temperature-dependent thermal conductivity can be input for each material property
{Q} can be constant or a function of temperature
-Temperature-dependent film coefficients can be input for convective boundary conditions

Fourier's Law provides the basis of the previous equation:
-This means that the thermal analysis Simulation solves for is a conduction-based equation.
-Heat flow within a solid (Fourier's Law) is the basis of [K]
-Heat flux, heat flow rate, and convection are treated as boundary conditions on the system {Q}
-No radiation is currently considered
-No time-dependent effects are currently considered
-Heat transfer analysis is different from CFD (Computational Fluid Dynamics)

-Convection is treated as a simple boundary condition, although temperature-dependent film coefficients are possible.
-If a conjugate heat transfer/fluid problem needs to be analysed, one must use ANSYS CFD tools instead.

2. Design and analysis of cylinder blocks:

For all the models there are only one material is used i.e cast iron[5] and their properties are given in table – 2.1

Table 2.1 Material properties

Thermal property	Cast iron
Isotropic thermal conductivity	8.3e-002 W/mm *c
Density	7.2e-009 tonne/mm^3

Also, the boundary condition[1] applied for all the models are same as shown in table 2.2

Table 2.2 Boundary conditions.

Parameter	specification
Base temperature	150°C (ramped)
Ambient temperature	35°C
Film coefficient	5e^{-6} w/mm^2 °C (step applied)
Convection matrix	Program controlled

2.1 Design and analysis of existing model

2.1.1 Modelling of existing model.

3D Model of existing cylinder block with fins is modified on 3D modelling software which is shown in fig 2.1

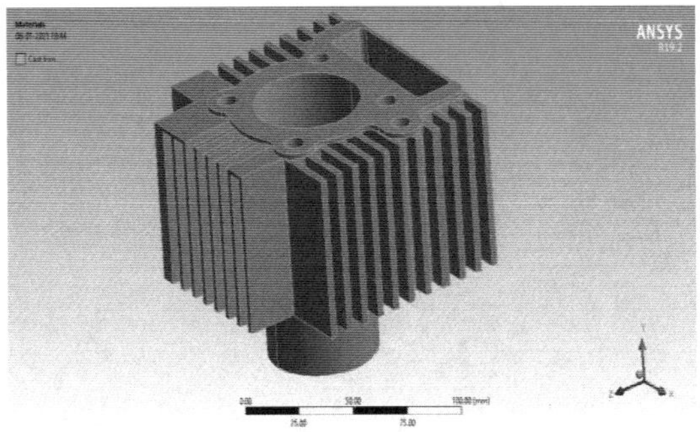

Figure 2.1 3D existing model.

2.1.2 Thermal analysis of existing model.
The existing 3D model of existing cylinder block is imported to the Parasolid format for the further thermal analysis.

A. Imported model

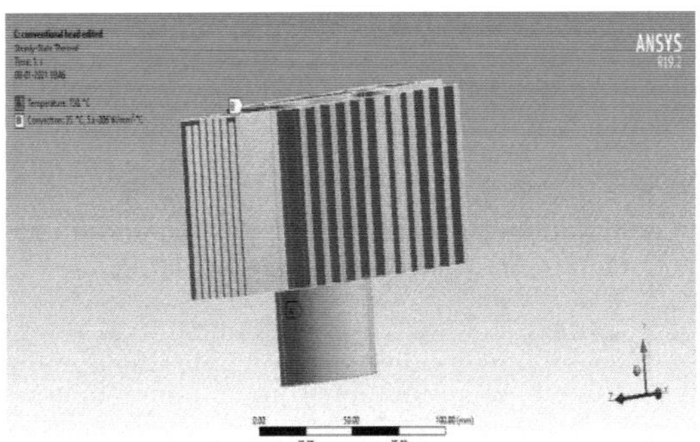

Figure 2.2 Imported existing model.

B. Meshing

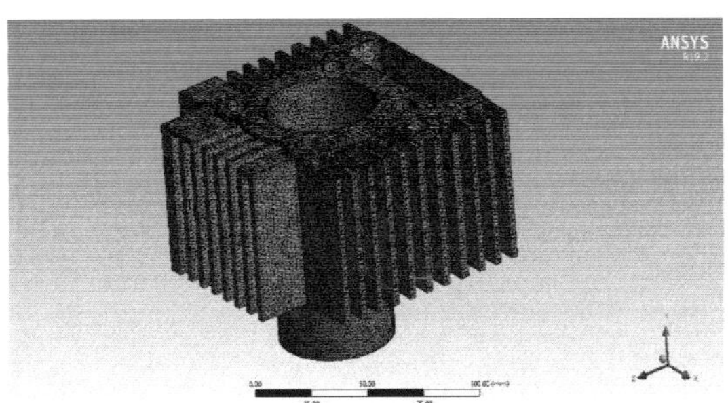

Figure 2.3 Meshing of existing model.

IOP Conf. Series: Materials Science and Engineering 1168 (2021) 012012 doi:10.1088/1757-899X/1168/1/012012

No. of nodes = 524151
No. of elements = 296649

C. Temperature distribution

Figure 2.4 Temp. distribution of existing model.

D. Total heat flux

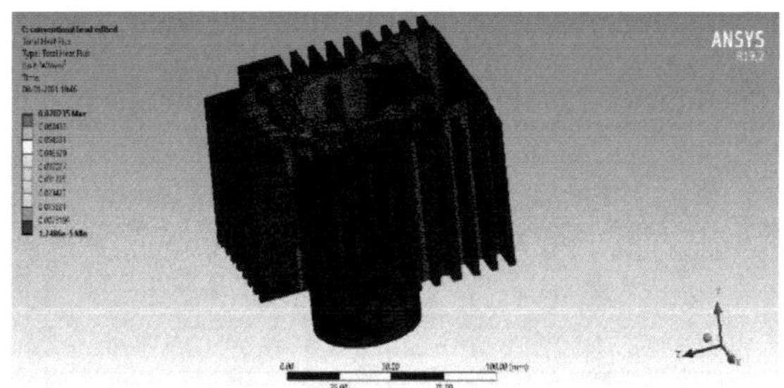

Figure 2.5 Existing model total heat flux.

E. Directional heat flux

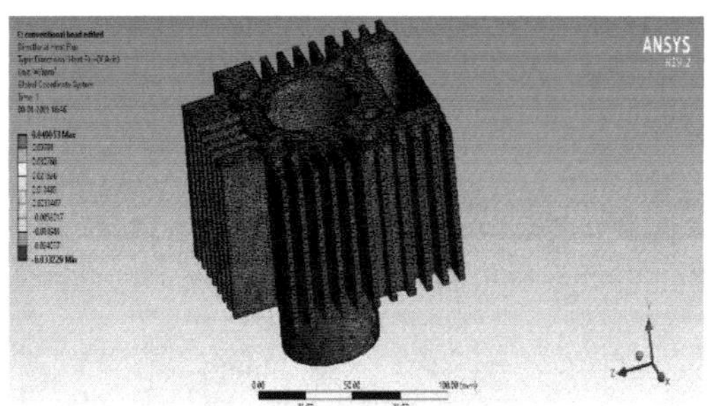

Figure 2.6 Existing model directional heat flux.

Results:
max temp: $150\ ^{O}C$, **min. temp:** $140.77^{O}C$
Total heat flux:
min: $1.74e^{-005}$ W/mm^2, **max:** $7.02e^{-002}W/mm^2$
Directional heat flux:
Max: $4.9e^{-002}W/mm^2$, **min:** $-3.32e^{-002}$ W/mm^2

IOP Conf. Series: Materials Science and Engineering 1168 (2021) 012012 doi:10.1088/1757-899X/1168/1/012012

2.2 Design and analysis of modified model-1

2.2.1 Modelling of modified model-1.
In this model no. of fins are increased by reducing the gap between fins of existing model.3D model of modified – 1 cylinder block with fins is modelled on 3D modelling software which is shown in fig 4.13.
No. of fins= 15-12-13, thickness = 3mm, gap=3mm

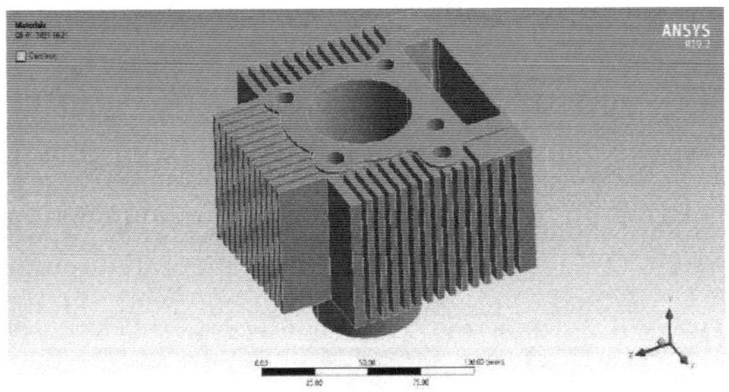

Figure 2.7 3D modified 1 model.

2.2.2 Thermal analysis of modified model– 1.

A. Imported model

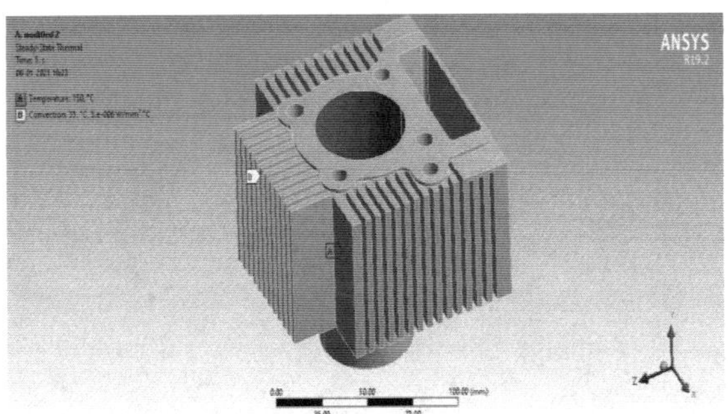

Figure 2.8 Imported modified model-1.

B. Meshing

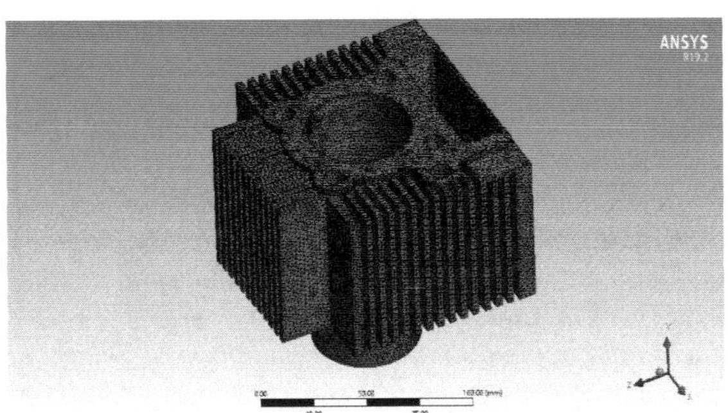

Figure 2.9 Meshing modified model-1.

No of nodes = 635363
No. of elements = 360479

C. Temperature distribution

Figure 2.10 modified model-1 temp. distribution.

D. Total heat flux

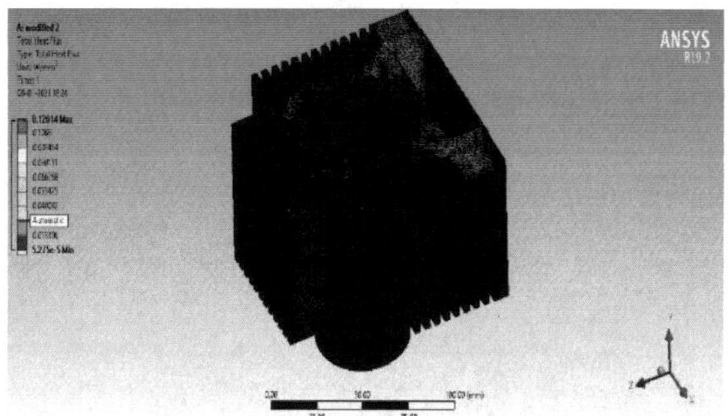

Figure 2.11 modified model-1 total heat flux.

E. Directional heat flux

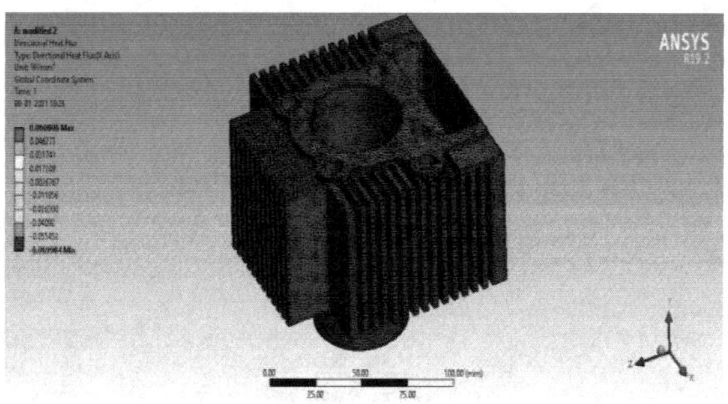

Figure 2.12 Modified model-1 directional heat flow.

Results:
Max temp. = 150^{0}C, **min temp.** = 130.24^{O}C
Total heat flux:
Max = 0.120w/mm^2, **min** = 5.2e^{-005}w/mm^2
Directional heat flux:
Max = 6.08e^{-002} w/mm^2, **min=** -6.99e^{-002}w/mm^2

IOP Conf. Series: Materials Science and Engineering 1168 (2021) 012012 doi:10.1088/1757-899X/1168/1/012012

2.3 Design and analysis of modified model-2

2.3.1 Modelling of modified model-2.
In this we have created circular penetration of diameter 4mm on the faces of fins of the conventional model and keeping all other parameters same as the conventional model and this became our modified model-2.The sketch of the circular perforation done on conventional model is shown in Figure 2.13.

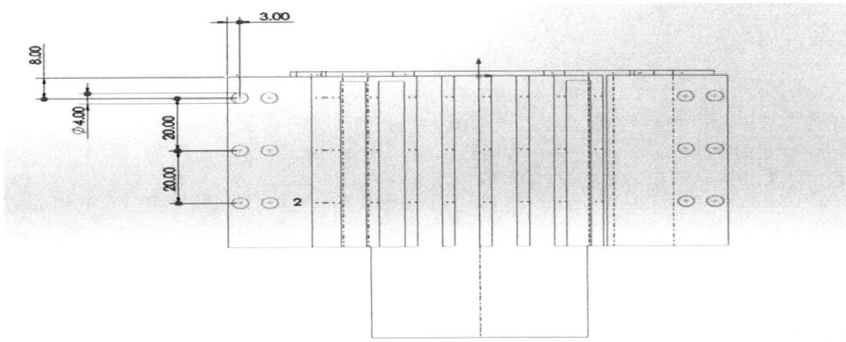

Figure 2.13 Sketch of circular perforation on fins modified model-2.

2.3.2 Thermal analysis of modified model-2.
The 3 D model of modified- 2-cylinder block is imported to the Parasolid format for the further thermal analysis.

A. Imported model

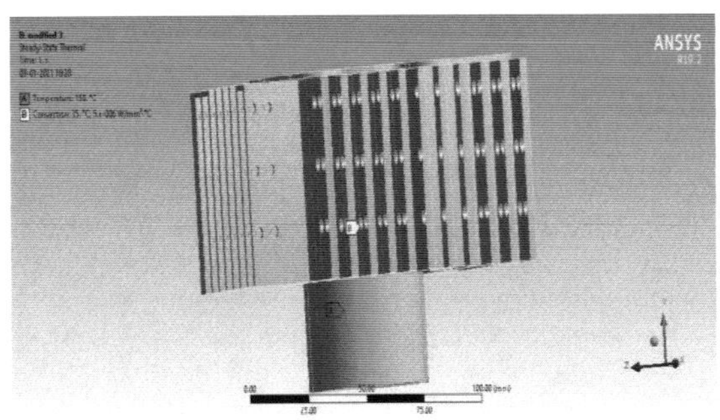

Figure 2.14 Imported modified model-2.

B. Meshing

Figure 2.15 Meshing of modified model- 2.

No. of nodes = 799456
No. of elements = 447266

C. Temperature distribution

Figure 2.16 Temp. distribution of modified model- 2.

D. Total heat flux

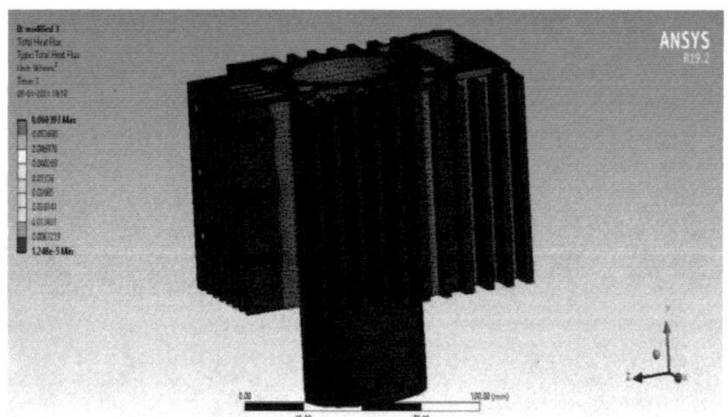

Figure 2.17 Total heat flux modified model-2.

E. Directional heat flux

Figure 2.18 Modified model- 2 directional heat flux.

Results:
Max temp. = 150°C, **Min. temp.** = 140.6 ^{O}C
Total heat flux
Max = 6.18e^{-002}w/mm^{2}, **min** = 1.24e^{-005}w/mm^{2}
Directional heat flux
Max = 5.93e^{-002}w/mm^{2}, **min** = -3.43e^{-002}w/mm^{2}

IOP Conf. Series: Materials Science and Engineering 1168 (2021) 012012 doi:10.1088/1757-899X/1168/1/012012

3. Results and discussion

In the above analysis of all the model we have got the result as temperature distribution over the fins, Total heat flux and directional heat flux in the ANSYS software. In which the modified model-1(no. of fins increased by decreasing the gap between the fins of exiting model) model gives us more cooling as compare to existing conventional model and modified model-2. Modified model-1 showed good results and it has more rate of heat transfer and it provide more cooling to the engine to work efficiently. The results of all the models obtained from ANSYS software are listed in **table 3.1** below

Table 3.1 Result of complete analysis of all the models.

Property	**Before modification**	**After modification**	
-	-	Modified-1	Modified-2
Material	Cast iron	Cast iron	Cast iron
No of fins	11-8-11	15-12-13	11-8-11
Circular perforation	Not used	Not used	Used
Fin gap	6 mm	3 mm	6 mm
Temp distribution max (^{0}C)	150	150	150
Temp distribution min (^{0}C)	140.77	130.24	140.60
Total heat flux max. (W/mm$^{2)}$	7.02e -2	12.0e -2	6.03e -2
Total heat flux min. (W/mm$^{2)}$	1.74e -5	5.2e -5	1.24e -5
Directional heat flux max. (W/mm$^{2)}$	4.9e -2	6.08e -2	5.93e -2
Directional heat flux min. (W/mm^2)	-3.32e -2	-6.99e -2	-3.43e -2
Maximum fin length	25mm	25mm	25mm
Minimum fin length	8mm	8mm	8mm

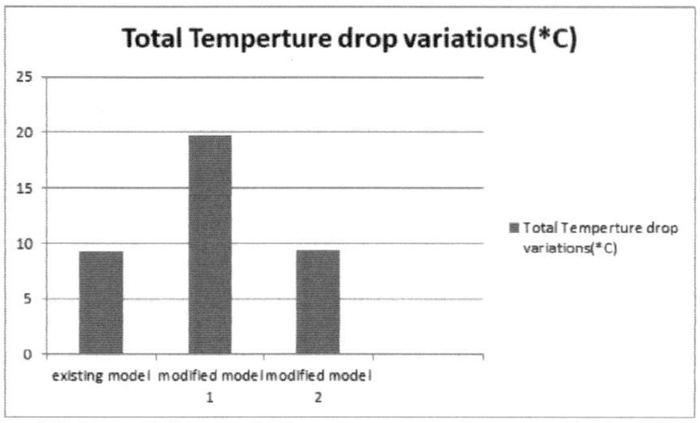

Figure 3.1 Total Temperature drop variations

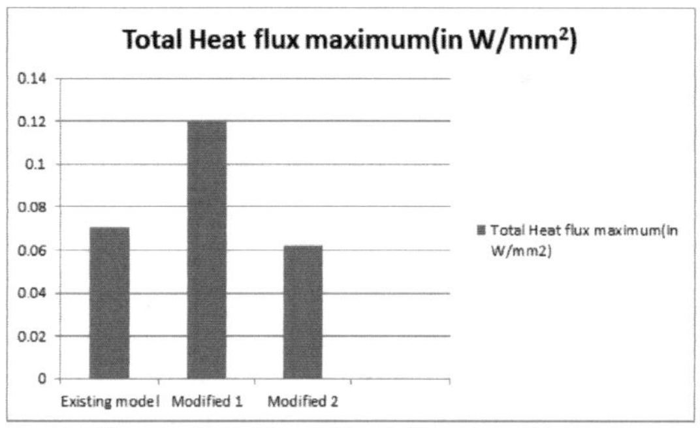

Figure 3.2 Total Heat flux maximum

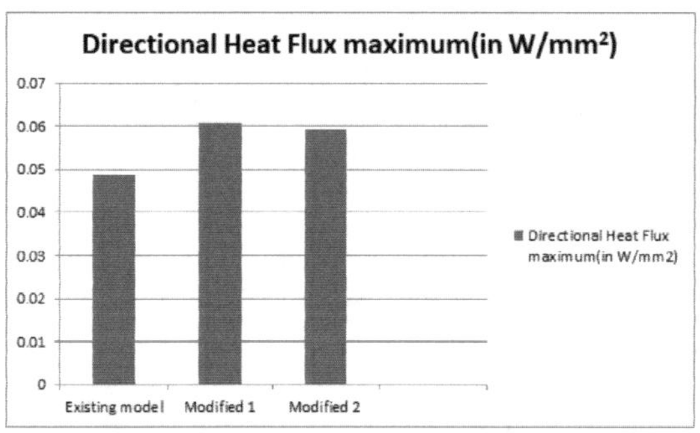

Figure 3.3 Directional Heat flux maximum

Table 3.2 Heat flux increase as compared to exiting model.

Type of Model	**HEAT FLUX (in w/mm^2**	**Increase in heat flux (in w/mm^2)**	**Percentage Increase in heat flux (in %)**
Modified - 1	.120	.0498	70.94
Modified - 2	.0618	-.0084	-11.96

IOP Conf. Series: Materials Science and Engineering 1168 (2021) 012012 doi:10.1088/1757-899X/1168/1/012012

4. CONCLUSION

Steady state thermal analysis of the two-wheeler engine cylinder block with fins with cast iron is carried out in this research. Analysis of existing model of cast iron fins in which the no of fins are 11-8-11 each side respectively and the gaps between them are 6mm. Modified-1 model of cast iron carried out in which we reduce the gaps between fins from 6mm to 3mm due to which no of fins increases from 11-8-11 to 15-12-13 each side respectively. Modified-2 model of cast iron is carried out in which we created two holes of 4mm diameter on the face of the fins. This analysis carried out in which Modifed-1 model give us more cooling as compare to Modified-2 model and conventional models due to increases in the surface area. Fins which will cause more heat transfer and cools engine more effectively. As we increase the area under the convection the cooling of engine improves and due to which engine cools effectively.

5. REFERENCES

[1] Iyengar M, Bar-Cohen A. Least-material optimization of vertical pin-fin, plate-fin, and triangular-fin heat sinks in natural convective heat transfer. InITherm'98. Sixth Intersociety Conference on Thermal and Thermomechanical Phenomena in Electronic Systems (Cat. No. 98CH36208) 1998 May 27 (pp. 295-302). IEEE.

[2] Jozaei AF, Ghafouri A. Optimization of Fin Type and Fin per Inch on Heat Transfer and Pressure Drop of an Air Cooler. International Journal of Aerospace and Mechanical Engineering. 2015 Sep 1;9(9):1657-60.

[3] Kang HS, Look Jr DC. Optimization of a trapezoidal profile annular fin. Heat transfer engineering. 2009 Apr 1;30(5):359-67.

[4] Mert Cuce P, Cuce E. Optimization of configurations to enhance heat transfer from a longitudinal fin exposed to natural convection and radiation. International Journal of Low-Carbon Technologies. 2014 Dec 1;9(4):305-10.

[5] Chandanshive A, Pise S M. Thermal analysis and optimisation of two – wheeler engine cylinder fins. International Research Journal of Engneering and Technology, 2019 May 1;6(5).

[6] Kundu B, Bhanja D. Performance and optimization analysis of a constructal T-shaped fin subject to variable thermal conductivity and convective heat transfer coefficient. International Journal of Heat and Mass Transfer. 2010 Jan 15;53(1-3):254-67.

[7] Bahadur R, Bar-Cohen A. Thermal design and optimization of natural convection polymer pin fin heat sinks. IEEE Transactions on Components and Packaging Technologies. 2005 May 23;28(2):238-46.

[8] Ahmed HE, Salman BH, Kherbeet AS, Ahmed MI. Optimization of thermal design of heat sinks: A review. International Journal of Heat and Mass Transfer. 2018 Mar 1;118:129-53.

[9] Chaitanya PS, Rani BS, Kumar KV. Thermal Analysis of Engine Cylinder Fin by Varying Its Geometry and Material. IOSR Journal of Mechanical and Civil Engineering (IOSR-JMCE). 2014 Nov;11(6):37-44.

[10] Sagar MV, Suresh N. Thermal Analysis of Engine Cylinder with Fins by using ANSYS Workbench. International Journal of Engineering Research & Technology (IJERT). 2017 Jun;6(06):2278-0181.

[11] Babu G, Lavakumar M. Heat transfer analysis and optimization of engine cylinder fins of varying geometry and material. IOSR Journal of Mechanical and Civil Engineering (IOSR-JMCE) e-ISSN. 2013:2278-1684.

[12] Tekhre D, Saini J. Design Modification and Thermal Analysis of IC Engine Fin. International Refereed Journal of Engineering and Science (IRJES) ISSN (Online). 2017:2319-1821.

[13] Alam MM, Sushmitha MD. Design and Analysis of Engine Cylinder Fins of Varying Geometry and Materials. International Journal of Computer Engineering In Research Trends, ISSN (O). 2016 Feb:2349-7084.

Most commonly used metallic biomaterials for plasma sprayed hydroxyapatite coatings

Sachin Solanke[1], Vivek Gaval[2*]

[1]Institute of chemical technology, Mumbai 400019, India.

vr.gaval@ictmumbai.edu.in

Abstract.
In load bearing implants metallic biomaterials have the first preference in orthopedic surgery because of their good mechanical properties which satisfied the prerequisite of human cortical bone. However metallic implants have limitations of poor wear resistance and release of metal ions resulting in to implant failure. To overcome these limitations hydroxyapatite based coatings on metal substrate can be used. The application of hydroxyapatite coating with the second reinforcement improve the fatigue strength of composite coatings. In this paper, commonly used metallic biomaterials for application of hydroxyapatite based composite coatings are discussed. Plasma spray coating technique with variable process parameters has been considered for this review. Presently plasma spray technique is commonly used for application of HA coatings in medical industry.
Keywords: Metallic biomaterials; Hydroxyapatite; Plasma spray technique; Process parameters.

1. Introduction:

Recently a lot of research is being carried out pertaining to biomedical engineering as far as artificial joint assembly is concerned. Orthopedic biomaterials have the largest market share as compared to all other biomaterials. The load bearing prosthesis such as hip-knee, Glenohumeral referred as shoulder and Ginglymus referred as elbow joint prosthesis is most active area of joint replacement in the current scenario. Near about five lakhs and more hip and knee joint surgery replacement are being performed per year in the world [1]. The possibility of this number is increases day by day as the population ages. Orthopedic biomaterial implants and fixing accessories are generally made of metals, ceramics or their composites. *In vivo*, longevity and performance of these load bearing units is still a big challenge for scientist, engineers and researchers. For reduction of wear rates and biocompatibility of orthopedic implants, ceramic surface modification process offers the excellent benefits. Bones are the load bearing parts of human body. Good mechanical and tribological properties are initial requirement of bone to prevent the failures like cracking, fracture, wear loss, aseptic loosening etc.

The artificial bones must have good fracture toughness on the application of high and cyclic loading during movement and action of the limbs. Poor fracture toughness results in poor wear resistance leading to wear loss during the movement and action of limbs. Generally, 80 % of bioimplant devices/ prosthesis in the medical industry are made from metal biomaterials [2, 3]. Nevertheless, the metallic biomaterials have the poor bio-function capability and the low adhesion strength between the substrate and coating layer [4]. Metallic biomaterials are especially used in the manufacturing of knee joint prosthesis, hip joint prosthesis, dentalimplants, surgical instruments, etc. Schematic of an orthopedic load bearing metallic hip and knee implant prosthesis is shown in figure 1 [5].

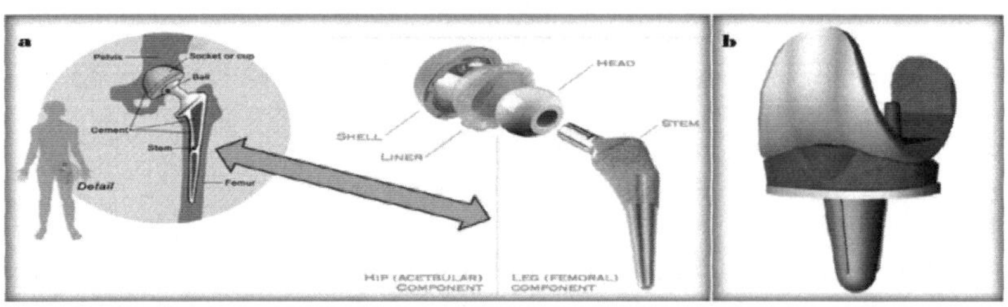

Figure 1. a) Hip Implant Prosthesis, b) Knee Implant Prosthesis [5]

1.1 Agreeable properties of artificial bioimplant material:

Following properties are essential for artificial orthopedic assembly to survive *in vivo* for 15 to 20 years in the living organism.

1. Good mechanical characteristic such as high fatigue resistance, hardness, elastic modulus, strength. Artificial

bioimplant material resist deformation when subjected to cyclic load during normal walking, jogging and stumbling. Natural bone assembly such as hip and knee joints sustain the three to eight time's load of the body weight of human being [6].

2. Good tribological properties such as high wear resistance, low friction and low surface roughness value for high surface finish. Low wear resistant biomaterial generates the wear debris when it slides against the bone tissue and articulating surface of bone assembly. This wear debris when comes in contact with living tissues or cells causes inflammation or allergy resulting in revision surgery [7].
3. High chemical corrosion resistance for bioactive, bioinertness and biodegradable artificial bioimplant material. Metal ions are released in case of low corrosion resistant biomaterials resulting in harmful and toxic reactions.
4. Good biocompatibility and Osseointegration properties. Biomaterials should give positive response to bone cells. Biomaterials should have good bone–tissue compatibility without any toxic effect. *In vivo* bone forming cells such as osteoblasts and fibroblasts shows excellent growth and proliferation on biomaterials. Implant biomaterials must prove its biocompatibility in terms of viability, proliferation and differentiation. Osseointegration ability is very decisive for orthopedic biomaterials for determining the life-time of the implants. Adhesion of fibroblast on biomaterials surface plays an important role in Osseointegration [8].

2. Types of metallic alloys for artificial bioimplant material:

The mechanical and tribological performance of metallic bioimplant make them more reliable and exceptional candidate for orthopedic applications. Generally metallic biomaterials and devices are used for the replacement of hard tissue like artificial knee and hip joint, bone screw, bone plates, dental implant and shoulder replacement. Several types of orthopedic implant biomaterials and alloys system with their physical and chemical characteristic have been studied in the research carried out in biomedical field. Out of this only a few biomaterials are presently used for orthopedic bioimplant surgery. This review paper is focused on selected biomaterials which are in great demand and are used in current scenario.

1. Stainless steel alloys:

Stainless steel was first biomaterial introduced in a total hip arthroplasty. This system has seen high generation of wear debris and high friction coefficient resulting in osteolysis and loosening of the implant. This system has number of variation over the years by modifying the surface properties for long longevity and service life. AISI 316L grade is most commonly used material in orthopedic application. The letter L in the designation signifies low carbon content. This grade is largely used in the manufacturing of surgical instruments, stents and fracture fixation parts. Some studies have reported allergic reactions due to presence of nickel content in the chemical composition of stainless steel. To avoid this allergic reaction, nitrogen ion implantation surface modification technique is employed. An Attempt has already been made by researchers to develop nickel free stainless steel without compromising on its physical properties [9].

2. Cobalt alloys:

Implant loosening or failure is a common factor affecting the total hip and knee replacement. Cobalt alloys possess high wear resistance, strength and ductility and its wear resistance is higher than stainless steel and titanium alloys. In artificial hip assembly, the femoral head is subjected to high cyclic load resulting in wear. Femoral head made from cobalt alloys exhibit good tribological properties when in contact with articulating liner. In our research study it was found that the specific wear rate and coefficient of friction of Co- Cr-Mo alloys is lower as compared to other metal alloys. Wrought cobalt alloys have high strength as compared to cast cobalt alloys so wrought cobalt alloys are in more demand. These alloys are used in dental implant, hip and knee assembly, heart valves etc. Carbides play an important role in strengthening of the cobalt alloys. Low carbon cobalt alloys have more wear resistance as compared to high carbon cobalt alloys. Modulus of elasticity of cobalt alloy is twice that of the titanium alloy. Elastic modulus of cortical bone is generally in the range of 15-25 GPa [10]. Mismatch in the modulus of elasticity between bone and implant surfaces may cause stress shielding resulting in revision of surgery.

3. Titanium alloys:

Over a past decade titanium alloys have been selected as a biomaterial for the replacement of hard tissue damaged like knee and hip joints, bone plates, fracture fixation etc. It is most commonly used in femoral stem part of hip prosthesis. The high resistance, good biocompatibility and Osseointegration of titanium alloys makes it suitable for this application. Torsional stiffness and modulus of elasticity of titanium alloys is closer to the human critical bone. This helps in reducing the stress shielding of bone. Titanium alloys are not suitable for femoral head because of their poor tribological properties. However, the strength to weight ratio of titanium alloys is higher as compared to other metallic alloys. The less weight of medical device or implant is one of most important criteria in design of artificial prosthesis in recent scenario. Several surface modification techniques were investigated to improve properties of titanium alloys. Thermal coating process is found to be most promising technique for coating titanium alloy bioimplant. In the present market, noncemented press fit stems are generally used for younger patients. This press fit stem is coated with bioceramic such as hydroxyapatite (HA) nano powder. Titanium press fit stem have good stability as compared to cemented stem. In cemented press fit stem, early failure or degradation of cement leads to poor stability resulting loosening of implants [11]. Table 1 summarizes the most commonly used metallic bioimplant materials for the orthopedic application.

Table 1. Comparison of orthopedic Metallic biomaterials in terms of their advantages, disadvantages, applications, Major alloying elements and density [11, 12, 13]

Orthopedic biomaterials	**Stainless steels**	**Cobalt alloys**	**Titanium alloys**
Advantage	1. Very low cost. 2. Commercially available. 3. Formed metal oxide sometimes act as lubricating agent.	1. High wear resistance. 2. High chemical resistance. 3. High Fatigue Strength. 4. Very Low coefficient of friction as compare to stainless steel and titanium alloys. 5. Manufacturing process (easily castable) is easy.	1. Strength to weight ratio is good as compare to stainless steel and cobalt alloys. 2. Good biocompatibility and Osseointegration. 3. Low stress shielding. 4. Light in weight (low density).
Disadvantage	1. High coefficient of friction. 2. Poor wear resistance. 3. Poor corrosion resistance. 4. Moderate biocompatibility and Osseointegration. 5. Wear generation leads to aseptic loosening. 6. High Elastic modulus. 7. Presence of Ni in some cobalt alloys causes allergy.	1. Elastic modulus is high chances to produce the stress shielding. 2. Micro wear generation leads to inflammation and allergy. 3. High density compare to stainless steel and titanium alloys. 4. Presence of Ni in some cobalt alloys causes allergy.	1. Moderate wear resistance. 2. Moderate corrosion resistance.
Application	1. Artificial Hip joint assembly (Femoral head of hip, stem part of hip prosthesis, acetabular cup). 2. Artificial knee joint assembly (Knee femoral component, Tibial component). 3. Medical surgical instruments. 4. Fracture plates and fixation.	1. Artificial Hip joint assembly (Femoral head of hip, acetabular cup). 2. Dental implants. 3. Artificial knee joint assembly (Knee femoral component, Tibial base plate).	1. Artificial Hip joint assembly (Femoral head of hip, stem part of hip prosthesis, acetabular cup) 2. Artificial knee joint assembly (Knee femoral component, Stemmed Tibial plate)
Major chemical contents/All oying elements	Fe-Iron (Balance) Cr- Chromium Ni-Nickel) Mn-Manganese Mo-Molybdenum C-Carbon)	Co-Cobalt (Balance) Cr-Chromium Mo- Molybdenum Fe-Iron Ni-Nickel	Ti –Titanium (Balance) Al-Aluminum V- Vanadim Nb-Nubian C-Carbon
Density (g/mm^3)	In between $7.98x10^{-3}$ to $10.5x10^{-3}$	In between $8.3x10^{-3}$ to $11.4x10^{-3}$	In between $4.43x10^{-3}$ to $6.5x10^{-3}$

Table 2. Comparison of mechanical properties of orthopedic metallic biomaterials with cortical bone and HA bioceramic material [14].

Material	Tensile Strength (MPa)	Compressive Strength (MPa)	Elastic Modulus (GPa)	Fracture toughness ($MPa. m^{0.5}$)
Ti-alloys	780 -1050	450 -1850	110	40 -70
Titanium (Ti-6Al-4V)	345	250-600	117	60
Stainless Steel	540 -1000	1000	200	55- 95
Cobalt alloys	600 - 1795	500 -1900	200 - 230	70- 90
Cortical Bone	50 -160	130 -180	15 - 25	2- 12
Hydroxyapatite (HA)	40 - 300	300 - 900	80 -120	0.6- 1

2.1 pie chart: % of bone mineral content:

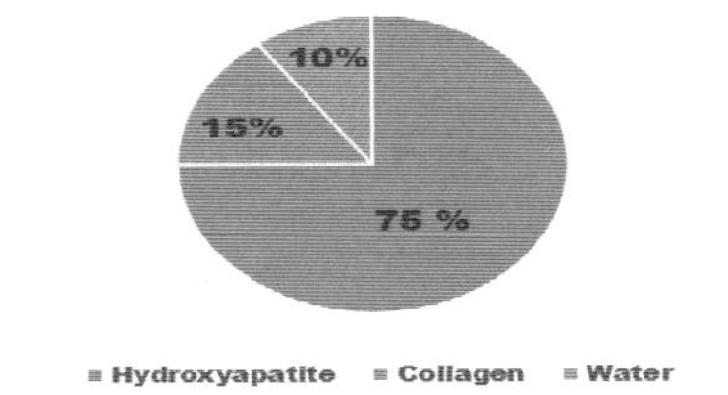

Figure 2. Bone Mineral Content.

To improve the Osseointegration and biocompatibility of bioimplant prosthesis HA ceramic has been used in last few years. The best suitable candidate for these coating are HA [$Ca_{10}(PO_4)_6(OH)_2$] due to its Osseointegration, biocompatibility and osteoconductive properties [15]. HA also shows the antimicrobial effect. HA is nothing but a calcium phosphate based ceramic and this offers physiological environment and surface for new bone growth and integration. HA is a clinically proved and acceptable ceramic whose chemical composition is same as human bones. It is largely accepted in the orthopedic industry by virtue of its inherent biological properties. Generally medical metallic implants are coated with HA. HA shows the brittle nature and low fracture toughness resulting in rapid wear and sometimes earlier fracture of artificial implant bone. From table no 2 it is evident that the fracture strength of HA is less than the fracture strength of human cortical bone. HA coating is applied to especially metal substrate for improving strength and fatigue resistance of implant coating in load bearing application. Hence there is a need to improve the mechanical and tribological properties of HA coated artificial implants without compromising on its biocompatibility. Various surface coating methods have been developed for deposition of HA on metallic surface. Thermal coating methods are mostly preferred because melting point of HA is very high approximately 1614^{0}C [16].

The most important factor during the surface modifications or consolidation is to maintain good dispersion of HA throughout the surface of implant. To control the porosity of HA coating is another important consideration because maintaining some degree of porosity and proper distribution of HA imparts good tribological, mechanical properties. Porosity helps in new bone formation (cell/tissue growth) and integration on implant surface. Appropriate degree of porosity is desirable property for orthopedic bioimplant material. Since 21st century, HA is used as a bioceramic material on orthopedic implant. Improvement of HA based coatings on the metal substrate is necessary for the future scope of human being real life application. Laser surface alloying was first technique used for HA coating on bioimplant metal substrates. Thermal spray HA coating is widely used to improve the Osseointegration and to control the healing process of metallic bioimplant prosthesis. The surface properties of base material is changed by using the surface modification technique. The surface modification can improve the tribological, mechanical and biological properties of bioimplant prosthesis. Table 3 shows the comparison of Weight % Constituents of cortical bone with HA bioceramic.

Table 3. Comparison of Weight % Constituents of cortical bone with HA bioceramic [17].

Weight % Constituents	Cortical Bone	Hydroxyapatite (HA)
Ca	24.5	39.6
P	11.5	18.5
Ca/P ratio	1.65	1.67
Na	0.7	Trace
K	0.03	Trace
Mg	0.55	Trace
CO_3^{2-}	5.8	-

Researcher recommended HA bioceramic powder not only because of its similar weight % of different elements with natural bone but also by considering its *in vivo* performance. Different studies were carried out on the *in vivo* performance of HA composite implant coatings [18].

2.2 Need of second reinforcement:

It was investigated that the HA coated bioimplant suitable for the load bearing prosthesis. Tsui et al. coated HA on the Ti6Al4V substrate by using the plasma spray technique. They have reported that HA composite shows excellent adhesion strength at the interface between the HA and Ti6Al4V surface. They have also reported good phase stability and chemical purity [19]. Fracture strength of HA 1 MPa.m$^{0.5}$ was reported which is less than the cortical bone 2 MPa.m$^{0.5}$. Hence to improve the fracture toughness of HA based bioimplant composites researcher suggested need of second phase reinforcement

3. Composite fabrication technique:

Various fabrication techniques have been investigated systematically to form the HA based bioimplant composites prosthesis. Due to high melting temperature of HA, most of processes have used elevated temperature for solidification. Challenge with these techniques is to control the dissociation of HA to other phases. During these processes HA powder experiences high temperature. These high temperature results in thermal decomposition of HA in to new phases such as tri-calcium phosphate (TCP), β-tri-calcium phosphate (β-TCP) and tetra-calcium phosphate (TTCP). For maintaining the mechanical strength of bioimplant composites it is necessary to control the dissociation of HA bioceramic. The phase purity of HA is one of the essential requirement of bioimplant coatings. As per biomedical BS ISO standard, 5% dissociation of HA in other non HA phases is allowed during the thermal composite technique. The composite fabrication technique is categorized in to two groups based on the application of bioimplant composite. The classification of composite is shown in figure 3 [20].

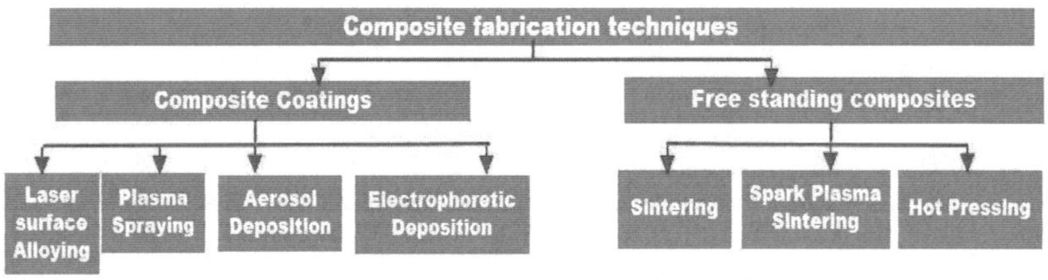

Figure 3. Classification of composite fabrication techniques [20].

Various composite fabrication process are ineffective and failed to satisfy the mechanical, tribological and biological properties of HA coatings. The main reason of failures are decomposition of HA powder particles, weak adhesion/ bonding strength between the implant and coating, overheating of metal substrates and Cytotoxicity related issues. Plasma spray process has ability to address all of the above issues and to produce coating satisfying all requirement such as purity phase of HA, good adhesion/ bond strength, good wear resistance, good biocompatibility, good bioceramic etc. [21].

3.1 Plasma spray technique:

Plasma spray technique is generally used for coating of metallic implants suchas stainless based alloys, Cobalt based alloys and titanium based alloys. For coating of metallic biomaterials plasma spray technique is US Food and Drug Administration (FDA) approved technique [22]. This process has been used in industry for many years for coating of bioceramic on metal substrates because of its high degree of flexibility, high production coating rates, low porosity and excellent bond strength. This process is commonly used for application of bioceramic HA coating. In plasma spray technique only bioceramic powders are exposed to high temperature not metallic substrate. Hence there is no chance of changing the physical and chemical properties of metallic substrate due to high heating. In the plasma spray process, the bioceramic powders in nanosize or micron size are fed in to hot plasma flame by the plasma carrier gas. This bioceramic powder is exposed to very high temperature and melted in a hot plasma. This melted bioceramic powder is driven towards the metallic substrates to from a coating. In plasma sprayed coatings generally the splats and lamellar microstructure are observed due to the layer by layer deposition of molten bioceramic powder on the metallic substrates. This technique is suitable to produce uniform thickness of bioceramic coating on the metallic substrates. Low plasma velocity in the plasma spray process results in to larger size of splat morphology whereas at high plasma velocity results in smaller size of splats morphology. In plasma spray process the degree of melting of bioceramic particle depends on the temperature of hot plasma flame, plasma velocity and size of bioceramic particle.

Figure 4. Plasma spray technique [20].

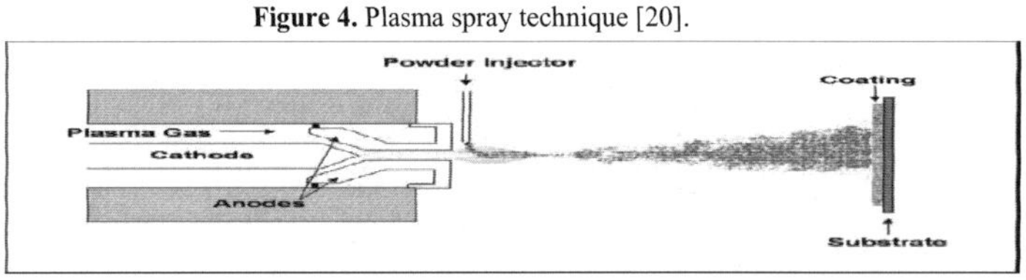

In a recent study HA coating with (002) orientation was fabricated on titanium alloy by using the atmospheric plasma spraying technique. For the coating 60 mm working distance was maintained with spray current 40 ampere.

The result demonstrates that (002) oriented coating improves the Osseointegration and mechanical properties [23]. Chou & Chang examined the microstructure of two different composite coatings on titanium alloy substrate. The composite coatings were plasma sprayed HA coating and plasma sprayed HA- 10 wt % ZrO2. They found that addition of second phase ZrO2 improves the wear resistance and fatigue strength of coatings. Various investigators closely examined the *in vivo* behaviour and its biocompatibility of plasma sprayed HA coating on titanium alloys. Plasma sprayed HA coating on Ti-6Al-7Nb alloy resulted in better osseointegration and osseoconduction ability of coating [24]. Thian et al. reported silicon-HA based composite coating can be used for artificial knee-hip implants. They deposited silicon-HA films on titanium alloy substrate and observed denser and crack free coating [25]. As compared to only HA coating, the plasma sprayed HA-CNT (Carbon nanotube) coating showed excellent improvement in terms of wear resistance, hardness and fracture strength [26]. Therefore, plasma spray technique is an efficient process for application of HA coating reinforced with second phase CNT on metal substrates. Kaleem et al. stated that reinforcement of ceramic as a second phase helps in improving mechanical properties of HA based bioimplant composite coatings. Addition of Al2O3 in HA based composite by using plasma spray showed tremendous improvement in bending strength and fracture toughness [27]. Reduction in decomposition of HA phases in to other phases has been observed on using HA-ZrO2 composite coatings with plasma spray technique. Jorge et al. reported that the hybrid composite (HA- Al_2O_3-CNT) showed threefold increase in the fracture strength than the HA coatings. These hybrid coatings result in to the lower adhesion (attachment) but have the higher proliferation rate [28]. Similar to hybrid composite coatings, ebrahimi et al. carried out research study using HA- Al2O3 and HA-SiO2 nano-composites on titanium alloy substrate. Results revealed that bi-layered plasma coating improved the cytotoxicity and proliferation rate with wettability of composites as compared to HA coating [29]. Figure 5 (a) and 5 (b) shows [30] the plasma sprayed HA coating and HA-CNT composite coating on Ti-6Al-4V substrate. Figures clearly indicate the typical splats and lamellar structure.

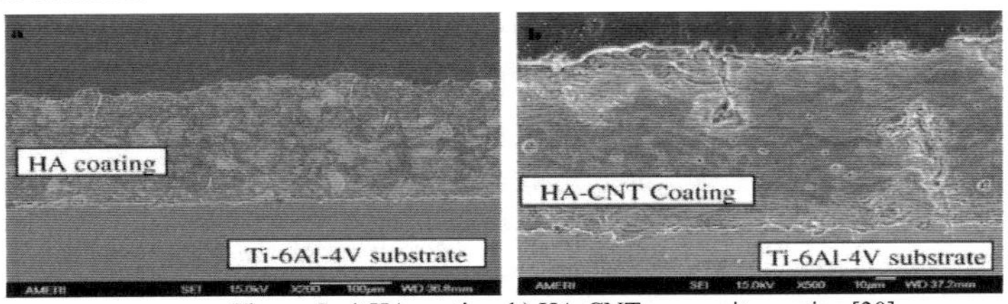

Figure 5. a) HA coating, b) HA-CNT composite coating [30].

Anstis equation was used for calculating fracture toughness. The equation [31] as follow:

$$k_{IC} = 0.016\left(\frac{E}{H}\right)^{0.5}\frac{P}{C^{1.5}} \quad (1)$$

Where K_{IC} is the fracture toughness, E is the modulus of elasticity, H is the harness of coating, P is the applied load and c is the radial crack length.

Evans and Marshall suggested the model for brittle ceramic material. This model computed the wear volume loss of brittle ceramic material. The wear volume loss is a combined function of the fracture strength, hardness and modulus of elasticity. The proposed equation [32] as follows.

$$V = P^{1.125}k_{IC}^{-0.5}\, H^{-0.625}\left(\frac{E}{H}\right)^{0\cdot 8} S \quad (2)$$

Where V is the wear volume, P is the normal applied load, H is the hardness, E is the modulus of elasticity and S is the total travelling distance by specimen on wear track.

3.2 preparation of metallic substrates for plasma spray process:

Researchers have used various surface techniques as a part of substrate preparation prior to plasma spray process. The processes are chemical etching, Micro roughing and grit blasting. Out of this grit blasting is a standard process and mostly used for preparation of bioimplant metallic substrates for HA coating. In plasma process grit blasted prepared metal surface is used for the good adhesion of implant coating. Generally, Al_2O_3 ceramic powder in micron size is used for grit blasting. Most preferred blasting angle was 750 and blasting pressure is in the range of 5 to 7 bar which has been identified by Amada and Hirose for grit blasting process [33]. The grit angle is responsible for the surface roughness of sample/substrates and this surface roughness improves the adhesion strength of plasma sprayed coatings. After the grit blasting, mostly ultra- sonic cleaning process was used to remove some of the grit particles that are embedded during the grit blasting.

3.3 Process parameters/ variables used for plasma process technique:

The quality of bioimplant coating is largely dependent on the selection of optimum process parameters in the plasma spray technique. Spraying process parameters influence the mechanical and morphological properties of bioimplant coating. User or operator can directly control the primary parameters before or during the plasma process. Secondary parameters depend on the primary parameters. The list of parameters for plasma spray process is mentioned in table 4.

Table 4. The list of primary and secondary parameters for plasma spray process [34].

Primary Parameters/ Variables for plasma spray process	Secondary Parameters/ Variables for plasma spray process
Bioceramic powder particle microstructure	Temperature of plasma flame
Bioceramic powder particle chemical composition	Velocity of plasma flame
Bioceramic powder injection angle used	Plasma flame dwell time
Plasma forming gas	Bioceramic powder particle velocity
Flow rate of plasma forming gas	Bioceramic powder particle melting
Plasma power /Current (Power = Current x Voltage)	Bioimplant substrate temperature
Carrier gas/ Carrier gas flow rate	Bioceramic particle quench rate/ Development of stress
Working spray distance	Plasma spray coating thickness

Considering the all above mentioned parameters in table 4 during the plasma spray process is not possible which is not economically accepted or plausible. So in the experimental design (DOE) it is required to select some specific important primary parameters. DOE is a desirable tool that furnish appropriate information on the basis of optimized number of experiments. Researcher have already shown the advantage of DOE while studying the plasma spray coatings technique for alumina, aluminum nitride and Zirconia materials. In few research studies, DOE tool was successfully implemented to HA coatings applied using plasma spray technique. Xiaomei et. al. recently investigated that the process parameters had significant effect on bioceramic HA coating. The morphology and phase purity of HA coatings is greatly influenced by varying spray distance and current in the plasma spray technique [35]. Larger spraying distance tend to increase the dwell time of inflight particle in the plasma flame. So due to this the velocity of particle reduces resulting in complete melting of HA powder particles leading to decomposition of HA phases. The parameters that have significant effect on the HA based bioceramic coatings during plasma process includes plasma power /Current, powder Particle Size, working spray distance, powder feed rate, plasma Forming Gases and powder Carrier Gas [36].

4. Conclusion/ summary:

The different types of most commonly used metallic biomaterials used in orthopedic surgery were reviewed in detail in this paper. The common deposition/ modification technique such as plasma spray was evaluated on the basis of process parameters. The different process parameters of plasma spray technique that affects the quality of final coating were discussed. The following conclusions are drawn from this review.

1. The metallic biomaterials such as stainless steel alloys (316L), Titanium alloys (Ti-6Al-4V), Cobalt alloys (Co-Cr-Mo) are used globally and extensively in the world due to their superior mechanical properties.
2. The mechanical and tribological properties of bioceramic coatings can be maintained by selecting the proper functional metallic biomaterials, proper fabrication coating technique and proper selection of process parameters.
3. Plasma spray technique is the commercially accepted method for fabrication of HA coating in orthopedic medical industry. A small change in process parameters can immensely affect the properties of coatings.
4. It researches out that the second phase reinforcement in HA based composite coating imparts better wear resistance, fracture toughness and biochemical properties as compared to HA coating.

References:

1. Crowninshield R., Rosenberg A., Sporer S.: Changing Demographics of Patients with Total Joint Replacement, Clin. Orthop. Rel. Research, 443: 266–272 (2006).
2. Solanke S., Gaval V.: Tribological Studies of Different Bioimplant Materials for Orthopaedic Application, ASM Science Journal, 13 (2020).
3. F. Baino F., Potestio I.: Orbital Implants: State-of-the-art review with emphasis on biomaterials and recent advances, Materials Science and Engineering: C, 69:1410-1428 (2016).
4. Hamidi M., Harun W., Samykano M., Ghani S.,Ghazalli Z., Ahmad F.: A review of biocompatible metal injection moulding process parameters for biomedical applications, Materials Science and Engineering: C, 78: 1263-1276 (2017).
5. Lahiri D., Ghosh S., Agarwal A.: Carbon nanotube reinforced hydroxyapatite composite for orthopedic application:A review, Materials Science and Engineering C, 32: 1727–1758 (2012).
6. Bergmann G., Graichen F., Rohlmann A.: Hip Joint Loading during Walking and Running, Measured in Two Patients, J. Biomech., 26 (8): 969–990 (1993).

7. Ramsden J.: The design and manufacture of biomedical surfaces. CIRP Ann.-Manuf. Technology, 56: 687–711 (2007).
8. Lahiri D., Benaduce A., Kos L., Agarwal A.: Quantification of carbon nanotube induced adhesion of osteoblast on hydroxyapatite using nano-scratch technique, IOP PUBLISHING (Nanotechnology), 22 (2011).
9. Hussein M., Mohammed A., Al-Aqeeli N.:Wear characteristics of metallic biomaterials: a review. Materials (Basel), 8 (5): 2749–2768 (2015).
10. Cuppone M., Seedhom B., Berry E., Ostell A.: The longitudinal Young's modulus of cortical bone in the midshaft of human femur and its cor- relation with CT scanning data, Calcif. Tissue Inter., 74: 302–309 (2004).
11. Paital S., Dahotre N.: Calcium Phosphate Coatings for Bio-Implant Applications: Materials, Performance Factors, and Methodologies. Mater. Sci. Eng., 66 (1-3), 1–70, 2009.
12. Wennerberg A., Albrektsson T., Jimbo R.: Implant Surfaces and Their Biological and Clinical Impact, Springer-Verlag: Berlin/Heidelberg, Germany (2014).
13. Saini M., Singh Y., Arora P., Arora V., Jain K.: Implant biomaterials: A comprehensive review. World Journal of Clin. Cases, 3(1): 52–57 (2015).
14. Ratner B., Hoffman A., Schoen F., Lemons J.: Biomaterials Science: An Introduction to Materials in Medicine (2nd Edition), Academic Press, USA, (2004).
15. S. Solanke, V. Gaval and S. Sanghavi.: In vitro tribological investigation and osseointegration assessment for metallic orthopedic bioimplant materials, Materials Today: Proceedings, 4 December (2020).
16. Teshima K., Lee S., Sakuri M., Kameno Y., Yubuta K., Suzuki T., Shishido T., Endo M., Oishi S.: Well-Formed One-Dimensional Hydroxyapatite Crystals Grown by an Environmentally Friendly Flux Method , Crystal Growth and Design, 9: 2937-2940 (2009).
17. Kuhn L., Grynpas M., Rey C., Wu Y., Ackerman J., Glimcher M.: A Comparison of the Physical and Chemical Differences Between Cancellous and Cortical Bovine Bone Mineral at Two Ages, Calcif Tissue Int., 83 (2): 146–154 (2008).
18. Ducheyne P., Qiu Q., Bioactive ceramics: the effect of surface reactivity on bone formation and bone cell function, Biomaterials, 20 (12), 2287-2303 (1999).
19. Tsui Y., Doyle C., Clyne T.: Plasma sprayed hydroxyapatite coatings on titanium substrates. Part 1: Mechanical properties and residual stress levels, Biomaterials, 19 (22): 2015-2029 (1998).
20. Lahiri D., Ghosh S., Agarwal A.: Carbon nanotube reinforced hydroxyapatite composite for orthopedic application: A review, Materials Science and Engineering C, 32: 1727–1758 (2012).
21. Ding C., Liu X., Zheng X.: Biocompatibility of plasma-sprayed biomedical coatings, Material Science Forum, 439: 30-33 (2003).
22. Sun L., Berndt C., Gross K., Kucuk A.: Material fundamentals and clinical performance of plasma-sprayed hydroxyapatite coatings: A review, J. Biomed. Mater. Res., 58: 570-592 (2001).
23. Liu X., He D., Zhou Z., Wang G., Wang Z., Wu X., Tan Z.: Characteristics of (002) Oriented Hydroxyapatite Coatings Deposited by Atmospheric Plasma Spraying, Coatings, 8: 252-258 (2018).
24. Massaro C., Baker M., Cosentino F., Ramires P., Klose S., Milella E.: Surface and biological evaluation of hydroxyapatite-based coatings on titanium deposited by different techniques, Journal of Biomedical Materials Research (Applied Biomaterials), 58: 651–657 (2001).
25. Thian E., Huang J., Best S., Barber Z., Bonfield W.: Magnetron co-sputtered silicon-containing hydroxyapatite thin films an in-vitro study, Biomaterials, 26 (16): 2947–2956 (2005).
26. S. G. Solanke, V. R. Gaval, Optimization of wet sliding wear parameters of Titanium grade 2 and grade 5 bioimplant materials for orthopedic application using Taguchi method, Journal of Metals, Materials and Minerals, 30(3), 2020.
27. Gautier S., Champion E., Assollant D.: Processing, Microstructure and Toughness of Al2O3 Platelet-Reinforced Hydroxyapatite, Journal of Eur. Ceramic Society, 17: 1361–1369 (1997).
28. Tercero J., Namin S., Lahiri D., Balani K., Tsoukias N., Agarwal A.: Effect of carbon nanotube and aluminum oxide addition on plasma-sprayed hydroxyapatite coating's mechanical properties and biocompatibility, Materials Science and Engineering C, 29: 2195–2202 (2009).
29. Ebrahimi N., Hossein Z., Vaezi M.: A new double-layer hydroxyapatite/alumina-silica coated titanium implants using plasma spray technique, Surface Coatings Technology, 352: 474–482 (2018).
30. Balani K., Anderson R., Laha T., Andara M., Tercero J., Crumpler E., Agarwal A.: Plasma- sprayed carbon nanotube reinforced hydroxyapatite coatings and their interaction with human osteoblasts in vitro, Biomaterials, 28 (4): 618-624, (2007).
31. Anstis G., Chantikul P., Lawn B., Marshall D.: A critical evaluation of indentation technique for measuring fracture toughness: I. Direct crack measurements, J. Am. Ceram. Society, 64 (9): 533-538 (1981).
32. Chiang S., Evans A.: Influence of a Tangential Force on the Fracture of Two Contacting Elastic Bodies, Journal of the American Ceramic Society, 66: 4-10 (1983).
33. Amada S., Hirose T.: Influence of grit blasting pre-treatment on the adhesion strength of plasma sprayed coatings: fractal analysis of roughness, Surface and Coatings Technology, 102:132-137 (1998).
34. Levingstone T., Ardhaoui M., Benyounis K., Looney L., Stokes J., Plasma sprayed hydroxyapatite coatings: Understanding process relationships using design of experiment analysis, Surface & Coatings Technology, 283: S0257-8972 (2015).
35. Lin B., Jean M., Chou J.: Using response surface methodology for optimizing deposited partially stabilized zirconia in plasma spraying, Applied Surface Science, 253 (6): 3254–3262 (2007).
36. S. G. Solanke, V. R. Gaval, R. K. Thakur, A. P. Pratap, Effect of varying stand-off distance on tribological and mechanical properties of plasma sprayed hydroxyapatite coated metallic substrates", Tribology in Industry, 2021.

Geometrical improvement in machined micro-hole using adaptive tool feed system in electrochemical based discharge machining

Viveksheel Rajput[1] Mudimallana Goud[2] and Narendra Mohan Suri[3]

[1-3] Production and Industrial Engineering Department, Punjab Engineering College, Chandigarh, India
Sheelrajput03@gmail.com

Abstract. Electrochemical based discharge machining (ECDM) removes the material by utilizing the thermal heating of the sparks followed by the chemical action. The tool feed method in ECDM controls the geometrical accuracies since gravity assisted and constant tool feed results in tool contacts that further deteriorates the micro-hole geometry. Moreover, high number of thermal cracks are produced with these tool feed methods. The present study investigates the adaptive tool feed system that withdraws the tool in upward direction when its contact with the work material is observed. Adaptive tool feed results in minimum tool contact and enhances the geometrical accuracies of the micro-holes. Hole circularity error (HCE) is taken as the response parameters for investigating the geometrical accuracies. The microscopy images reported a significant improvement in the geometry of the micro-holes machined with adaptive tool feed when compared to other tool feed methods. An improvement of 43.7% in HCE is acquired with the application of adaptive tool feed at 40 V. Moreover, the effect of input parameters on geometrical characteristics in terms of HCE is also discussed. The present study successfully reports the geometrical improvement of micro-holes in ECDM.

Keywords: Adaptive tool feed, tool contacts, Micro-holes, ECDM.

1. Introduction

The demand of geometrical accuracies during the fabrication of micro-features on advanced materials is increasing over the past decades ease because of increased applications in Micro-electro-mechanical system (MEMS). It further lays down the platform for investigating the new methods of machining these materials with improved geometries. ECDM is one such method that have the potential to machine with better geometrical accuracies. It is based on the principle of thermal melting and chemical dissolution to remove the material. ECDM is known to be a blend of popularly known methods i.e., electric discharge machining. (EDM) and electrochemical machining (ECM). The principle schematic diagram of the process is shown in Figure 1., consists of two electrodes dipped in aqueous electrolyte solution along with the work material. The tool electrode is built as cathode while auxiliary electrode is built as anode, NaOH and KOH are the generally preferred electrolytes for machining with ECDM. With the supply of applied voltage, there is a development of small hydrogen and oxygen bubbles at the tool electrode and auxiliary electrode respectively. The hydrogen bubbles start joining each other in order to form a big size bubble with the increase in voltage supply. These big size bubbles form an isolating gas film at the tool electrode surrounding; thereby constricting the flow of current within the circuit. Thereafter, the sparks are produced due to the electric breakdown of the film (consisting immense electric field). The spark striking is the reason that material gets removed from the work material followed by the chemical dissolution. The work material needs to be placed under the tool tip at a very small distance known as a machining gap. The process is first reported by Kurafuji and Suda [1] during glass drilling with the electric discharge. The spark mechanism in ECDM was briefly described by Basak [2,3] during micro-machining with ECDM process and highlighted those critical values of both applied voltage and current is required to initiate the sparks. Jain et al. [4] successfully described the thermal heating phenomena alongside chemical action during machining micro-holes. Wuthrich et al. [5-6] made various contributions in the field of ECDM discussing critical parameters such as tool feed, gas film etc.

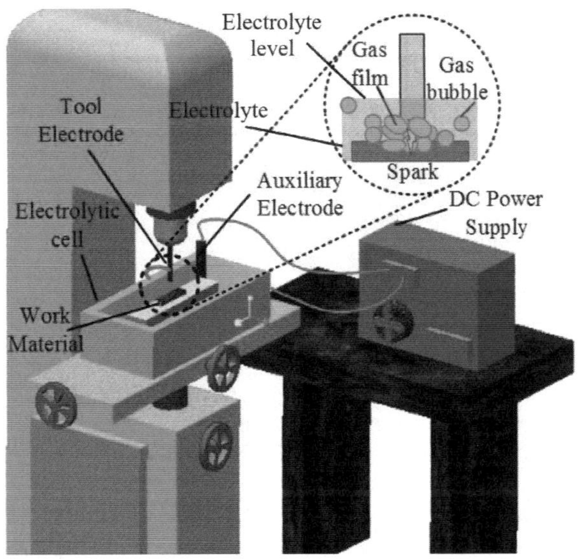

Figure 1. Principle diagram of ECDM process [7].

Rajput et al. [7] described the use of gravity tool feed for machining micro-holes on glass materials and emphasized that higher number of thermal cracks were formed with gravity tool feed method. It produces poor circularity of the micro-holes. Ziki et al. [8] demonstrated the application of constant tool feed method and elaborated that it resulted into the tool stick and jump phenomena with the material. It produces poor geometrical characteristics in terms of hole circularity error (HCE). It is observed that tool feed higher than the mean material removal causes unnecessary tool contacts and deteriorates the micro-holes characteristics. Rajput et al. [9] optimizes the ECDM parameters with tool contacts (TC) as one of the response parameters and concluded that the combined increase of applied voltage and concentration alongside low level of tool feed rate results in a smaller number of tool contacts. However, many numerical and experimental studies have been reported covering different aspects of the process and exploring different areas to improve its performance [10-11].

1.1. Problem formulation

Despite having various experimental studies, very rare studies concerning the tool contacts with the material is performed. Additionally, controlling tool feed according to the material removal is rarely described in the literature studies. Improving geometrical accuracies or HCE of the micro-holes is still a challenging task. This present study investigates the application of adaptive tool feed to improve the geometrical characteristics of the micro-holes. It monitors the tool feed with respect to the material removal rate and controls tool feed accordingly. The prevention of unnecessary tool contacts is provided to enhance the geometrical accuracies and HCE of the micro-holes. Additionally, the effect of input parameters on HCE is also observed under different machining conditions.

2. Methodology and Measurement

The experiments are conducted on a developed experimental setup using one factor at a time (OFAT) approach. The soda lime glass is selected as a work material for performing the micro-drilling operations. The cylindrical shaped stainless-steel tool of 1 mm diameter is considered as a tool electrode. Table 1 gives the machining conditions utilized for studying the geometrical characteristics of the micro-holes. The machined holes are assessed using a stereo microscope (model: Stemi-305 Trino, make: Carl Zeiss installed with Zen blue software) from the hole entrance side in order to study the geometrical characteristics. Applied voltage (V), tool feed rate (mm/min) and electrolyte concentration (wt.%) are chosen as an input parameter while geometrical accuracies in terms of HCE is chosen as a response characteristic.

Table 1. Machining conditions used in the present study.

Parameter	Range/Value	**Parameter**	Range/ Value
Tool electrode	Stainless steel	Applied Voltage	35 -50 V
Auxiliary electrode	Graphite	Electrolyte concentration	15 wt.%-30 wt.%
Electrolyte	NaOH	Tool feed rate	3-5 mm/min
Machining time	2 minutes	Inter-electrode gap	40 mm

2.1. Experimental setup

The adaptive tool feed experimental setup is fabricated in the laboratory and attached to the vertical milling machine as shown in Figure 2. It consists of force sensor or load cell that senses the presence of tool force when its contact with the material takes place. It works on the principle of closed loop system and controls the tool feed accordingly i.e., tool monitoring system. The load cell develops a signal once its contact is detected and send it to the controller. The controller further removes the tool contact by moving it in opposite direction. As a result, the tool contacts are prevented during micro-drilling process.

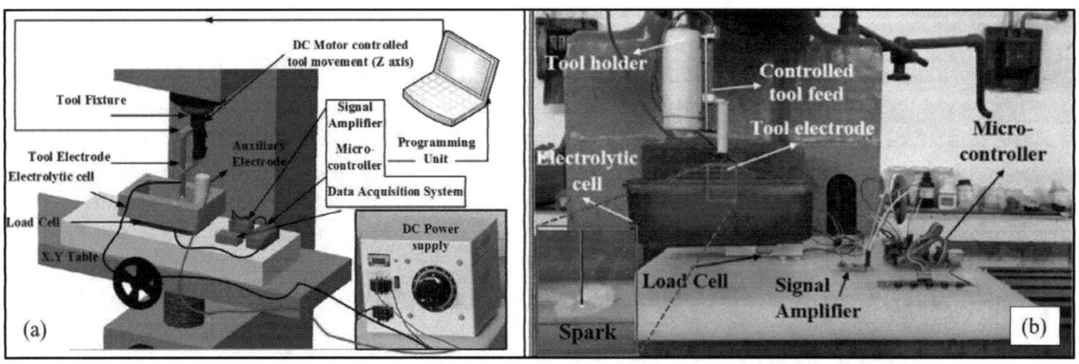

Figure 2. Developed experimental setup of adaptive tool feed system [9].

2.2. Hole circularity error (HCE)

It is described as the out of roundness error and indicates the hole geometrical accuracies. It is measured as described in Figure 3.

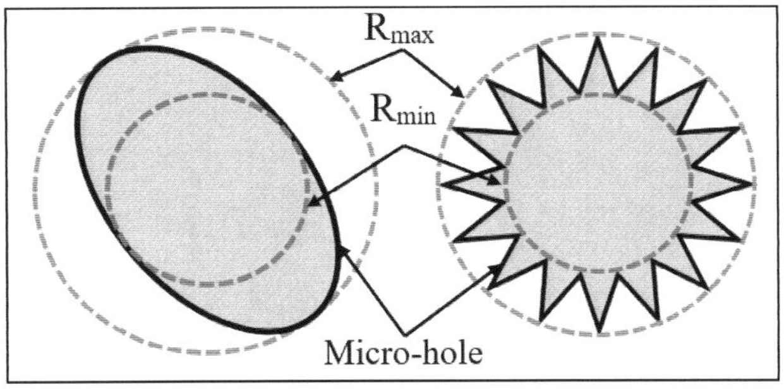

Figure 3. Schematic diagram of hole circularity error [7].

3. Results and Discussions

The microscopy images of the micro-holes are analyzed for comparing the geometrical accuracies with and without adaptive tool feed. Figure 4 shows the comparison of micro-holes machined at 35 V and 15 wt.% using 3 mm/min tool feed rate. It is seen that the micro-holes machined with adaptive tool feed system exhibits better geometrical accuracies in terms of roundness (Fig. 4(b & d)) when compared to geometrical accuracies obtained without adaptive tool feed system (Fig. 4(a)). Adaptive tool feed results in lesser number of tool contact that prevents the production of unnecessary thermal cracks. Moreover, adaptive tool feed helps in attaining a constant machining gap that enables the availability of the electrolytes under the tool for a constant and consistent formation of the gas film. As a result, more channelized sparking takes place and better geometrical accuracies are obtained. On the contrary, a permanent contact is seen in gravity tool feed that causes more thermal cracks at the hole boundary. Also, no gas film is observed under the tool tip due to its contact and hence, poor geometrical characteristics are observed (Fig.4(c)). Table 2 presents the experimental results of HCE obtained during the study. Figure 5 shows the plot of circularity error with and without application of adaptive tool feed at different applied voltages when machined with 15 wt.% NaOH concentration and 3 mm/min tool feed rate. It is depicted that the circularity error improves with the utilization of adaptive tool feed as discussed. An improvement of 43.7 % in circularity error is obtained with the application of adaptive tool feed at 40 V (Fig.5).

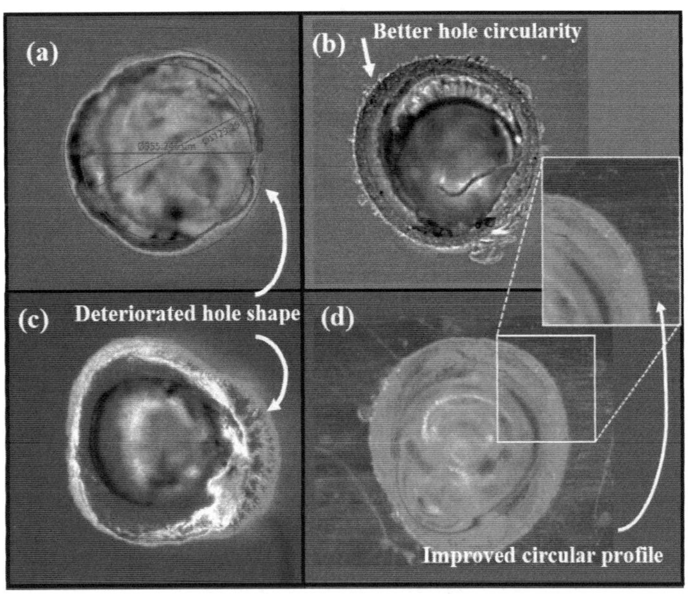

Figure 4. Comparison of micro-holes drilled with different tool feed method (a) Without adaptive tool feed (b) adaptive tool feed (c) Gravity tool feed (d) adaptive tool feed.

Table 2. Experimental data obtained in the present study.

Hole Circularity Error (HCE)								
A (V)	**Without Adaptive Tool feed**	**Adaptive Tool feed**	**B (wt.%)**	**Without Adaptive Tool feed**	**Adaptive Tool feed**	**C (mm/min)**	**Without Adaptive Tool feed**	**Adaptive Tool feed**
35	0.0331	0.029	15	0.0533	0.019	3	0.0533	0.019
40	0.0427	0.024	20	0.068	0.015	4	0.068	0.025
45	0.0533	0.019	25	0.071	0.011	5	0.078	0.034
50	0.0786	0.028	30	0.083	0.018			
55	0.0813	0.037	35	0.089	0.021			
60	0.0832	0.048	*A: Applied Voltage, B: Electrolyte concentration, C: Tool feed rate*					

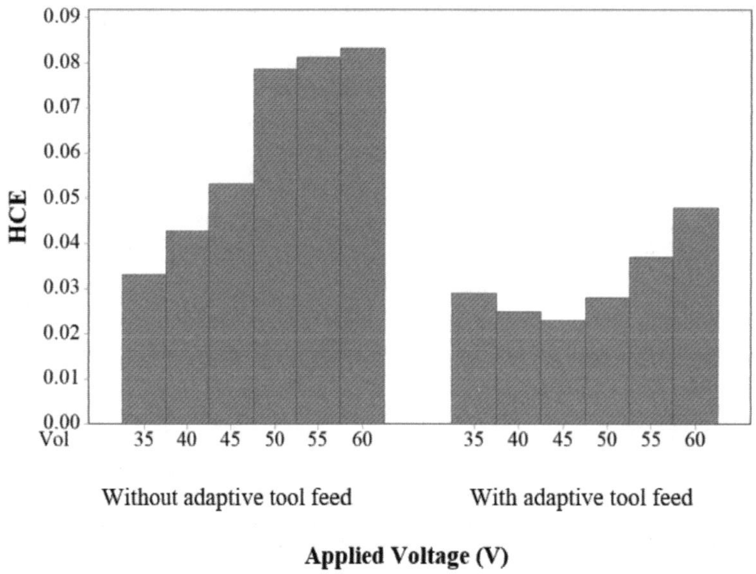

Figure 5. Comparison of HCE at different applied voltage with and without the application of adaptive tool feed system.

3.1. Effect of applied voltage on HCE

It is known fact that the level increase in applied voltage causes the increase in formation rate of small bubbles inside the electrolytes also increases that further give rise to the increase in sparking action [12]. Thus, an increase in material removal occurs that further leads to the prevention of the tool contacts. As a result, it is expected to obtain better geometrical characteristics or improved HCE since less tool contacts are obtained. But there is also a high possibility of side sparking from the tool sides that may causes a deterioration of the geometrical accuracies. Hence, poor circularity is developed. Figure 5 shows the trend of HCE with respect to applied voltage with and without the application of adaptive tool feed system. It is seen that without adaptive tool feed system, there is an increase in HCE with the increase in applied voltage due to high side sparking and high tool contacts. On the contrary, HCE improves with the level increase in applied voltage up to 45 V due to decrease tool contacts but an increase in HCE is observed with the further increase in applied voltage above 45 V at a constant tool feed rate of 3 mm/min. Gravity tool feed extensively produces a high HCE due to the tool permanent contact with the work material (Fig.4(c)).

3.2. Effect of electrolyte concentration on HCE

The effect of electrolyte concentration on HCE is observed similar as seen in case of applied voltage. HCE decreases with the increase in electrolyte concentration up to 25 wt.% and then an increase in HCE beyond 25 wt.% is observed at 45 V. Increase in electrolyte concentration leads to the enhanced frequency of the sparks that further removes the material faster. It reduces the contact of the tool with the work material. A further increase in concentration may leads to side sparking from the tool sides; as a result, poor circularity is observed [13]. Figure 6 shows the trend of HCE with respect to electrolyte concentration with and without adaptive tool feed system.

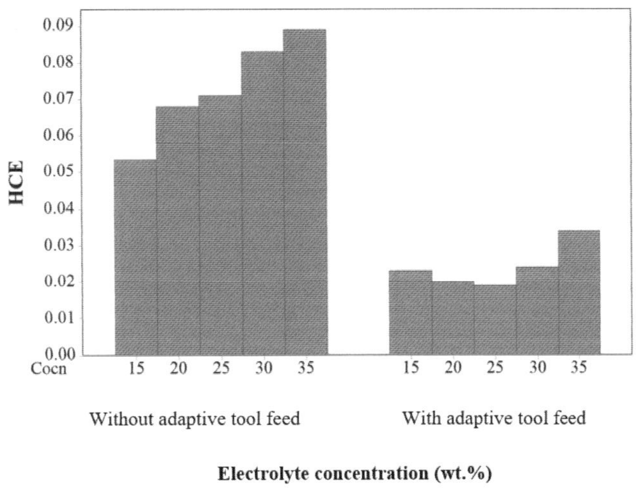

Figure 6. Comparison of HCE at different electrolyte concentration with and without the application of adaptive tool feed system.

3.3. Effect of tool feed rate on HCE

Tool feed significantly controls the geometrical accuracies of the micro-holes since too high tool feed rate results in high tool contacts. As a result, may deteriorates the hole circularity. Any level increase in applied voltage and electrolyte concentration produces more sparks over the work material. Thus, a combination of higher level of these two alongside low levels of tool feed rate results in better hole circularity. Figure 7 shows the trend of HCE with respect to tool feed rate with and without adaptive tool feed system.

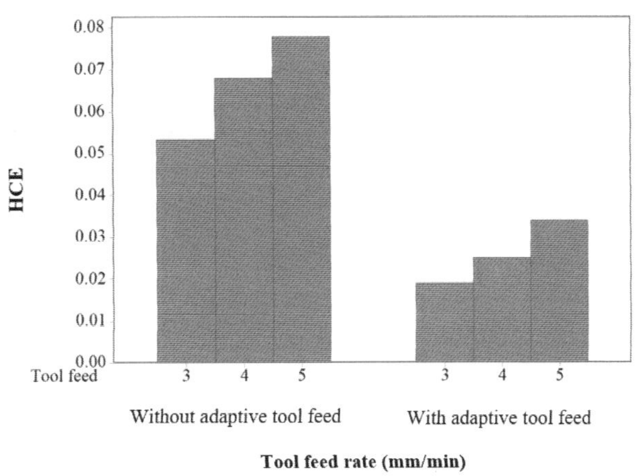

Figure 7. Comparison of HCE at different tool feed rate with and without the application of adaptive tool feed system.

4. Conclusions

The present article investigates the use of adaptive tool feed system for improving the geometrical accuracies of the micro-holes in terms of hole circularity error (HCE). The influence of input parameters on HCE is also discussed. The conclusions are given below:

- The adaptive tool feed-based system is a successful method in improving the geometrical accuracies of the micro-holes.
- Adaptive tool feed minimizes the tool contacts with the material and improves the formation of gas film under the tool tip.
- The combined effect of tool feed rate, electrolyte concentration and applied voltage control the contact of the tool with the material. Higher level of electrolyte concentration and applied voltage alongside low tool feed rate are preferable for controlling the geometrical accuracies. It produces better hole circularity during micro-drilling operation.
- An improvement of 43.7 % in circularity error is observed with the application of adaptive tool feed system in ECDM at 40 V.
- An increase in the level of applied voltage and electrolyte concentration results in better circularity up to some extent, thereafter a deterioration in circularity error is seen due to increase in side sparking.

References

[1] Kurafuji H and Suda K 1968 Electrical discharge drilling of glass *Ann CIRP* **16** 415–419.
[2] Basak I and Ghosh A 1997 Mechanism of spark generation during electrochemical discharge machining a theoretical model and experimental verification *J. Mater. Process. Technol.* **62** 46-53. DOI: 10.1016/0924-0136(95)02202-3
[3] Basak I and Ghosh A 1997 Mechanism of material removal in electrochemical discharge machining a theoretical model and experimental verification *J. Mater. Process. Technol.* **71** 350-9. DOI: 10.1016/S0924-0136(97)00097-6
[4] Jain V K, Dixit P M and Pandey P M 1999 On the analysis of the electrochemical spark machining process *Int. J. Mach. Tools Manuf.* **39** 165–186.
[5] Wuthrich R, Hof L A, Lal A, Fujisaki K, Bleuler H, Mandin P H and Picard H 2005 Physical principles and miniaturization of spark assisted chemical engraving (SACE) *J. Micromech. Microeng.* **15** S268-S275.
[6] Wuthrich R, Fujisaki K, Couthy P, Hof L A, Bleuler H (2005) Spark assisted chemical engraving SACE in micro factory. *J. Micromech. Microeng.* 276–280. DOI: 10.1088/0960-1317/15/10/S04
[7] Rajput V, Pundir SS, Goud MM and Suri NM 2020 Multi-Response Optimization of ECDM Parameters for Silica (Quartz) Using Grey Relational Analysis. *Silicon.* https://doi.org/10.1007/s12633-020-00538-7
[8] Ziki J D A and Wüthrich R 2013 Forces exerted on the tool-electrode during constant-feed glass micro-drilling by spark assisted chemical engraving *Int. J. Mach. Tools and Manuf.* **73** 47-54. https://doi.org/10.1016/j.ijmachtools.2013.06.008
[9] Rajput V, Goud MM and Suri N M 2021 Three-Dimensional Finite Element Modeling and Response Surface Based Multi-response Optimization During Silica Drilling with Closed-Loop ECDM *Silicon*. https://doi.org/10.1007/s12633-020-00867-7
[10] Rajput V, Goud MM and Suri NM 2020 Finite Element Modeling for Comparing the Machining Performance of Different Electrolytes in ECDM, *Arab. J. sci. technol.* https://doi.org/10.1007/s13369-020-05009-0
[11] Antil P 2020 Modelling and Multi-Objective Optimization during ECDM of Silicon Carbide Reinforced Epoxy Composites *Silicon* **12** 275–288. DOI:10.1007/s12633-019-00122-8

[12] Rajput V, Goud MM and Suri NM 2021 Review-Electrochemical Discharge Machining: Gas Film Electrochemical Aspects, Stability Parameters, and Research Work *Journal of the electrochemical society* **168 (1).** DOI: 10.1149/1945-7111/abd516
[13] Rajput V, Goud MM and Suri NM 2021 Performance analysis of closed-loop electrochemical discharge machining (CLECDM) during micro-drilling and response surface methodology based multi-response parametric optimisation *Advances in materials and processing technologies.* DOI: 10.1080/2374068X.2020.1860494

Acknowledgments
Authors wishing to acknowledge the assistance and support given by Punjab Engineering College (PEC), Chandigarh, India.

Energy storage using NaOH phase change material for solar thermal application at constant heat flux

B Kumar[*,1], M K Das[2] and J N Roy[1]

[1]School of Engineering Science and Engineering, Indian Institute of Technology Kharagpur - 721302, West Bengal, India
[2]Department of Mechanical Engineering, Indian Institute of Technology Kharagpur - 721302, West Bengal, India,

*Corresponding Author; First Author, School of Engineering Science and Engineering, Indian Institute of Technology Kharagpur - 721302, West Bengal, India, Tel: +91 7679554218, brajeshkumar998@gmail.com

Abstract. There is a mismatch between energy demand by consumers and supply from the sun. Therefore, we store Solar Energy in the phase change material (PCM) for storage energy. We used medium temperature range salt, Sodium hydroxide (NaOH), melting temperature at 591 K and heat of fusion 160 kJ/kg. We did the numerical simulation using COMSOL Multiphysics®, we kept radially variable surface flux at the bottom of the container. Linearly variable flux at the centre is 78 kW/m^2 and 18 kW/m^2 at the periphery. For numerical analysis, we used 3D enthalpy based model. We found that the bulk temperature of NaOH reaches about 800 K after 6 hours and total energy stored by the PCM is about 21 MJ. This amount of energy is very useful for domestic and small industrial process.

Keywords: Heat flux, Phase change material, Enthalpy based model, Numerical simulation.

1. Introduction

Solar energy has the potential to meet the energy requirements of domestic and industrial processes, but there is a time mismatch between solar energy supply and energy demand by the process. In this case, thermal energy storage allows the use of solar energy without the presence of solar radiation. Application of solar thermal energy with the industrial process will eliminate CO_2 emissions and fossil fuel consumption. However, the industry has problems to use solar energy due to ample space requirement. Phase change material (PCM) can resolve this issue and can supply heat energy at a constant rate for different thermal applications. Work reported on PCM heat storage by various authors is listed in table 1.

Table 1. Summary of work done on latent heat storage by various authors.

Methods	Procedure	Reference
1D enthalpy	Gauss-Seidel /Experimental	[1]
1D and 2D enthalpy	Analytical	[2]
2D fixed-grid enthalpy	Finite volume approach	[3]
2D fixed-grid enthalpy	Gauss-Seidel iteration	[4]
-	Experimental	[5]
Review	-	[6]
-	Experimental	[7]

There are many PCM available for thermal applications but among all the PCM, NaOH is medium-temperature range PCM having cost-effective, and good heat storage capacity. Therefore, NaOH is choosen as area of the present work.

2. Mathematical Model

The heat transfers in container and PCM are assumed to be transient and three-dimensional. Body forces and convection are neglected. The container and PCM interface are uniform. Purely conductive and radiative heat transfer is allowed for all domains. For modelling, a three-dimensional heat transfer 3-D enthalpy based model has been used. Many authors have used enthalpy based model in 1D or 2D [1-4]. In this work, we have used 3D enthalpy model in which all three dimensions are considered for simulation.

Thermophysical property of the container and phase change material which is used during numerical simulation is listed in table 2.

Table 2. Thermophysical property of the container and PCM.

Thermophysical property of Material	Density ρ (kg/m^3)	Specific heat c_p (kJ/kg-K)	Thermal conductivity k (W/m^2-K)	Heat of fusion λ (kJ/kg)	Melting temperature T_m (K)
Stainless steel (Polished) [8]	7609	0.59	23	-	1740
NaOH [8]	2130-1780	2.01-2.09	0.92-0.7	160	591

2.1 Governing equation

The governing equations are:

$$\rho\frac{\partial H}{\partial t}+\nabla.\mathbf{q}=\dddot{Q} \tag{1}$$

Where $\mathbf{q} = -k\nabla T$

For solid container

$$H = c_p T \tag{2}$$

For Solid-phase PCM

$$H = c_p T \; : \; \left(T \le T_m - \varepsilon\right) \tag{3}$$

$$H = c_p T + \frac{\lambda}{2\varepsilon}\left(T - T_m + \varepsilon\right) \; : \; \left(T_m - \varepsilon \le T \le T_m + \varepsilon\right) \tag{4}$$

For Liquid phase PCM

$$H = c_p T + \lambda \; : \; \left(T \geq T_m + \varepsilon \right) \tag{5}$$

Nondimensional Stefan number (Ste)

$$Ste = \frac{c_p \left(T_m - \varepsilon - T_b \right)}{\lambda} \tag{6}$$

Melt fraction (Φ)

$$\Phi = \left(1 - Ste \right) \tag{7}$$

where ρ (kg/m^3) is density, c_p (kJ/kg-K) is specific heat capacity, H (kJ/kg) is specific enthalpy, k (W/m-K) is thermal conductivity, $\dddot{Q}$ (W/m^3) is volumetric heat generation rate, T_b (K) is bulk temperature of PCM, T_m (K) is melting temperature, *Ste* is Stefan number, ε (K) is transition interval between solid to the liquid phase [9], Φ is solid-phase PCM, λ (kJ/kg) is the heat of fusion.

2.2 Boundary layout and grid generation

3-D enthalpy based model is computed using COMSOL 5.3a Multiphysics®. The simulation domain and grid generation are shown in figure 1 and 2 respectively. Various assumptions have been taken for the simulation setup.

- The solid container and PCM interface is uniform.
- The solid-liquid interface of PCM is progressed uniformly.
- Transport properties of the solid-liquid phases of the PCM are negligible.
- Super-heating and sub-cooling effects are neglected.
- Clearance effect between PCM and container are neglected
- The backward differentiation formula (BDF) is used for numerical simulation.

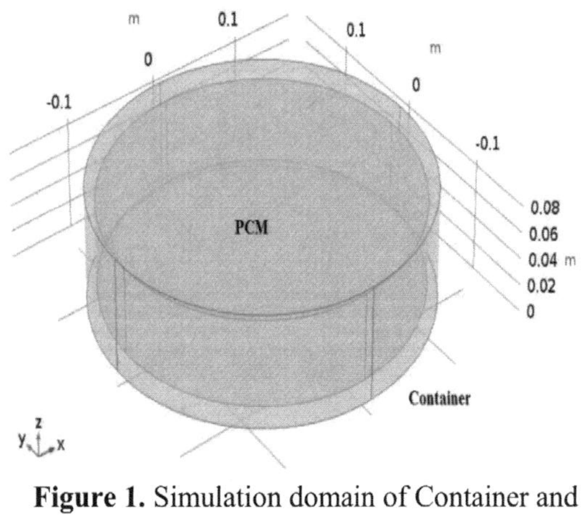

Figure 1. Simulation domain of Container and PCM.

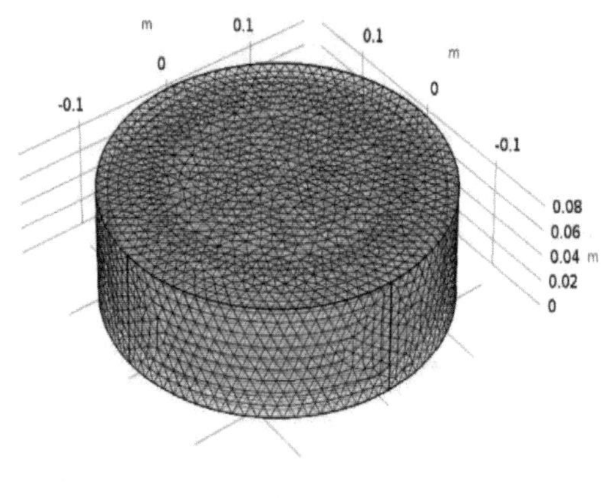

Figure 2. Grid generation of Container and PCM.

2.3 Grid independence test and validation

Three different sets of grids are generated using the mesh generation module. First set of grids 44004, the second set of grids 88106 and third set of grids 176012 are taken for simulation. Grid independence test of temperature at the center vs. time shown in figure 3. Grid set of 88106 and 176012 show identical temperature vs. time distribution at the center of the PCM. Therefore, an optimal set of grid 88106 is used for further computation.

The present computational model is validated with the results of Siyabi et al. [10]. Figure 4 shows the validation of the present work with the benchmark. The temperature variation vs time shows good agreement of the present simulation setup with the published benchmark results.

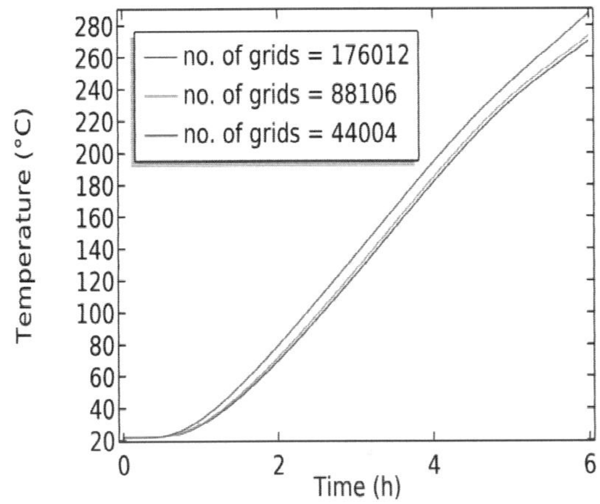

Figure 3. Grid independence test of present work concerning temperature vs. time at the center of the PCM.

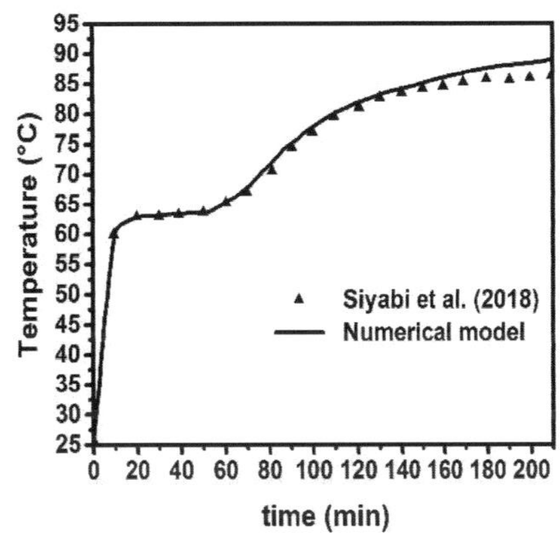

Figure 4. Validation of this work with Siyabi et al (2018) [10].

3. Results and discussions

Stainless steel polished container [9] having the outer radius of 16.7 cm, the height of 16.2 cm, and thickness 10 mm are filled with a NaOH phase change material (PCM). Mass of the PCM is selected based on receiver volumetric capacity and density of PCM. Radially variable constant inward heat flux is taken for the bottom surface, and all surface is exposed at a constant surface emissivity of 0.1 for radiation heat transfer. Heat flux on the receiver surface is shown in Table 3.

Table 3. Heat flux variation in the radial direction (R) at the bottom surface (taken from ray optics simulation for 1 kW solar parabolic dish collector system for six hours a day on receiver surface by Kumar et al (2020) [11]).

R (m)	Heat flux q_0 (kW/m^2)	R (m)	Heat flux q_0 (kW/m^2)
0.003	78.70	0.970	55.74
0.026	73.38	0.115	49.65
0.048	68.36	0.138	44.03
0.670	63.34	0.154	35.96
0.087	58.63	0.166	18.02

Initial and boundary condition:
when $t \leq 0$, T = 295 K; and when t>0, $\mathbf{q} = q_0$.

Transverse temperature contour of NaOH at 2nd, 4th and 6th hour is shown in figure 5. The temperature at bottom surfaces gradually increases from 850 K to 1025 K and temperature at the top surfaces increases from 370 K to 460 K. Figure 6 shows the temperature contour on a horizontal plane at the top, middle and bottom surface of NaOH at 6th hour. At the bottom surface temperature at the center is high in comparison with the periphery but in case of middle and top surface temperature at the center is low in comparison with the periphery. This is due to heat flux at the center of the receiver is high at the bottom surface but in case of middle and top surface temperature is dominating inward direction due to container periphery.

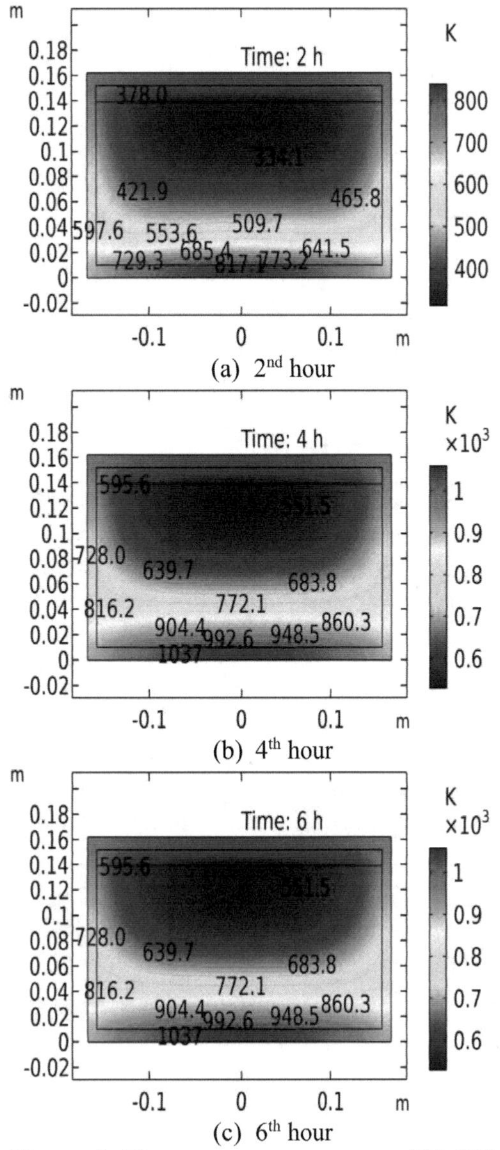

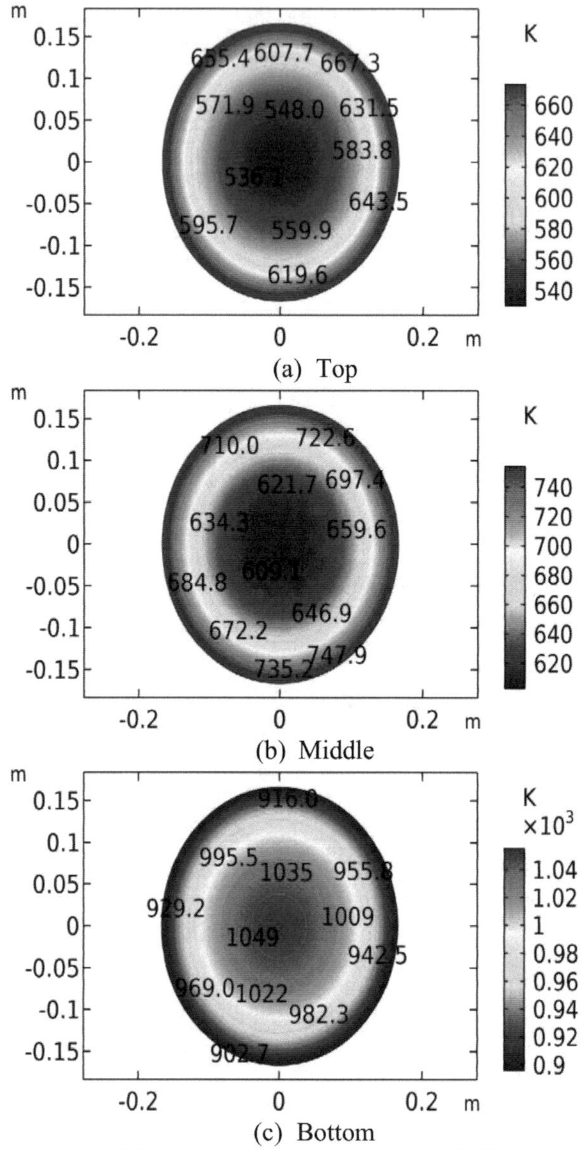

Figure 5. The temperature contour of NaOH PCM along a vertical plane at 2nd, 4th and 6th hours.

Figure 6. The temperature contour of NaOH along a horizontal plane at 6th hour on the top, middle and bottom surface of PCM.

The temperature distribution profile of NaOH along diameter at middle of PCM is shown in figure 7. Initially, the temperature inside the complete domain is 295 K and gradually temperature increases. Temperature gradient at the periphery is high and at the center is negligible but after some time the temperature gradient at center starts increasing. Figure 8 shows the temperature distribution profile of NaOH along diameter at the top, middle and bottom surfaces of PCM at 6th hour. From figure 8 it is clear that temperature at the bottom surface is high at the center, and gradually decreases up to periphery, but for the top and the middle plane temperature at the center is higher and it reduces towards the periphery.

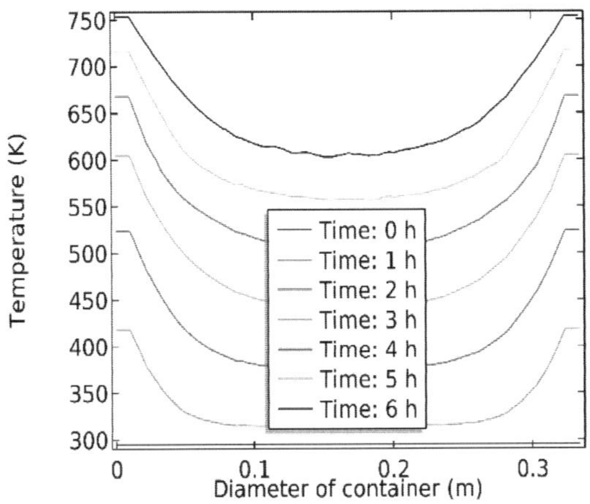

Figure 7. The temperature distribution profile of NaOH along diameter at middle of PCM after different time intervals.

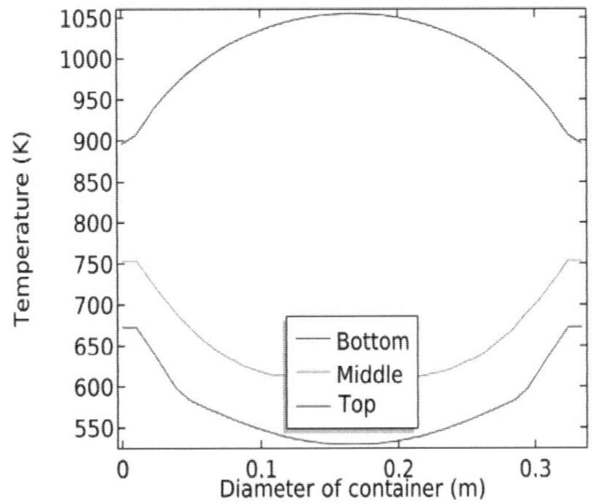

Figure 8. The temperature distribution profile of NaOH along diameter at the bottom, middle and top surface of PCM at 6th hour.

The bulk temperature and total energy storage by PCM is calculated using volumetric integration at each time intervals shown in equation 8.

$$\langle \psi \rangle = \frac{1}{\Delta V} \sum_{k=1}^{n} \int_{\forall} \psi(x,y,z,t)\mathrm{d}V \tag{8}$$

where ψ is a function, $\langle \psi \rangle$ is average of function, ΔV is control volume, $k = 1$ *to* n is time interval, (x,y,z) is spatial and t is time coordinate. Figure 9 shows the bulk temperature of NaOH phase change material. From figure 9 it is clear that as time increases, bulk temperature of the PCM increases. Initially upto 3rd hour, bulk temperature gradient with time is higher but after 3rd hour temperature gradient is slightly less, it shows that after 3rd hour melting starts. Figure 10 shows total energy storage by NaOH PCM. From figure 10 it is clear that total energy stored by the NaOH phase change material is about 21 MJ.

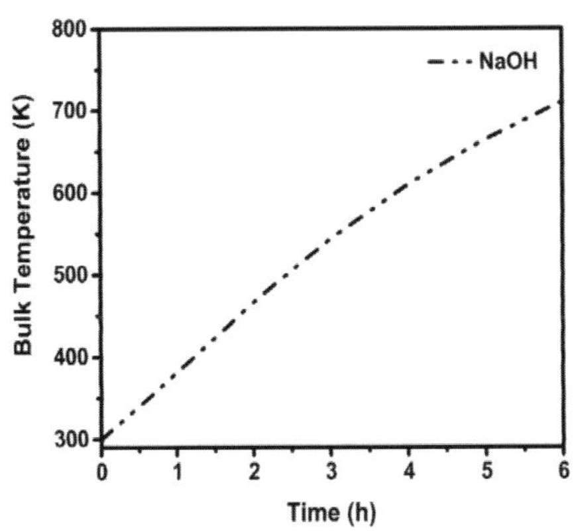

Figure 9. Bulk temperature of NaOH, PCM at a constant heat flux

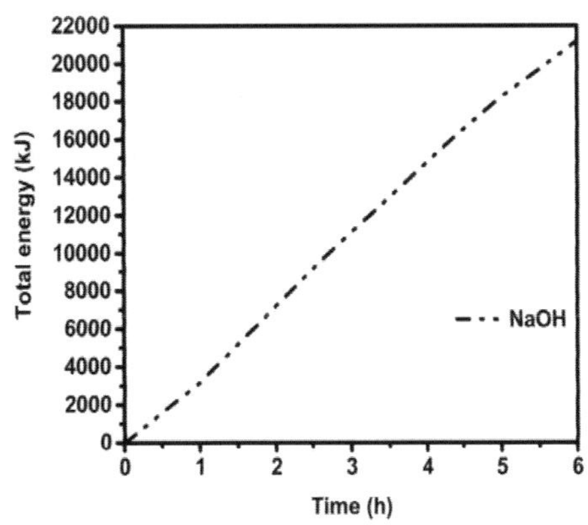

Figure 10. The total energy stored by NaOH, PCM at constant heat flux.

4. Conclusions

The temperature of the bottom surface reaches about 1050 K and top surface about 550 K. The huge temperature gradient is because of incomplete melting of the PCM. The bulk temperature of NaOH reaches about 800 K at 6th hours and total energy stored by the PCM is about 21 MJ. This amount of energy is very useful for domestic and small industrial process.

5. References

[1] Zivkovic B and Fujii I 2001 An analysis of isothermal phase change of phase change material within rectangular and cylindrical containers. *Solar energy*, **70**(1) 51-61.

[2] Manohar S S, Matthias A E, Sabharwall P, and Phil S 2010 Engineering database of liquid salt thermophysical and thermochemical properties. Technical Report, Idaho National Laboratory (INL), **10** 18297.

[3] Dacheng L, Yulong D, Peilun W, Shuhao W, Hua Y, Jihong W, and Yun H 2019 Integrating two-stage phase change material thermal storage for cascaded waste heat recovery of diesel-engine-powered distributed generation systems: A case study. *Energies,* **12**(11) 1-20.

[4] Kamal A R I and Mabruk M A 2000 Performance of a thermal storage system of the vertical tube type. *Energy Conversion and Management*, **41**(11) 116-1190.

[5] Riffat S, Mempouo B and Fang W 2015 Phase change material developments: a review. *International Journal of Ambient Energy*, **36**(3) 102-115.

[6] Elias C N and Stathopoulos V N A 2019 Comprehensive review of recent advances in materials aspects of phase change materials in thermal energy storage. *Energy Procedia*, **161** 385-394.

[7] Motahar S and Jahangiri M 2020 Transient heat transfer analysis of a phase change material heat sink using experimental data and artificial neural network. *Applied Thermal Engineering*, **167** 114817.

[8] Mills K C 2011 *Thermophysical Properties of Selected Commercial Alloys* (Woodhead Publishing Limited Cambridge England).

[9] Ahmet S and Kamil K 2002 Thermal and heat transfer characteristics in a latent heat storage system using lauric acid. *Energy Conversion and Management*, **43**(18) 2493-2507.

[10] Siyabi I A, Khanna S, Mallick T, and Sundaram S 2018 Multiple phase change material (PCM) configuration for PCM-based heat sinks-an experimental study. *Energies*, **11**(7) 1629.

[11] Kumar B, Das M K and Roy J N 2020 Heat energy storage using parabolic dish collector in $LiNO_3$ phase change material. *Journal of the Indian Chemical Society* Vol. **97**(10a),1-5. (Accepted)

Acknowledgements
The work is supported by the Indian Institute of Technology Kharagpur. The first author also acknowledges the financial support of the Ministry of Education Academic Research Fund for Doctoral Students.

Assessment of Electrical and Thermal Performance of Photovoltaic Thermal Air Collector in the Climate of Pantnagar, Uttarakhand

Tarun Singh Samant*, L Varshney, P Maheshwari, P K Vishnoi and P S Gautam

Department of Mechanical Engineering, College of Technology, G. B. Pant University of Agriculture and Technology, Pantnagar, Udham Singh Nagar, Uttarakhand 263145, India

E-mail: tarun.samant@gmail.com

Abstract. The photovoltaic thermal collector (PVT) produces electricity and extracts heat from a common area exposed to solar radiation. The present work provides a methodology to assess the potential of PVT at any location. The climate of Pantnagar (a town situated in the Terai belt of Uttarakhand) is selected for the study. An unglazed PVT air heating collector is modelled, taking account of both electrical and thermal aspects of design. The performance of the collector is then simulated on four days, representing average weather of four seasons observed in North Indian Plains, in correspondence with the typical meteorological year (TMY) data of 2006-2015 for the location. The diode parameters and characteristic curves of PV module selected for PVT are obtained for reference and operating conditions. The operating temperature of solar cell and heated air are plotted after convergence of numerical solution at available data points. Finally, electrical and thermal power output of the PVT with associated optical and thermal losses are obtained. The highest electrical power output of 223 W (136 W/m^2) is observed at 12:20 hours (IST) on equinox of September while highest thermal power output of 411 W (251 W/m^2) is obtained at 12:20 hours (IST) on summer solstice.

1. Introduction

The photovoltaic thermal (PVT) collector is a promising hybrid technology with the benefit of simultaneous electrical and thermal production from a common land or roof area. It is beneficial when the area available for installation is limited. It has been proved that the PVT collector gives increased total yield compared to that provided by photovoltaic module and thermal collector, installed side-by-side on the same area [1]. The increased yield from the same area thus reduces the payback time of the investment. Despite major developments in PVT systems around the world, India is lagging far behind in developing a PVT market with only 7 m^2 of flat-plate PVT water collectors (total installed capacity of 4 kW_{th} thermal and 1 kW_p electrical) and 255 m^2 of PVT concentrators (total installed capacity of 132 kW_{th} thermal and 27 kW_p electrical) installed till 2019 when compared with the total of 1,166,888 m^2 of installed PVT collector area (total installed capacity of 605,910 kW_{th} thermal and 207,655 kW_p electrical) worldwide [2]. A major reason behind low acceptance of this technology in India is a lack of supporting studies which demonstrate the PVT potential in Indian climate.

The earliest studies on PVT collectors appeared in the second half of the 1970s [3–5]. Many researchers contributed in the development of the performance prediction models for PVT collectors and systems in the intermediate period [6–10]. Major developments in the PVT field have been witnessed in the past decades [11–12]. Recently, task 60 of Solar Heating and Cooling Programme addressed the issue of low awareness about PVT in heating, ventilation and air conditioning (HVAC) industry with progress in the development of international standards for PVT collectors [13].

The purpose of the study is to simulate the performance of an unglazed PVT air collector with the climatic parameters typically observed in Pantnagar (29.02°N 79.48°E), a town located at an elevation of 229 m above sea level in Terai region of Uttarakhand. The air heated by PVT may be used for space heating during cold months or drying of agricultural produce in any season and helps in achieving renewable targets for a sustainable future. The climate of Pantnagar is hot and humid in summer and monsoon while cold and dry/foggy during winter. The average temperature peaks up to a maximum of 40.6°C in May and drops down to a minimum of 1.2°C in January [14]. Four days (solstices and equinoxes) from the 2006–2015 typical meteorological year (TMY) data of the location are chosen as representative of the average climate of four seasons typically observed in northern India. The data of these days also include weather uncertainties as observed in real time satellite images to simulate dynamic behavior of PVT in actual conditions. For example, there is a fair probability of fog in morning during winter season in Pantnagar which reduces the irradiance as reflected in data.

2. Methodology

2.1. Meteorological Data

The accurate climate data of any place can be recorded by a weather station. Due to costs and time associated with collection of real data on the site, data from satellites is often used in prediction studies. The meteorological data is obtained from PVGIS database in present study [15]. The solar insolation data is available in hourly, daily, monthly, and TMY formats. The TMY of 2006–2015 obtained from PVGIS for the selected location use data of 2006 for March (Spring equinox) and June (Summer solstice) while data of 2014 for September (Autumn equinox) and data of 2008 for December (Winter solstice). The days corresponding to the occurrence of these events in the respective month of the year are selected for study. The data of the corresponding days are plotted concisely in figure 1.

The dew point temperatures, plotted in figure 1, are obtained from CoolProp thermophysical property library [16] with atmospheric pressure, ambient temperature and relative humidity as input parameters. The sky temperature (T_s) is calculated from equation (1) with ambient temperature (T_a), dew point temperature (T_{dp}) and hour from midnight (t) as input parameters (units in parentheses) [17] :

$$T_s(\mathrm{K}) = T_a(\mathrm{K})\left[0.711 + 0.0056T_{dp}(°\mathrm{C}) + 0.000073T_{dp}^2(°\mathrm{C}^2) + 0.013\cos 15t(h)\right]^{1/4} \tag{1}$$

2.2. PVT Design Data

The PV module selected to design PVT is JJ-M660 (Jain Photovoltaic) with a peak power of 245 W at standard test conditions (STC). The electrical characteristics of PV module, design and thermal characteristics of materials selected to construct the integrated photovoltaic thermal collector (PVT) and operating conditions are specified in table 1. A conceptual line diagram to visualize the layers of unglazed PVT air collector investigated in present study is shown in figure 2.

2.3. Electrical Model

The five parameter single diode model is used to simulate the $I-V$ curve from the Manufacturer's specification data. The initial points are obtained by analytical approach while the solution is obtained by numerical approach [18]. The well-known five parameter, single diode equation of current (I) and voltage (V) is given as [9]:

$$I = I_L - I_o[\exp\{(V + IR_s)/a\} - 1] - (V + IR_s)/R_{sh} \tag{2}$$

where a is ideality factor, I_L is light current, I_o is reverse saturation current, R_s is series resistance and R_{sh} is shunt resistance. These are the five unknown parameters which need to be solved simultaneously from the equations which are obtained by applying open circuit, short circuit and maximum power conditions to equation (2). A set of translation equations are then used to obtain these parameters at operating conditions [10]. The electrical efficiency is then calculated at the maximum power point of the operating $I-V$ curve, obtained by solving equation (2) with operating parameters.

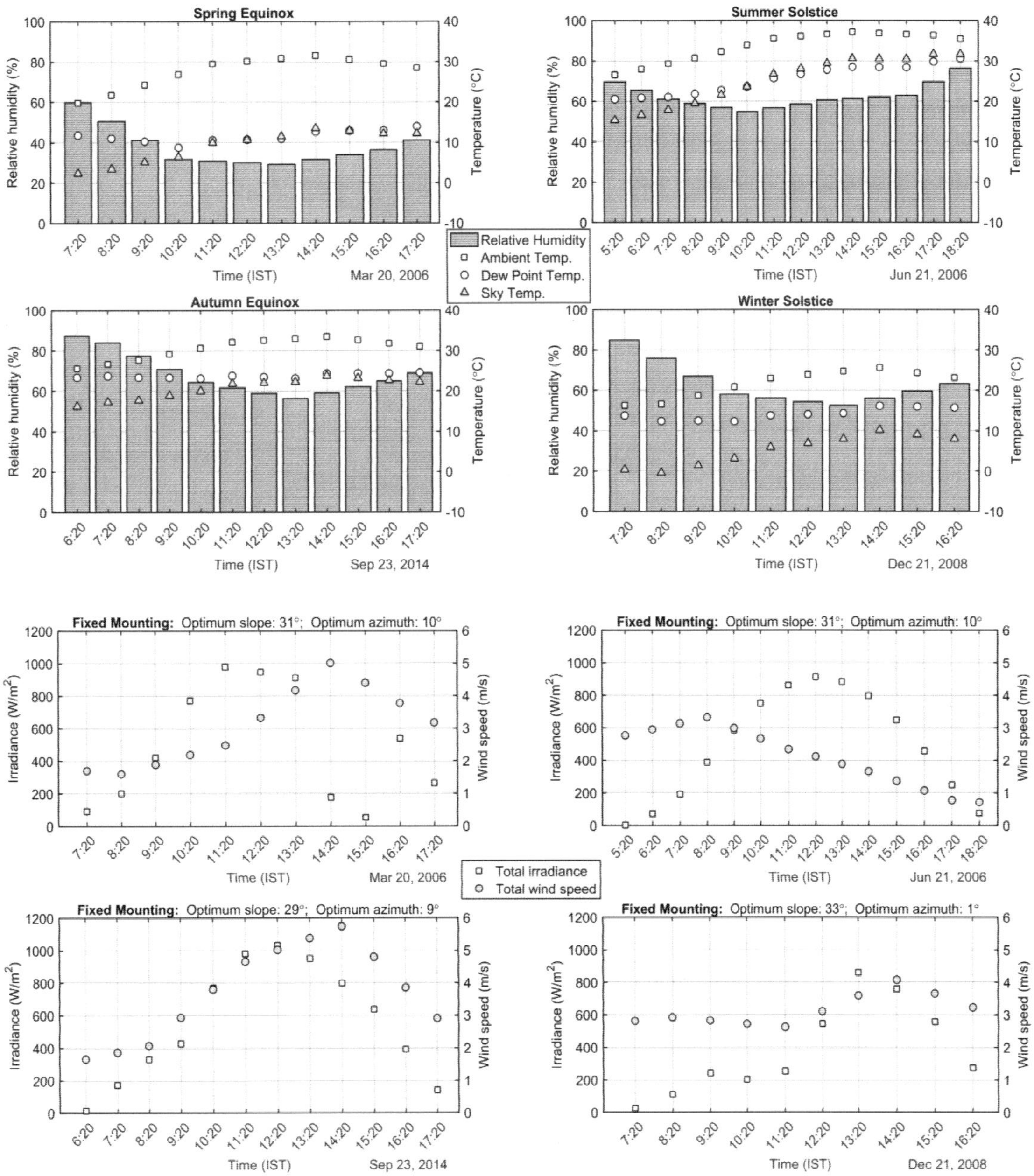

Figure 1. Meteorological data of Pantnagar, corresponding to TMY 2006-2015 [15].

IOP Conf. Series: Materials Science and Engineering 1168 (2021) 012016 doi:10.1088/1757-899X/1168/1/012016

Table 1. Design and operating parameters of PVT air collector simulated in the study.

Symbol	Parameter	Value
L	Length of collector (= Length of PV module)	1643 mm
W	Width of collector (= Width of PV module)	996 mm
H	Thickness of collector (= Thickness of PV module frame)	40 mm
δ	Depth of air duct	20 mm
V_{oc}	Open circuit voltage of PV module	36.75 V
I_{sc}	Short circuit current of PV module	8.87 A
V_{mp}	Maximum power point voltage of PV module	29.31 V
I_{mp}	Maximum power point current of PV module	8.36 A
α	Temperature coefficient of PV module current	4.08×10^{-3} A/K
β	Temperature coefficient of PV module voltage	– 0.121 V/K
$NOCT$	Nominal operating cell temperature of PV module	45 ± 2°C
α_c	Absorptance of solar cell	0.9
α_T	Absorptance of TPE backsheet	0.5
β_c	Cell packing factor in PV module	0.89
τ_g	Transmittance of glass (Low iron, tempered)	0.917
ε_g	Emittance of glass	0.88
ε_d	Emittance of duct inner surfaces	0.95
L_g	Thickness of glass	3.2 mm
k_g	Thermal conductivity of glass	1.7 W/m K
L_E	Thickness of EVA	350 ± 50μm
k_E	Thermal conductivity of EVA	0.235 W/m K
L_c	Thickness of solar cell (Polycrystalline silicon)	200 ± 30μm
k_c	Thermal conductivity of solar cell (Silicon)	148 W/m K
L_T	Thickness of TPE backsheet [25 μm EVA + 250 μm PET + 25 μm PVF (Tedlar)]	300 ± 5 μm
k_T	Thermal conductivity of PVF (Tedlar)	0.16 W/m K
k_P	Thermal conductivity of PET	0.15 W/m K
L_i	Thickness of insulation	20 mm
k_i	Thermal conductivity of insulation (Polyurethane foam)	0.035 W/m K
v_a	Velocity of air at inlet (Vary in 0.25 m/s steps; Peak at noon)	1 – 2.5 m/s

2.4. Thermal Model

The thermal behavior of PVT is simulated by energy balance equations with following simplifying assumptions:

1. At a particular data point, the collector is in steady-state with the environment.
2. The thermal inertia of collector material is considered negligible.
3. The heat conduction is one-dimensional and perpendicular to the layers of PVT.

4. The temperature variation along width of a particular layer of PVT collector is neglected.
5. The heat losses from the edges of the PVT collector frame are neglected.
6. The ohmic losses in cell due to bad contacts, current leakage and hotspots are not present.
7. The optical properties of PV module are independent of temperature.
8. The ambient temperature along the collector front and back is same.
9. The flow of air inside duct is uniform and fluid properties are evaluated at bulk temperature.
10. The wind flows at a constant speed on top and bottom surface with no directional effects.

The thermal model of present study is developed taking Amori et al. [10] as reference with several improvements like good convergence of parameters in the electrical model with analytical guess values for numerical solution [18], improved material property definitions and fluid property evaluations from reference tables [16]. An interactive program for coupled solution of electrical and thermal model is developed in MATLAB. The flow of information and conditional statements in the program are coded according to the flowchart scheme of the computer program developed by Amori et al. [10] and validated by replication of design parameters in test run which shows excellent similarity between results of both programs. The collector characteristics are calculated at each step of iteration from the Hottel-Whillier-Bliss model which is omitted here for brevity [17]. The mean fluid temperature can be obtained from approximate solution of these parameters which is then used to calculate the fluid properties at next iteration using CoolProp library [16].

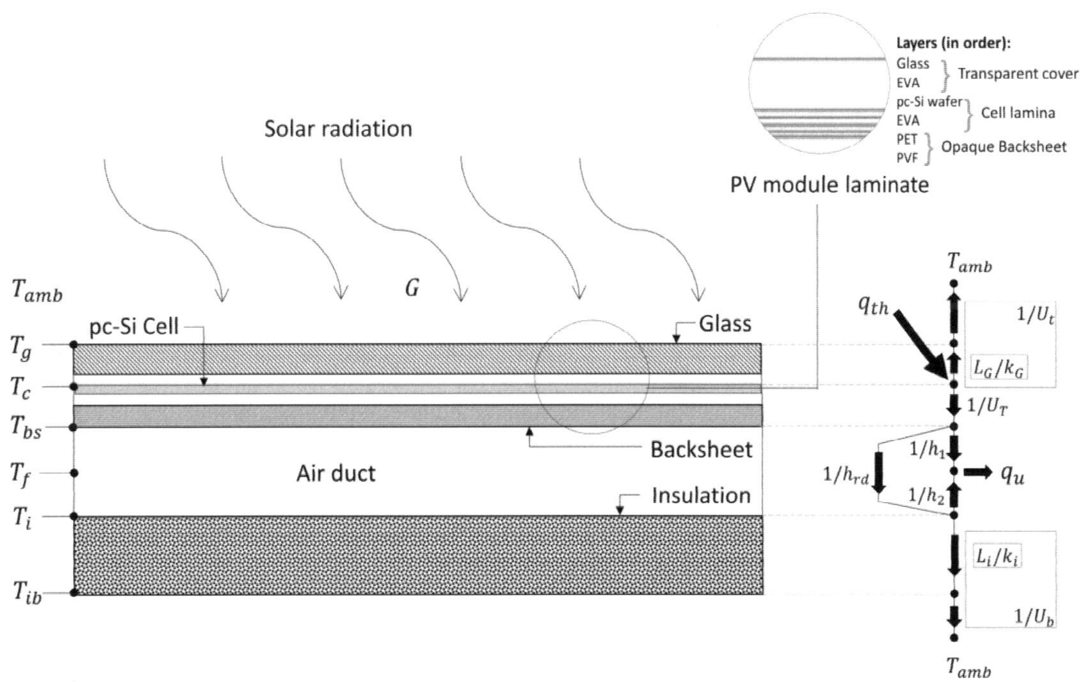

Figure 2. Cross-section of unglazed PVT air collector with corresponding thermal network: Labels denote thermal resistance between nodes of temperature while arrows show the direction of heat flow.

3. Results and Discussion

3.1. PV Characteristic Curves at Reference and Operating Conditions

The solution of the five parameter single diode model for the selected module (245 W_p) at STC conditions (1000 W/m^2, 25°C and 1.5 AM) results in an ideality factor (a) of 1.6048, light current (I_L) of 8.87 A, reverse saturation current (I_o) of 1.005×10^{-9} A, series resistance (R_s) of 0.34 Ω and shunt resistance (R_{sh}) of 9098 Ω. The $I-V$ curve with maximum power point (MPP) located with the help of $P-V$ curve at reference and operating conditions are plotted in figure 3. The drop in the PV power output at operating conditions is due to higher cell temperature as PV module voltage has a negative temperature coefficient (see table 1).

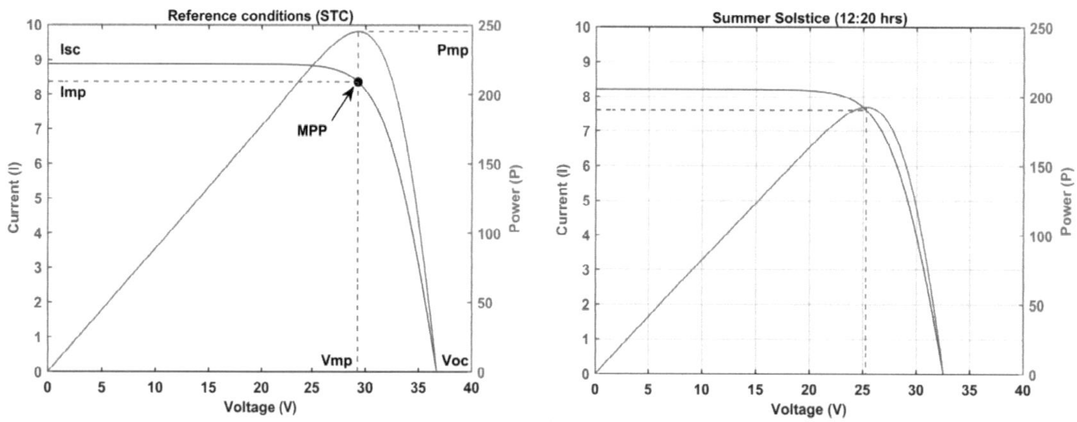

Figure 3. $I-V$ and $P-V$ characteristic curves of PVT at reference and operating conditions.

3.2. *Temperature of Solar Cell and Heated Air*

The solar irradiance, ambient temperature, wind speed and velocity of air in duct has a direct impact on the operating temperature of the solar cell in the PVT collector. The meteorological parameters are uncontrollable and vary with season, the operating parameters can be optimized for high yield through the day and throughout the year. The PVT is fixed at the optimum slope and azimuth for the year in consideration as suggested by PVGIS for providing in-plane irradiance. The velocity of air in the duct is varied each hour in steps of 0.25 m/s with peak (2.5 m/s) occurring at noon so that high yield is obtained at high level of irradiance. The hourly simulated temperatures of solar cell and heated air are plotted in figure 4. The highest cell temperature of 57°C is obtained at 13:20 hours for the day of summer solstice and the highest outlet air temperature of 45°C is attained an hour later. The inlet air velocity is 2.25 m/s at the time of highest cell temperature and 2 m/s for highest outlet air temperature.

3.3. *Power Output and Losses*

The accounting of electrical and thermal power output with associated losses in the collector is presented in figure 5. It provides a clear view of PVT potential at the selected location. The fraction of in-plane solar irradiance that reaches the solar cell is determined by optical efficiency of the transparent cover material. The electrical power output is determined from electrical conversion efficiency of the solar cell at the operating conditions. The amount of thermal power that can be extracted from the PVT depends on the maximization of heat transfer from solar cell to working fluid and minimization of heat losses to the environment. The maximum power that can be obtained from PVT primarily depends on the level of irradiance that is available on the plane of the collector. As the tilt of the collector is optimized for whole year operation by fixing it facing the median of the paths of the sun in the sky during year, the angle of incidence on the plane of PVT at noon in equinoxes is lower than that in solstices. The operating temperature of solar cell is also higher in summer solstice due to higher ambient temperature in summer. The irradiance on the plane of PVT and temperature of solar cell are lowest in winter solstice hence lowering the thermal yield but the electrical yield is still higher than in summer due to better cell efficiency at low temperature. The highest electrical power of 223 W is observed at 12:20 hours (IST) on autumn equinox while the highest thermal power of 411 W is obtained at 12:20 hours (IST) on summer solstice. These results are normalized with the area of collector to determine the highest PVT electrical and thermal power potential of 136 and 251 W/m^2, respectively, that is available in the typical climate of the location during the median and extremes of the sun in the sky. The highest thermal and overall (electrical + thermal) efficiencies are 27.56% and 40.46%, respectively, both observed at 12:20 hours (IST) on summer solstice while the highest electrical efficiency of 15.08% is observed at 9:20 hours (IST) on winter solstice.

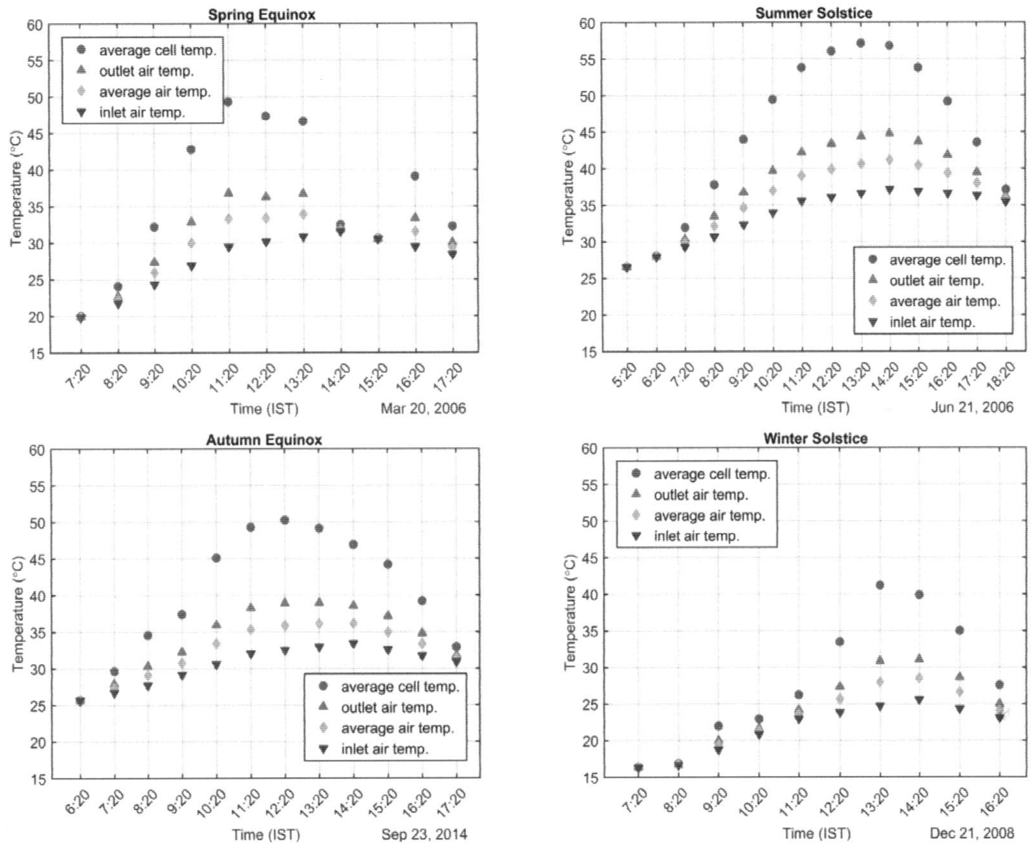

Figure 4. Simulated air and average cell temperature in PVT for different seasonal days in Pantnagar.

4. Conclusions

The power produced by the PVT during the days selected in the present study are instantaneous values at corresponding points of time, obtained from irradiance data from the satellite images and climate data from weather monitoring services. These data do not represent the true ultimate potential of PVT at the location but only provide a snapshot of the possible yields that can be obtained from PVT. If these values are assumed as a constant for the hour, the energy yield from the PVT at the midpoint of different seasons can be estimated. The calculations with this assumption gives a total daily yield of 1.76, 2.57, 2.22 and 1.20 kWh/m^2 with energy conversion efficiency of 32.9%, 37.5%, 33.4% and 31.3% on spring equinox, summer solstice, autumn equinox and winter solstice, respectively. The ratio of thermal and electrical yield during these days are 1.39, 1.83, 1.45 and 1.18, respectively. The simulation results show the potential and utility of PVT for heating air with electricity generation in Pantnagar.

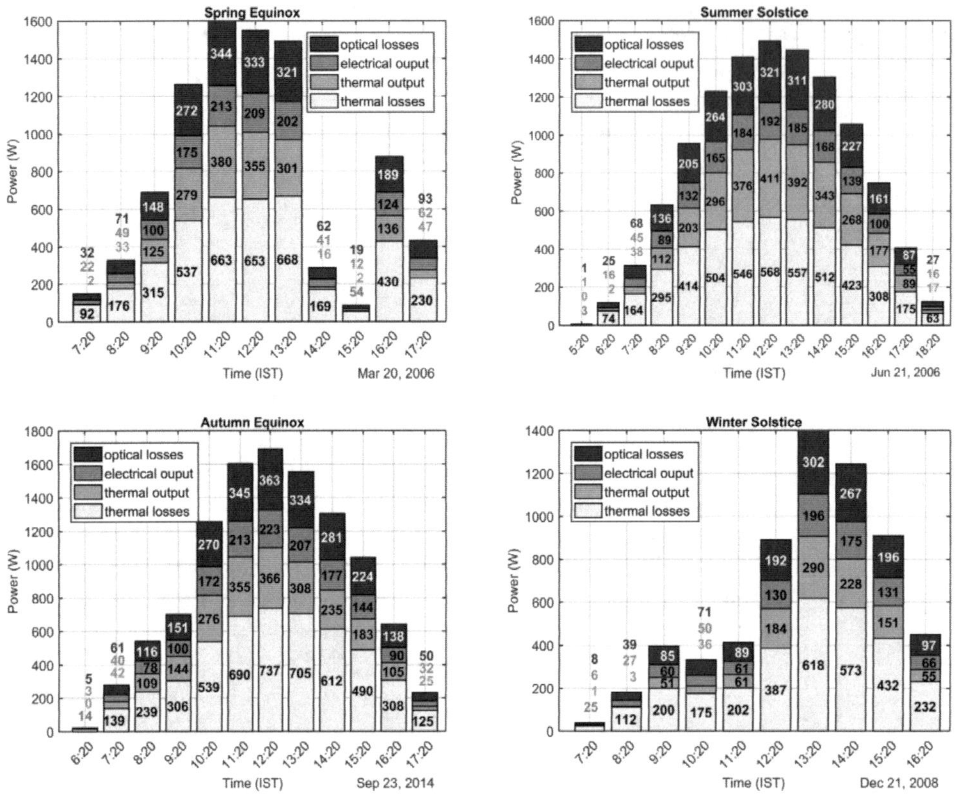

Figure 5. Distribution of irradiance available on PVT plane for different seasonal days in Pantnagar.

References

[1] Lämmle M, Oliva A, Hermann M, Kramer K and Kramer W 2017 *Sol. Energy* **155** 867–79
[2] Weiss W and Spörk-Dür M 2020 *Solar Heat Worldwide* (Gleisdorf: AEE INTEC) pp 27–31
[3] Wolf M 1976 *Energy Conversion* **16** 79–90
[4] Kern Jr E C and Russell M C 1978 *Proc. The 13th IEEE Photovoltaic Specialists' Conf.* 5–8 June 1978 (Washington DC) pp 1153–7
[5] Florschuetz L W 1979 *Sol Energy* **22** 361–6
[6] Garg H P, Agarwal R K and Joshi J C 1994 *Energy Convers. Manag.* **35** 621–33
[7] Dubey S, Sandhu G S and Tiwari G N 2009 *Appl. Energy* **86** 697–705
[8] Joshi A S, Tiwari A, Tiwari G N, Dincer I and Reddy B V 2009 *Int. J. Therm. Sci.* **48**, 154–64
[9] Sarhaddi F, Farahat S, Ajam H, Behzadmehr M and Adeli M M 2010 *Appl. Energy* **87** 2328–39
[10] Amori K E and Al-Najjar H M T 2012 *Appl. Energy* **98** 384–95
[11] Zondag H A 2006 *IEA SHC Task 35 Report DA2-1 – Commercially available PVT products.*
[12] Collins M and Zondag H 2009 *IEA SHC Task 35 Report DB2 – Recommended Standard for the Characterization and Monitoring of PV/Thermal Systems*
[13] Hadorn J C 2020 *IEA SHC Task 60 Report – PVT Systems. 2020 Highlights*
[14] Rathore L S 2015 *Climatological Normals 1981–2010* (Pune: India Meteorological Department) pp 589–90
[15] Huld T, Müller R and Gambardella A 2012 *Sol. Energy* **86** 1803–15
[16] Bell I H, Wronski J, Quoilin S and Lemort V 2014 *Ind. Eng. Chem. Res.* **53** 2498–508
[17] Duffie J A, Beckman W A 2013 *Solar Engineering of Thermal Processes* (New Jersey: John Wiley and Sons)
[18] Hejri M, Mokhtari H, Azizian M R and Söder L 2016 *Int. J. Sustain. Energy* **35** 396–410

EXPERIMENTAL EVALUATION OF DIESEL ENGINE OPERATING WITH A TERNARY BLEND (BIODIESEL-DIESEL-ETHANOL)

Pradeep Kumar Vishnoi[1,2], Puneet Singh Gautam[1], Peeyush Maheshwari[1], Tarun Singh Samant[1], V. K. Gupta[3]
[1]Research Scholar, Department of Mechanical Engineering, G. B. Pant University of Agriculture and Technology, Pantnagar, Uttarakhand, India,
[3]Professor, Department of Mechanical Engineering, G. B. Pant University of Agriculture and Technology, Pantnagar, Uttarakhand, India,

Abstract: The present investigation aims to study the future potential fuel of a non-edible tamanu oil (calophyllum inophyllum). The raw oil of tamanu oil was converted into biodiesel in two-step esterification process and different test fuel blends with biodiesel were prepared with conventional diesel and ethanol as an additive to evaluate its effect on engine characteristics such as BP, BTE, and BSFC, EGT and smoke density on a 4.4 kW rated power, single-cylinder four-stroke of constant rpm C.I. engineat different loads (25%, 50%, 75%, and 100% of rated power). The fuel blends were prepared with biodiesel in the ratio of 40%,100% coded as B40 and B100 respectively along with diesel fuel and ethanol as an additive in the fixed ratio of 10%, 15%, and 20%. It was found that B40E20D40 showed a reduction in smoke density and BSFC of 14.28% and 5.9% respectively, 18.3% increase in BTE and 3.73% reduction in exhaust gas temperature at full load whereas B100 showed a maximum reduction in smoke density of 15.97% with 4.53% increase in BSFC and a marginal increase in EGT of 0.53%. B40E20D40 gave a better performance in a diesel engine from the perspective of minimum BSFC, lower exhaust gas temperature and higher BTE among all other blends

Keywords: Biodiesel, Exhaust emissions, Ethanol, Engine performance

1. INTRODUCTION

The Energy utilization per capita is 33% of the worldwide average, with the growing development in Indian's economy will drive interest for energy throughout many sectors. Thus, access to sufficient and solid wellsprings of energy ends up noticeably imperative; especially when one-fourth of the population doesn't have access to electricity and reliance on sources of energy of non-renewable either imported or produced locally keeps on increasing. The latter meets around three-quarters of India's energy demand [1].

The energy consumption on an average total of India out of which 44 % originates from coal. Energy-rich sources, for example, petroleum and biomass constitute 22 % each. The transport and industry sectors are the biggest consumer of energy in India. The petroleum product usage will keep on growing due to development in the transport area, especially road transport, which represents a significant share of automobiles. The road transport share as a percent of freight and passenger traffic is assessed above 60 % and 90 %, expecting 10 % vehicle growth every year. As of now, 46 % of demand for transportation fuel is alone fulfilled by diesel whereas 24 % is fulfilled by gasoline. It is anticipated that in further next ten

[2] Corresponding author;
Email Id: pradeepvishnoi08@gmail.com

years, the demand on average for transport fuels will go up from an expected 134000 million liters in the year 2015 to 225000 million liters in the year 2026 [1].

The above data suggest that consumption of fossil fuels at the present rate would cause irreversible energy crisis along with a sharp increase in carbon footprints and other emissions which would not only cause environmental hazards but also deteriorate human health. The growth in transportation and continuous increase in demand on crude oil has led to life-threatening pollution level leading to stricter emission norms that have forced the engine manufacturers to comply with the norms which cost company money and time.

Biofuels are considered to provide energy security which is eco-friendly, recyclable, and biodegradable. Biodiesel and alcohol are considered to be potential renewable energy sources [2,3]. Based on many types of research on biodiesel it has been found that diesel fuel can be replaced to a greater extent but diesel engine running on biodiesel produces higher Nox emission as compared to diesel fuel. But the addition of alcohols with biodiesel reduces the Nox emission to some extent found by several authors. Renewable energy sources or oxygenated sources that are widely being researched nowadays provides similar physio-chemical fuel properties when compared to fossil fuels. There exists about more than 350 vegetable oil-bearing crops which have been studied by numerous authors, are considered among the most suitable vegetable oils. The most widely used vegetable oilssuch as jatropha curcas, tamanu oil, mustard oil, rapeseed oil, palm oil soybean oil sunflower oil, etc are consideredfuture potential fuel for diesel engines [3].

Tamanu oil is a non-edible oiltree found among the Clusiaceae family. Tamanu oil tree are found in Africa, Asia, and Pacific areas. These plants are also called by the name as Alexandrian Laurel, Tamanu, Pannay Tree, Sweet Scented Calophyllum, Punnai, etc. Usually, these trees are available in seashore areas, which are planted to neutralize soil breaking. They grow around 2- 3 m in length with a thick trunk having broken barks. The leaves are strong and round perfectly which transmits sweet lime-like smell. They bloom twice a year and yield different round seed found in clusters. Each seed has a smooth epidermis layer and is covered by a hard layer and has a diameter of about 50 mm, encasing a light-yellow piece of around 25 mm in length weighing generally 7g [5,6]. The word "Calophyllum" means "Beauty leaf" from Greek word.

This seed has numerous physio-chemical properties that might be utilized as an alternate future renewable energy source for automotive vehicles running on diesel engines that are not only searched for transportation services but also industrial and power sectors as well [6].

Tamanu oil has been studied by several researchers for alternative fuel on diesel engine by analyzing engine characteristics curve such as performance and emissions out of which few of them are discussed below [7-10].

Ong et al**.** [7] conducted a comparison on biodiesel production among palm oil, jatropha seed oil, and tamanu oil and showed that tamanu oil has the similar properties to conventional diesel fuel and could be the potential future fuel to replace depleting fossil fuel.

Deepan [8] invesigated on tamanu oil biodiesel on performance and emission characteristics of CI engine and analyzed that biodiesel mixed with diesel fuel coded as B15, B25, B50, and B100 respectively, performed from on diesel engine no load to full load resulted in higher BTE of 30% for B15, compared with diesel and lowest BSFC at all loads. With 100 % biodiesel, mechanical efficiency was higher at all loads and B100 produced higher smoke at all loads along with higher NOx emission from the engine. Krishnan et. al. [9] performed the experiment on tamanu oil biodiesel and studied the effect on engine performance and emission characteristics on a single-cylinder multi-fuel engine. Blends of biodiesel i.e., 10%, 20%, 40%, and 60% with diesel showed the brake power reduced when compared with diesel fuel. NOx emission from the blend B40 resulted in a lower amount, while at full load NOx emission increased as compared to reference diesel fuel. The CO emission was also found to be higher for

B40 fuel blend. Arun [10] studied performance of CI engine running on tamanu oil -diesel blends with the conclusions that BSFCconsumption was lower for B10 and B20 than the diesel fuel. Blends with 30% biodiesel reduced CO_2 emissions along with decrement in HC, and NOx emissions. BTE (brake thermal efficiency) slightlyincreased when load was increased when compared to diesel fuel and recommended that upto 30% biodiesel could be used in an unmodified CI engine.

Pugazhvadivu[11] experimented on the performance and emission characteristics of a diesel engine with ethanol in the Pongamia oil biodiesel-diesel blend. The ethanol addition in B100 (pure biodiesel) and B50 (50% biodiesel+50% diesel) showed a reduction in BTE and NOx emission reduced at full load. B100 and B50 fuel blends showed reduction in the smoke opacity. The ethanol blend with B100 and B50 showed better results when compared to the engine running on diesel fuel. Mofijur [13] studied the comparison of performance and emission characteristics between biodiesel–diesel and ethanol–biodiesel–diesel blends showed a reduction in CO emissions due to presence of oxygen in biodiesel and concluded that the exhaust gas emission with 5–10% of ethanol with 20–25% biodiesel could be added to diesel fuel and further investigation on biodiesel from second generation plantsarerequired. The work on tamanu oil discussed above showedthat further investigations are required by esterification process and addition of ethanol as an additive.

Thus, in this investigation, the engine performance parameters such as BSFC, BTE, and emission parameters on 4.4 kW rated power, single-cylinder four-stroke CI engine of constant rpm on Tamanu oil biodiesel and diesel with ethanol addition was studied. The other engine parameters such as compression ratio, injection timing, inlet temperature, and pressure were kept constant.

2. Materials and Methods

The raw tamanu oil was purchased fromTamil Traders, Suramangalam Salem, Tamilnadu. The raw tamanu oilpossesshigher viscosity and has a free fatty acid (FFA) value found to be more than 3% [5]. The FFA content in vegetable oils, if found in higher amount require whether the oil could be utilized in pure form or it requires an esterification process depending on the FFA content. If FFA is found to be more than 3% in vegetable oil, a two-step esterification process is required, for a single step esterification process would convert into soap rather than biodiesel when reacted with alkali catalysts having FFA content more than 3%. The first step known as the pre-esterification process requires the raw vegetable oil to get reacted with an acid catalyst such as sulphuric acid (H_2SO_4), thereby reducing FFA content to some extent which is again processed in the second step known as transesterification process where this low FFA oil obtained in the first step, is get reacted with base catalysts such as sodium hydroxide (NaOH) or potassium hydroxide (KOH) [5]. The chemical process of soap formation if the percentage of FFA is higher (> 3%) and the pre-esterification process is shown in **Fig. 1** and **Fig. 2**

$$CH_3 - CH_2 - CH_2 --- CH_2 - CH_2 - OOH + CH_3 - OH \xrightarrow[\text{Base catalyst}]{\text{KOH/NaOH}} CH_3 - CH_2 - CH_2 --- CH_2 - CH_2 - OO - Na^+ \text{ or } K^+ + H_2O$$

Free Fatty Acid | Alcohol | Soap | Water

Fig. 1 Soap formation in high FFA oils [5]

$$CH_3 - CH_2 - CH_2 ------ CH_2 - CH_2 - OOH + CH_3 - OH \xrightarrow[\text{Acid catalyst}]{H_2SO_4} CH_3 - CH_2 - CH_2 ------ CH_2 - CH_2 - OOCH_3 + H_2O$$

Free Fatty Acid | Alcohol | Methyl Ester | Water

Fig. 2 Acid pre-esterification [5]

All the parameters for esterification were set according to the investigation of Jahirul *et al.*[5,6]. The temperature in the pre-esterification process was set at 75°C, acid catalyst sulphuric acid (H_2SO_4) with

10% by weight of oil and methanol to oil ratio of 30:1 of the mixture was prepared and the mixture was continuously stirred for 2 hours. After the pre-esterification chemical process was completed in the reactor, the mixture was transferred into separating funnel from where two layers were found after settling: -methanol and low FFA Tamanu oil as shown in **Fig. 4(a).** The upper layer (methanol) was removed and low free fatty acid tamanu oil was again put on the reactor for base-catalyst transesterification process for the second step or transesterification process in which the temperature set was at 55°C in reactor setup, 7.5:1 methanol to oil ratio and 1% by weight NaOH as base catalyst was prepared and the mixture was stirred continuously for 1 hour. Once it was completed and again poured into separating funnel where after settling, two layers were visible: -biodiesel (upper layer), glycerol (bottom layer) as shown in **Fig.4(b).** Glycerol was removed from the bottom layer and the upper layer of crude biodiesel was processed for water wash to remove the excess methanol and impurities remained during the esterification process. The water wash was done in successive steps with hot water to avoid trapping of air bubbles for at least 4 times to have completely clean water visibility in successive steps and due to higher density, it was collected at the bottom part as shown in **Fig.4(c).** The conversion efficiency was obtained by approximately 89%. Eventually, the finished biodiesel was heated at about 50°C in an oven for 15 minutes to remove moisture trapped into it during a water wash.

Further, the fuel blends were prepared with final processed biodiesel with the reference fuel i.e. diesel and with ethanol as an additive that was purchased from the local market of the technical grade of 99.9% purity. The tamanu oil biodiesel was mixed with diesel fuel by 40% and 100% coded as B40, and B100 respectively. Afterward, ethanol was mixed in the ratio of 10%, 15%, and 20%. The composition and nomenclature of each blend fuel are shown in **Table 2.**

The important fuel properties of different fuel blends were carried out as per the ASTM standards. The important fuel properties - Density, cloud and pour point, flash and fire point kinematic viscosity, and calorific value for all the blends were determined and are tabulated in **Table 3.** The test was carried out on a 4.4 kW/6HP, single-cylinder four-stroke diesel engine at various load (25%, 50%, 75%, and 100% load of rated torque) at a constant speed of 1500 rpm. The engine specification is shown in **Table 1** and the schematic diagram of an engine test rig is shown in **Fig. 3.**The investigations were studied on engine characteristics and emission characteristics of these different blends

Table 1 Specification of engine test rig

Engine Characteristic	Specifications
Make/Model	Kirloskar TAFI
Maximum Power	4.4 kW @1500 RPM
Injection type	Direct Injection
Number of cylinders	Single
Cylinder bore/Stroke	87.5 /110 mm
Compression ratio	17.5.:1
Fuel Injection timing	-24 CAD a TDC
Fuel Injection pressure	170 to 250 bar

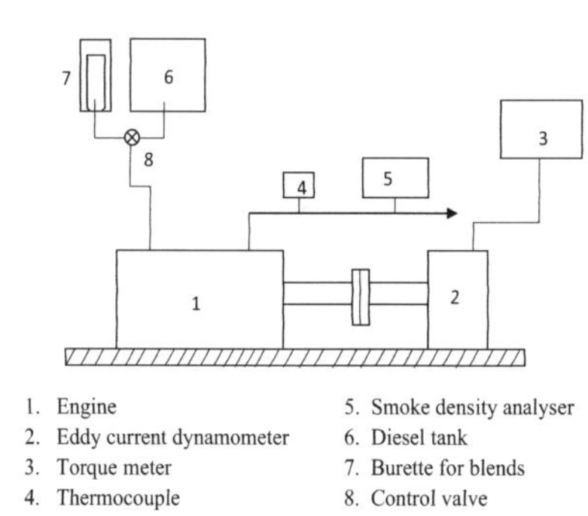

Fig. 3 Engine test rig

Fig. 4. A photographic view of esterification product (a) pre-esterification acid catalyst product (b) transesterification base-catalysts product (c) Water wash

Table 2 Proportions and nomenclature of fuel blends

Sl.No	Biodiesel (%)	Diesel (%)	Ethanol (%)	Nomenclature
1.	0	100	0	Diesel
2.	40	60	0	B40D60
3.	40	50	10	B40E10D50
4.	40	45	15	B40E15D45
5.	40	40	20	B40E20D40
6.	100	0	0	B100

Table 3 Fuel Properties of all the Blends

Fuels samples	Density (gm/cc) @15°C	Cloud Point (°C)	Pour Point(°C)	Kinematicviscosity (mm²/s) @ 40°C	Flash Point(°C)	Fire Point(°C)	Calorific Value (MJ/Kg)
ASTM	0.860-0.900	-3 to 12	-15 to 10	1.9-6.0	-	-	-
Diesel	0.823	-2	-10	2.51	60	63	45.86
B40D60	0.842	2	-2	3.62	78	84	43.82
B40E10D50	0.84	1	-2	3.37	40	51	42.37
B40E15D45	0.837	3	-1	3.20	30	40	41.68
B40E20D40	0.834	1	-1	2.82	28	32	41.20
B100	0.88	10	5	5.28	180	193	40.09
Ethanol	0.794	-15*	-	1.66	17	36	27.42

* = reference

3. Results and Discussion

3.1 Brake Specific fuel Consumption

Fig. 5 shows the variation of brake specific fuel consumption versus load (%) for various blends fuels. The brake specific fuel consumption (BSFC) decreases as the percentage of load increases for all the test fuels due to better combustion and lower heat losses. The brake specific fuel consumption for diesel were obtained as 0.544, 0.386, 0.330, and 0.322 kg/kWh. The observation with B40D60 the BSFC at 25, 50, 75, and 100% load were 0.554, 0.384, 0.344, and 0.326 kg/kWh respectively which indicated increase at all loads except 50% load by 4.22%. While the addition of ethanol with B40D60 the BSFC was increased for all the blends, with the highest percentage reported as 6.16% at 50% load for B40E15D45 as shown in **Fig. 5**. The BSFC with tamanu oil methyl ester (B100) at all loads were 0.565, 0.397, 0.351, and 0.337 kg/kWh increased by 3.75%, 2.93%, 6.36% and 4.4% respectively. This may be due to lower calorific value, higher viscosity and poor atomization of B100 and addition of ethanol further increase BSFC due to reduction in calorific value of the blend, lower heat release rate, and more energy consumption.

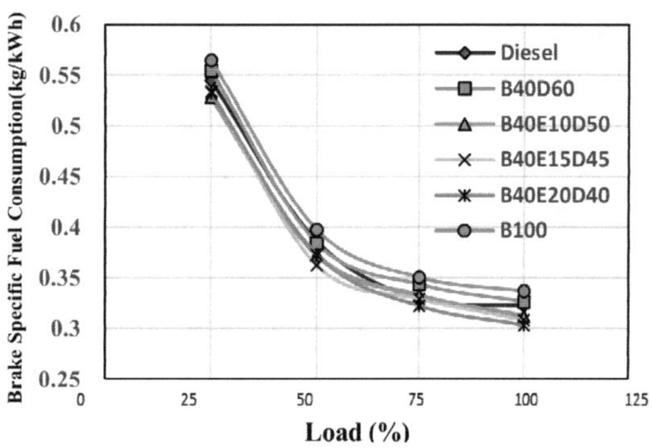

Fig. 5 Variation of brake specific fuel consumption of Diesel, B40D60 and its ethanol blends

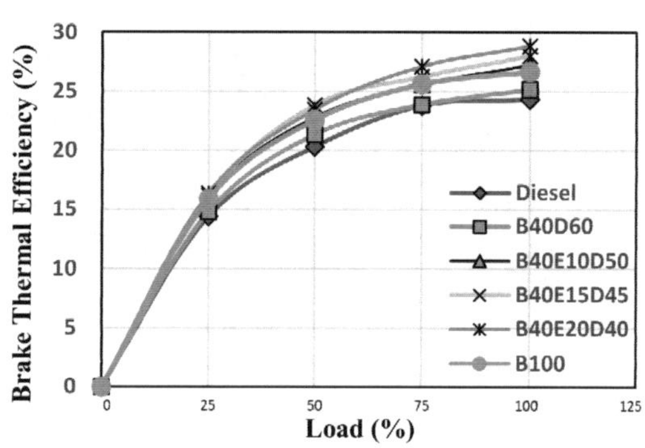

Fig. 6 Variation of brake thermal efficiency of Diesel, B40D60 and its ethanol blends

3.2 Brake thermal efficiency

Fig. 6 shows the comparison of brake thermal efficiency (BTE) versus load (%) for all different test fuels. For all the test fuels, the brake thermal efficiency was increased as the load was increased due to lower heat losses at higher loads. It was observed that the brake thermal efficiency (BTE) for, B40D60, B100, and Diesel was about 25.17%, 26.67%, and 24.35% respectively at full load and the percentage increase was 4.2%, 1.79%, and 9.56% with respect to diesel. The addition of ethanol to tamanu oil biodiesel reduces the viscosity of the fuel, increases volatility and the inherent oxygen (i.e. 32%) in ethanol improves the combustion phenomenon. The percentage of ethanol in blend reduces the calorific value of fuel which intakes more amount of fuel to develop the same power, and the highest BTE observed was 28.82% at 100% load for B40E20D40 increased by 18.41% with respect to diesel. The BTE with B100 was found to be 15.90%, 22.60%, 25.61%, and 26.67% at 25, 50, 75, and 100% load respectively with a percentage increase of 10.25%, 11.12%, 7.55%, and 9.56% with respect to diesel.

3.3 Exhaust gas temperature

Fig. 7 shows the variation in exhaust gas temperature for various test fuels versus load (%). The exhaust gas temperature (EGT) increases as the percentage of load increases for all the tested fuels. This increase in EGT was because, at higher load, an extra amount of fuel is required to be injected to develop more power. The exhaust gas temperature for Diesel, B40D60, and B100 was 375°C, 361°C, and 377°C at full load. With B40D60 there is a decrement in the exhaust temperature while it was further reduced with various ethanol blends with the highest percentage reduction of 3.94% for B40E15D45 at 75% load as compared to diesel. The reason for higher EGT might be due to poor atomization of vegetable oil because of higher viscosity which causes slow combustion thus producing higher temperatures at the exhaust. But the addition of ethanol caused lower exhaust gas temperature because of inherent oxygen content present in the ethanol which causes complete combustion with blends and due to heat of vaporization and lower heat losses in exhaust gases. The exhaust gas temperature with B100 was found to be marginally increased as compared to diesel fuel with the highest increase of 1.71% at 75% load.

3.4 Smoke density

Fig. 8 represents the variation of smoke emission with respect to load. The smoke emission increased as the percentage of the load was increased due to more requirement of fuel at higher load to develop the power. The smoke density (HSU %) with Diesel, B40D60 and B100 at full load was found to be 39.9% and 33.52% respectively that indicated an increase in smoke as the percentage of biodiesel was increased which might be because of higher viscosity and higher volatility thereby leading to poor combustion due to improper atomization of fuels. The smoke density was lowered as the percentage of ethanol blending with biodiesel-diesel was increased to diesel due to the complete combustion process with oxygenated ethanol, lower viscosity, and reduced fuel-rich regions in the combustion chamber.

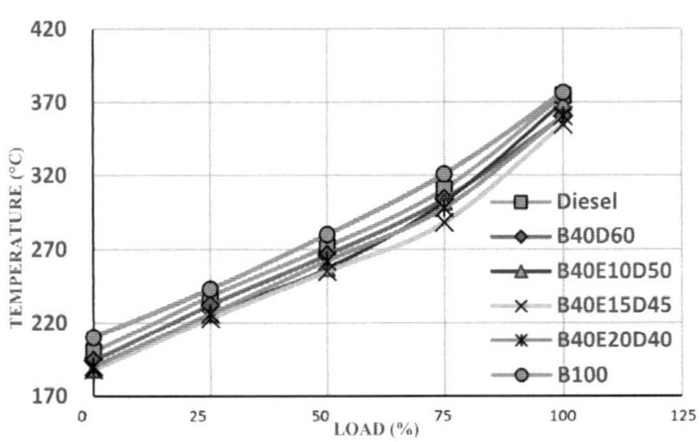

Fig. 7 Variation of exhaust gas temperature of B40D60 and its ethanol blends

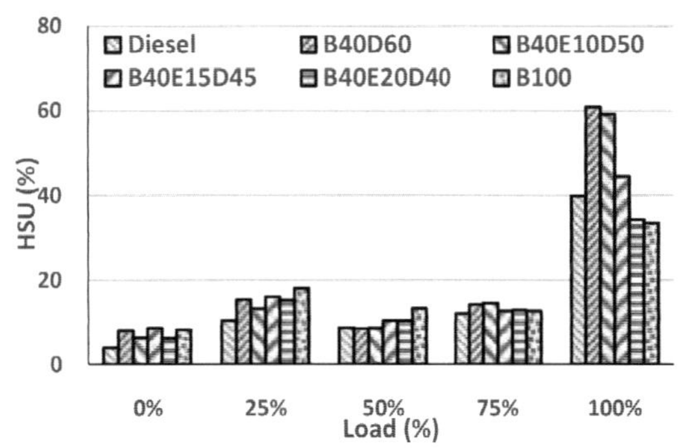

Fig. 8 Variation of smoke density with B40D60 and its ethanol blends

4. CONCLUSION

The performance evaluation of diesel engine on selected fuel blends was found acceptable based on brake power, brake specific fuel consumption, and brake thermal efficiency. The engine was able to develop similarly and accepted close results of brake power for all fuel blends as compared to diesel. It shows that the engine running on Tamanu oil biodiesel with ethanol as an additive was found to be potential fuel. Even 10 to 20% addition of ethanol, improves the properties, performance as well as emission characteristics of the. But in the present investigation where ethanol was added in diesel-biodiesel blends, it was observed that at 20% ethanol blend showed influential results with 40% of tamanu biodiesel by its smoke density reduction of the fuel at selected load condition and reduction in the exhaust gas temperature which is considered to be the main cause of NOx formation.

Acknowledgments

The author would like to express his acknowledgment to G. B. Pant University of Agriculture & Technology for experimenting in the internal combustion engine laboratory of College of Technology and expresses a deep sense of reverence and heartfelt gratitude to project guide Dr. V. K. Gupta, Professor, Department of Mechanical Engineering for his invaluable guidance, constant encouragement, and constructive suggestions throughout the investigation.

References

[1] India. Ministry of New & Renewable Energy. National Policy on Biofuels. 20018. National Policy on Biofuels. New Delhi, 18p.

[2] Altin R, Cetinkaya S, Yucesu HS. 2001. The potential of using vegetable oil fuels as fuel for diesel engines. *Energy Conversion and Management;* 42(5):529–38

[3] Demirbas A., 2008. The importance of bioethanol and biodiesel from biomass. *Energy Sources Part B-Economics Planning and Policy;* 3(2):177–85

[4] Bandyopadhyay, K.R. 2015. Biofuel promotion in India for transport: exploring the grey area. The Energy and Resource Institute TERI.

[5] Jahirul, M.I., Brown, R.J., Senadeera, W., Ashwath, N., Rasul, M.G., Rahman, M.M., Hossain, F.M., Moghaddam, L., Islam, M.A. and O'Hara, I.M., 2015. Physio-chemical assessment of beauty leaf (Tamanu oil) as second-generation biodiesel feedstock. *Energy Reports*, *1*, pp.204-215.

[6] Ong, H.C., Mahlia, T.M.I., Masjuki, H.H. and Norhasyima, R.S., 2011. Comparison of palm oil, Jatropha curcas and Tamanu oil for biodiesel: a review. *Renewable and Sustainable Energy Reviews*, 15(8), pp.3501-3515.

[7] Deepan K., G., 2015. Experimental Investigation of Performance and EmissionCharacteristics Of Tamanu Oil As Alternative Fuel In CI Engine. *International Journal of Engineering and Computer Science*, 4(6), pp.12471-12475

[8] Krishnan, P.N. and Vasudevan, D., 2015. Performance, Combustion and Emission Characteristics of Variable Compression Ratio Engine Fuelled with Biodiesel. *International Journal of ChemTech Research*, *7*(1), pp.234-245.

[9] Arun, B., 2014. A study on production, performance and emission analysis of Tamanu oil-diesel blends in a CI engine. *International Journal of Science and Research*, *3*, p.2319A7064.

[10] Pugazhvadivu, M., 2009. Studies on the effect of ethanol addition to biodiesel: Performance and emissions of a diesel engine. *Indian Journal of Science and Technology*, *2*(11), pp.23-26.

[11] Mofijur, M., Rasul, M.G., Hyde, J., Azad, A.K., Mamat, R. and Bhuiya, M.M.K., 2016. Role of biofuel and their binary (diesel–biodiesel) and ternary (ethanol–biodiesel–diesel) blends on internal combustion engines emission reduction. *Renewable and Sustainable Energy Reviews*, *53*, pp.265-278.

Experimental analysis and theoretical validation of C.I. engine performance and combustion parameters using zero-dimensional mathematical model fuelled with biodiesel and diesel blends

Puneet Singh Gautam[1,2] **Pradeep Kumar Vishnoi**[1], **Peeyush Maheshwari**[1], **Tarun Singh Samant**[1], **V. K. Gupta**[3]

[1] Research Scholar, Department of Mechanical Engineering, G. B. Pant University of Agriculture & Technology, Pantnagar, Uttarakhand, India
[3] Professor, Department of Mechanical Engineering, G. B. Pant University of Agriculture & Technology, Pantnagar, Uttarakhand, India

Abstract: The present work aims to analyze the experimental and numerical investigation on performance such as brake power (BP), brake specific fuel consumption (BSFC), brake thermal efficiency (BTE), and combustion parameters (in-cylinder pressure, heat release rate) on a 4.4 kW single-cylinder four-stroke C.I. engine running on biodiesel extracted from tamanu oil and diesel. A two-zone, zero-dimensional mathematical model was developed by using thermodynamic laws and cylinder volume, Wiebie's co-relation, in-cylinder pressure, heat transfer model, engine friction model was developed to validate the above engine performance parameters and analyze the combustion parameters comprising of in-cylinder pressure, in-cylinder temperature, net heat release rate at different loads with single varying-parameter i.e. equivalence ratio while other parameters engine speed, compression ratio, injection timing were kept constant. It was found that the theoretical results were in good agreement with the experimental results for straight diesel and its blends with biodiesel.

Keywords: Biodiesel, Diesel engine combustion, Two-zone combustion model, Heat release rate

Nomenclature

A	Instantaneous heat transfer area	$\frac{dT}{d\theta}$	Rate of change in temperature
B	Cylinder bore (mm)	m_a	Mass of air
a	Crank radius (mm)	m_f	Mass of fuel
hc	Convective heat transfer (W/m²-K)	V_c	Clearance volume (mm³)
hr	Radiative heat transfer coefficient (W/m²-K)	V_d	Swept volume (mm³)
s	Distance Between Crank, Piston Axes	W	Work transfer
Up	Mean piston speed (m/s)	χ	Mass fraction burned
Tw	Cylinder wall temperature (K)	η_c	Combustion efficiency
Tb	Burned temperature (K)	Υ	Specific heat ratio
l	Connecting rod length (mm)	λ	Excess air coefficient
$\frac{dP}{d\theta}$	Rate of change in pressure	θ	Crank angle (in degree)

[2] Corresponding author;
Email Id: psgautam51@gmail.com

$\frac{dU}{d\theta}$	Rate of change in internal energy	θ_d	Burn duration (in degree)
$\frac{dQ}{d\theta}$	Rate of heat transfer	θ_i	Start of combustion (in degree)

1 Introduction

Modeling in science and engineering might be for the most part viewed as the way toward depicting the physical phenomena in a specific system with the assistance of mathematical equations (with reasonable assumptions) and having a solution by either discretizing or by any other method to see more about such complex phenomena. As a rule, building models help in designing better devices by seeing more about the complex physical processes happening in that [1,2]. In such a manner, there is great importance in modeling engine combustion processes. zero-dimensional models are the simplest and easy to use and observe the variation of the theoretical results in the engine operating parameters on accumulated heat release rates, cylinder pressure, cylinder temperature, etc. The zero-dimensional models mean that they don't include the flow field dimensions, rather all the processes are instantaneous. Single zone models could be considered a thermodynamic system in which combustion takes place in the engine which exchanges heat (energy) and mass transfer with the surroundings and the first law of thermodynamics could be applied to determine heat transfer from the system [1]. These models have been approached used in two different directions: Both of these models could be used to anticipate the cylinder pressure with respect to crank angle by following energy release or Wiebe's mass fraction burned profile. Another utilization with these models could be in determining the heat transfer rate with respect to crank angle by determining in-cylinder pressure data experimentally [1,2]. There has been a large focus by the researcher in recent decades in modeling the engine combustion for designing better engines that produce less pollution [3-9]. In this present investigation first, the traditional model is used because there was no pressure trace device available in the engine set up hence in-cylinder pressure data is predicted with the help of a single-zone mathematical model thereafter the pressure data was used with a two-zone model to further obtain burned and unburned mass and numerical results such as heat release rate power, torque, work done, etc.

Many researchers conducted experiments on mathematical modeling either single-zone or two-zone zero-dimensional models to validate the experimental results or analyzing the effects of engine parameters such as compression ratio, ignition timing (injection pressure, start of injection), equivalence ratio, etc. out of which few of them are discussed below.

Patil studied mathematical modeling for the prediction of performance parameters on CI engines running on biodiesel and biodiesel-diesel blends. The predicted BTE and peak pressure were in closer approximation with that of experimental results [8]. Potdukhe et al. studied a modeling and energy analysis, zero-dimensional single zone combustion model simulation to predict the single-cylinder constant speed diesel engine performance. The theoretical model obtained the pressure and temperature variations inside the cylinder by using the combustion correlations which were in closer simulated results with experimental results. Heat loss could also be studied with that simulations model. The results of the present models were well in agreement with the experimental result [9]. Hariram et al. developed a comprehensive zero-dimensional model to estimate the in-cylinder pressure, the rate of heat release, and rate of pressure rise in a single-cylinder four-stroke CI direct injection engine using conventional diesel and blends of beeswax biodiesel. In this study, the thermodynamic approach was utilized to anticipate the in-cylinder pressure, and Wiebe's and Wolfer's relation was applied to investigate the heat release correlations and ignition delay respectively. At no load, the cylinder pressure for conventional diesel was found to be 48 bar and 53 bar by experimental and simulated studies respectively. The heat release rate for conventional diesel, 10% and 20% beeswax biodiesel (BWB10) at 0%, 50%, and 100% load shows an increasing trend between the values of 20 J/°CA and 58 J/°CA which was also supported by the simulated results with a deviation of 7.2%. The rate of pressure rise obtained from simulation of conventional diesel, 10%, and 20%

beeswax biodiesel indicated 8.3% higher as compared to the experimental data throughout all loading conditions [10].

In this present work a two-zone, zero-dimensional mathematical model was developed by applying the first law of thermodynamics and discretizing the same and empirical correlations with a forward difference method and solving the same in matrix laboratory (MATLAB) to validate the performance parameters such as brake specific fuel consumption, brake thermal efficiency with the experimental results running on 4.4kWsingle cylinder four-stroke C.I. engine. Other important theoretical results such as bulk in-cylinder pressure, in-cylinder temperature, heat release rate, power, torque, work done, etc. at different loads with varying equivalence ratios are analyzed. The engine parameters that was constant like engine speed, compression ratio, injection timing, inlet temperature, and pressure.

2 Materials and methods

In this investigation, biodiesel was obtained from tamanu oil also known as Calophyllum Inophyllum, and blends were prepared with conventional diesel. The investigation of these different blends has been studied by measuring its properties and performing these blends on a 4.4 kW/6HP, single-cylinder four-stroke diesel engine of which specification is shown in **Table 2** and schematic diagram of the engine test rig is shown in **Fig. 2.** As there were no measuring devices available such as pressure transducer, dynamometer, crank angle sensor, a data acquisition device, fuel flow transmitter, airflow transmitter, exhaust gas analyzer, etc. Thus in-cylinder pressure was predicted using a single zone model thereafter heat release rate, power, torque, work done, etc. were analyzed. The blends were prepared once finished biodiesel was obtained with diesel. The blends B20D80 (20% biodiesel and 80% diesel) were prepared and the important fuel properties for the blend were determined and is tabulated in **Table 1.**

2.1 Mathematical Modelling of Engine Combustion

In this investigation mathematical modeling of zero-dimensional was developed with the help of empirical relation suggested by many authors and researchers [10,11,12] to determine cylinder pressure, cylinder temperature, heat release rate, etc. theoretically that was further utilized after discretizing the empirical relations to solve it through MATLAB and validate the results obtained experimentally.

2.1.1 Cylinder volume

The fundamental geometry of the reciprocating piston engine is shown in **Fig. 1.** Using the engine kinematics model volume every crank angle (θ) can be calculated by solving [1, 2]

$$V(\theta) = V_c + \frac{\pi B^2}{4}[l + a - s] \qquad \text{-------------} (1)$$

Where, V_c= clearance volume, B = cylinder bore, l = connecting rod length, a = crank radius, s = distance between the crank axis and the piston pin axis and is given by

$$s(\theta) = a\cos(\theta) + (l^2 - a^2\sin(\theta^2)^{\frac{1}{2}}$$

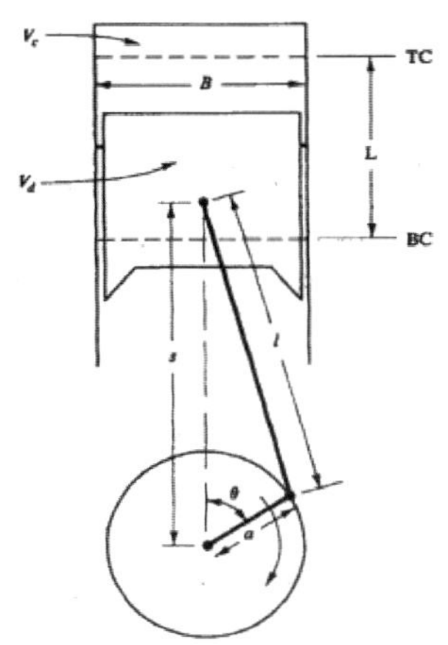

Fig. 1 An engine kinematics model [12]

Table 1 Fuel Properties of all the Blends

Fuels samples	Density (gm/cc) @15°C	Cloud Point (°C)	Pour Point (°C)	Kinematic viscosity (mm^2/s) @ 40°C	Flash Point (°C)	Fire Point (°C)	Calorific Value (MJ/kg)
ASTM standards	0.860-0.900	-3 to 12	-15 to 10	1.9-6.0	-	-	-
Diesel	0.823	-2	-10	2.51	60	63	45.86
B20D80	0.833	-2	-7	3.24	68	73	44.34

Table 2 Specification of the engine test rig

Engine Characteristic	Specifications
Make/Model	Kirloskar TAFI
Maximum Power	4.4 kW @1500 RPM
Injection type	Direct Injection
Number of cylinders	Single
Cylinder bore/Stroke	87.5 /110 mm
Compression ratio	17.5.:1
Fuel Injection timing	-24 CAD a TDC
Fuel Injection pressure	170 to 250 bar

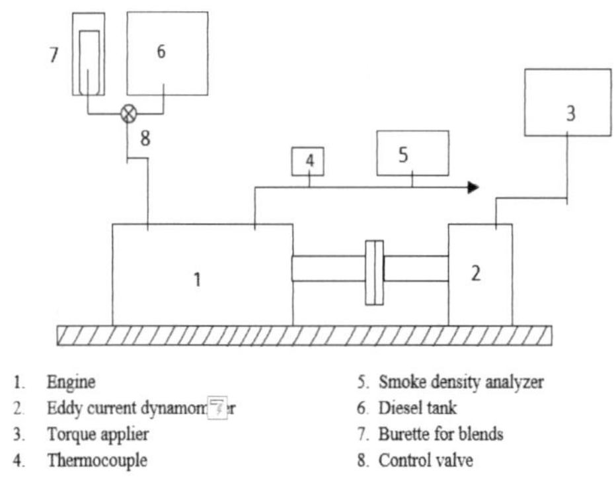

Fig. 2 A schematic diagram of engine set-up

2.1.2 Wiebie co-relation

Single-zone models typically use the Wiebie co-relation to represent the chemical, or gross, energy release as a function of crank angle [3,4]. The Wiebie co-relation has a characteristic "S-shape" and is defined as follows

$$\chi(\theta) = 1 - \exp\left[-a\left(\frac{\theta-\theta_i}{\theta_d}\right)^{(k+1)}\right] \tag{2}$$

Where a and k are adjustable constants (k= 5 and 2 are commonly used), a=6.908 b=3 was used in this case, θ = instantaneous crank angle, θ_i = start of combustion, and θ_d = burn duration. The burn profile is engine-specific, and the constants and k can be adjusted to tune the profile to a specific engine or application.

The ideal gas law is defined as:

$$\mathrm{PV} = \mathrm{mRT} \tag{3}$$

Where, P= pressure of an ideal gas, V= volume of the gas, m= mass of the gas, R= universal gas constant and T= mean gas temperature

Differentiating equation (3) with respect to θ

$$\frac{dP}{d\theta} = \left(\frac{P}{T}\right)\left(\frac{dT}{d\theta}\right) + \left(-\frac{P}{V}\right)\left(\frac{dV}{d\theta}\right) \tag{4}$$

Where, P, T, and V are instantaneous values that are modeled with respect to the engine's crank angle.

2.1.3 In-cylinder temperature

The same process can be applied to the first law of thermodynamics, for the four-stroke engine energy balance [2] which is expressed as:

$$\Delta U = Q - W \tag{5}$$

Where, Q = total energy transferred into the system, W = work transferred out of the system, ΔU=is the change in internal energy within the system. In differentiating equation (5) with respect to $d\theta$, equation (6) can be obtained

$$\frac{dU}{d\theta} = \frac{dQ}{d\theta} - \frac{dW}{d\theta} = mC_v\frac{dT}{d\theta} \tag{6}$$

Where, C_v =the specific heat of the combustion chamber gas.

And by applying $\gamma = \frac{C_p}{C_v}$ and $C_p - C_v = R$, and $\frac{dW}{d\theta}$= P$\frac{dV}{d\theta}$ which gives us the change in temperature as a function of crank angle:

$$\frac{dT}{d\theta} = \frac{T}{PV}(\gamma - 1)\frac{dQ}{d\theta} - \frac{T}{V}(\gamma - 1)\frac{dV}{d\theta} \tag{7}$$

2.1.4 In-cylinder pressure

Using η_c(the combustion efficiency) and LHV (the lower heating value of the supplied fuel) equation (8) was obtained as

$$\frac{dQ}{d\theta} = \frac{dQ_{in}}{d\theta} - \frac{dQ_{loss}}{d\theta}$$

$$\frac{dQ}{d\theta} = \left(\eta_c \times m_f \times LHV \times \frac{d\chi}{d\theta}\right) - \frac{dQ_{loss}}{d\theta} \tag{8}$$

Lastly, the change in pressure was obtained by substituting equation (7) into equation (8)

$$\frac{dP}{d\theta} = \frac{\Upsilon - 1}{V}\left(\frac{dQ}{d\theta}\right) - \Upsilon\frac{P}{V}\left(\frac{dV}{d\theta}\right) \tag{9}$$

The equation (9) is the basis of the model which was used for engine simulation.

2.1.5 Heat transfer model (Annand's heat release model)

This model incorporates all the processes taking place inside the cylinder for heat transfer calculations, i.e. air motion in the cylinder, fuel spray growth, and mixing sprays impact on the walls, turbulence, droplets evaporation, fuel ignition delay, and combustion process. Annand and Ma 1971 formulated the co-relation for heat transfer [1]

Annand's heat release model

$$\frac{dQ_{loss}}{d\theta} = (h_c(\theta) + h_r(\theta))A(\theta)(T_\theta - T_w)\left(\frac{1}{\omega}\right) \tag{10}$$

Where, ω = woschni's factor (60/ (360*RPM), and r denotes the reference, $h_c(\theta)$ = convective heat transfer coefficient and $h_r(\theta)$ = Radiation heat transfer coefficient

2.1.6 Modeling engine friction

Friction losses vary significantly from engine to engine and can be introduced through bearing components and pistons, along with the process of driving engine accessories [2,3,5]. Engine friction losses could be very difficult to model without known engine data and can vary based on engine coolant and oil temperatures, ambient conditions, and engine speed and throttle settings. Heywood suggested that friction mean effective pressure (FMEP) could be approximated for a CI engine as:

$$FMEP = C + 48 \times \left(\frac{RPM}{1000}\right) + 0.4 \times {U_p}^2 \tag{11}$$

Where, C=75 kpa, U_p=Mean piston speed

In choosing a PC program to execute the requirement of a two-zone model, Matrix Laboratory (MATLAB) was considered. It was considered due to its simplicity of reproduction and speed of validation. In light of keeping all limitations add up, the script was created in MATLAB for future utilization. The MATLAB code was set up using the script and all the script is separated into a few subsections. In the first section, the engine geometry and ambience inputs of known engine information sources such as bore, stroke, length of connecting rod, number of cylinders, compression ratio, and other operating parameters such as the timing of opening and closing of the valves in crank angle, the angle at the start of combustion, combustion duration, etc. was mentioned to the program as inputs for the validation. Given the information sources script would determine the area of the cylinder, clearance volume, cross-sectional piston area, and cylinder head piston area. The air inputs were set such as the ambient temperature 300K (room temperature) and atmospheric pressure as 1 atm. It was followed by Fuel input parameters, for example, the mass of the fuel, Calorific value, Lower Heating Values, and air-fuel ratio, equivalence ratio were taken according to the experimental model of the engine. The air-fuel ratio and lower heating value (LHV) to calculate the actual air-fuel ratio. Excess air coefficient or inverse of equivalence ratio was varied from 1.25 to 0.83. In the further section pre-allocation of arrays was done. This kept MATLAB from having to resize arrays or matrices in no. of iteration, reducing the overall computational time and space.

Keeping in mind the end goal that the fundamental program has separated into two sub-loops wherein the main loop with a predefined index (i =0:360, for compression and expansion only) computed instantaneous engine properties as said above. Notwithstanding the properties of immediate properties like volume, pressure, and temperature it has additionally scripted for to ascertain work done in a total cycle, indicated power, friction power, brake power, the correction factor for variable specific heat ratio were used. Variable specific heat ratio model was used from the derivation of the polynomial method.

3 Results and discussion

All the blends were performed on a constant speed 4.4 kW single-cylinder four-stroke of constant speed diesel engine and the engine performance test such as brake specific fuel consumption, brake power, brake thermal efficiency, exhaust gas temperature, and smoke density were determined.

3.1 Modeling Results

All the numerical results showed in this investigation were calculated by varying equivalence ratios at different loads (50%, and 100%) because assuming actual fuel-air and stoichiometry fuel-air ratio as equal that is Φ =1 might or might not fit the experimental results obtained for diesel, biodiesel and diesel-biodiesel blends. Hence for the blends running on biodiesel and diesel require that it is to be observed by varying equivalence ratio (Φ) from 0.8 to 1.2 with the other parameters kept as constant which is nothing but the ratio of actual fuel-air ratio and stoichiometry fuel-air ratio for closer approximation with the experimental results. The method to calculate the molecular weight of selected blends was taken from a reference [12]. Because of the complexity to approximately find the molecular weight and lower heating value (LHV) with the addition of ethanol thus in calculating the equivalence ratio so its ethanol blends were not considered. The experimental results that were validated were Diesel, B20D80.

3.1.1 Validation of brake specific fuel consumption and brake thermal efficiency

Fig. 3 shows the validated results of brake-specific fuel consumption and brake thermal efficiency versus load (%) that were calculated theoretically using MATLAB. The brake-specific fuel consumption was obtained numerically and the brake thermal efficiency was calculated with varying equivalence ratios from 0.8 to 1.2 to best fit the results obtained experimentally. It was observed that theoretical results were found to be closer to the experimental results by varying the equivalence ratio from 0.8 to 1.2. It was also evident from the figures that the equivalence ratio for diesel fuel was found to be greater than one while as the percentage of biodiesel was increased it was approaching towards the equivalence ratio other than one. That line plotted from the theoretical model for brake specific fuel consumption which is closer to the experimental line has been shown by the bold black dotted line of corresponding equivalence ratio and accordingly for brake thermal efficiency has been shown by bold black dotted line for the same equivalence ratio as from brake specific fuel consumption as shown in **Fig. 4**.

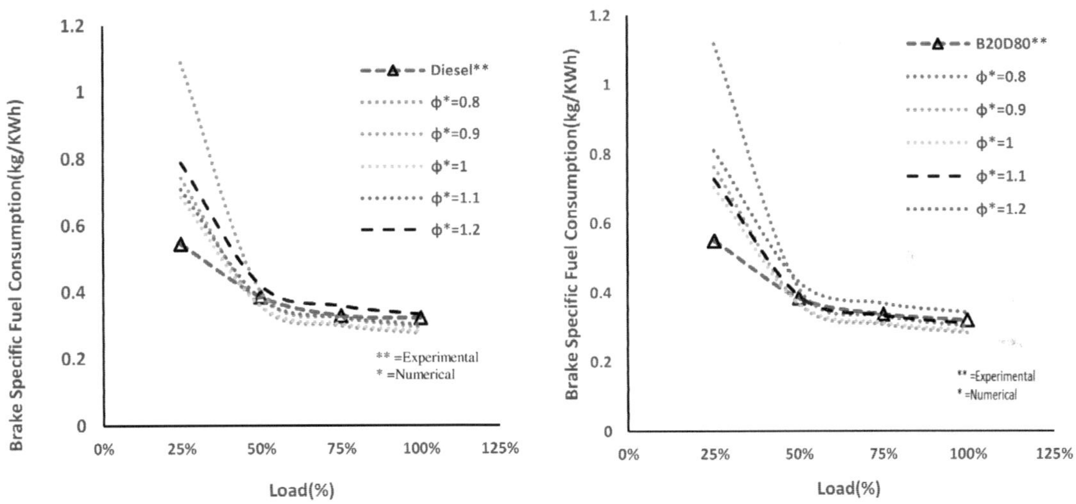

Fig. 3 Experimental and numerical brake specific fuel consumption with varying equivalence ratio for Diesel, B20D80

3.2 Numerical results

Other results plotted which were obtained numerically included cylinder pressure, cylinder temperature as a function of crank angle. The temperature of burned and unburned masses, heat release rate, torque, power, and work done calculated as a function of the crank angle at different loads (50%, and 100% load).

3.2.1 In-cylinder pressure

Fig. 5 represents cylinder pressure as a function of the crank angle at various loads with a variation of equivalence ratio. As the equivalence ratio was varied from 0.8 to 1.2 the pressure was also increased at all the loads. At 50% load the peak pressure was 42, 45, 47 and 47.5 bar for Φ=0.8, 0.9, 1, 1.1, and 1.2 respectively. The same fashion was followed for other loads. The maximum peak pressure of 95 bar was observed at an equivalence ratio of 1.1 at 100% load.

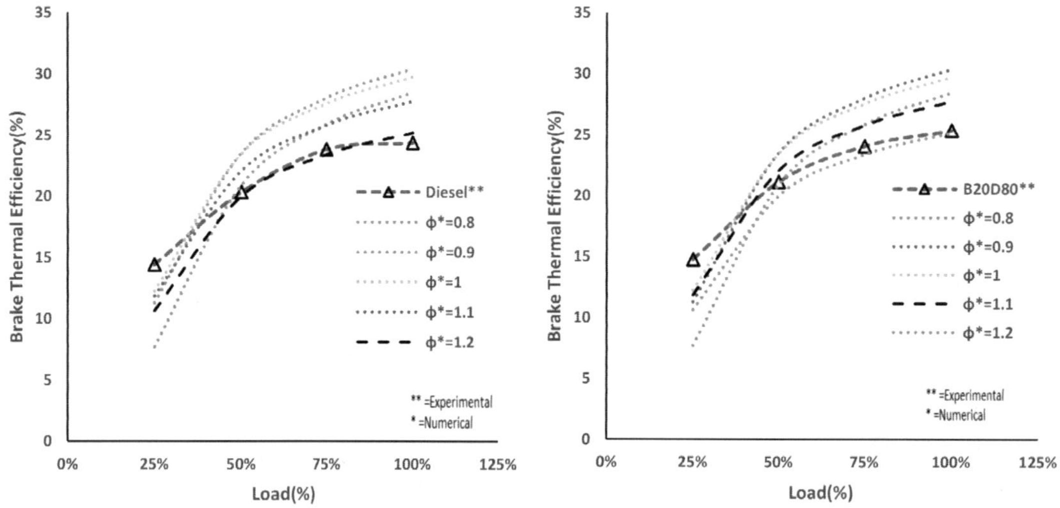

Fig. 4 Experimental and numerical brake thermal efficiency with varying equivalence ratio for Diesel

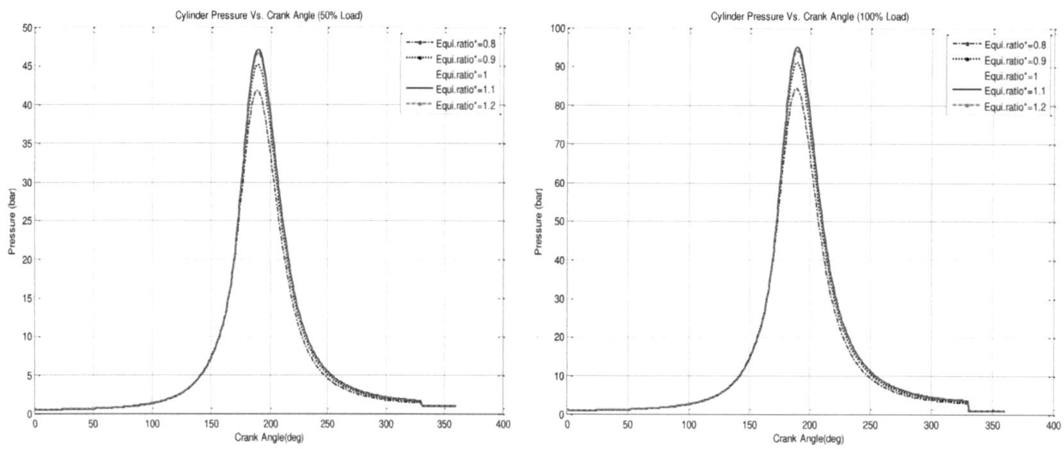

Fig. 5 Variation of in-cylinder pressure as a function of the crank angle at 50% and 100% load

3.2.2 Heat transfer (kW)

The heat transfer (or cumulative heat transfer) to the wall or heat loss is shown in **Fig. 6** as a function of the crank angle at various loads with variation in equivalence ratio. The heat transfer at lower load and equivalence ratio was found to be lower. The heat transfer was increased as the equivalence ratio increased at all the loads. As soon as the compression process ends there was a sudden rise of heat release and reached a peak point during the combustion period due to heat being produced from the fuel thereby started decreasing from the point where the expansion process started taking place. The lowest heat transfer at the peak was 10.5 kW at 50 % load at Φ=0.8 and was found to be highest at 100% load as 21 kW.

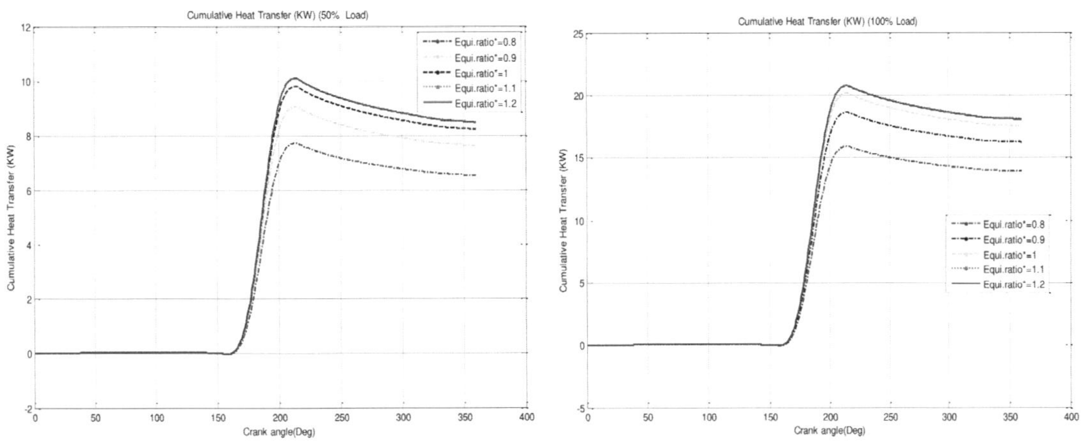

Fig. 6 Variation of heat transfer as a function of crank angle with variation in equivalence ratio at 50% and 100% load

4 Conclusion

The performance evaluation of diesel engines on selected fuel blends was found acceptable based on brake power, brake specific fuel consumption, and brake thermal efficiency. The validation with the present model showed a closer approximation with the experimental results such as brake specific fuel consumption and brake thermal efficiency even with biodiesel blends with diesel with varying equivalence ratios. Other results obtained such as in-cylinder pressure, heat release rate, burned and unburned temperature, etc. from the two-zone zero-dimensional model showed closer approximation with the practical results. Hence this two-zone model could be used for the prediction of performance characteristics of CI engine when there are fewer data available for any type of hydrocarbon fuels. This model could also be used for optimization purposes such as by varying compression ratio, inlet pressure, inlet temperature, fuel injection timing, etc.

Acknowledgments

The author would like to express his acknowledgment to G. B. Pant University of Agriculture & Technology and expresses a deep sense of reverence and heartfelt gratitude to project guide Dr. V. K. Gupta, Professor, Department of Mechanical Engineering for his invaluable guidance, constant encouragement, and constructive suggestions throughout the investigation. My abstruse regards go to the ministry of Human Resource and Development, New Delhi for financial support by awarding GATE Fellowship during this investigation.

References

[1] Ganesan V., 2000. "*Computer Simulation of Compression Engine*" 1st edition.

[2] Heywood, J.B., 1988. *Internal combustion engine fundamentals* (Vol. 930). New York: Mcgraw-hill.

[3] Blair, G.P., 1999. Design and Simulation of Four-Stroke Engines. *Technology, 2011*, pp.10-18.

[4] Cuddihy, J.L., 2014. *A User-Friendly, Two-Zone Heat Release Model for Predicting Spark-Ignition Engine Performance and Emissions* (Doctoral dissertation, University of Idaho).

[5] Hareesh, K., N Teja Rohith. , Reddy, B.Konda. 2014. Computer Simulation of Compression Ignition Engine through MATLAB. *International Journal for Research in Applied Science & Engineering Technology (IJRASET)*.2: 2321-9653

[6] Krieger, R. B. & Borman, G. L. 1966. "The Computation of Apparent Heat Release for Internal Combustion Engines," ASME paper 66-WA/DGP-4,

[7] Patil, S., 2013. Thermodynamic modelling for performance analysis of compression ignition engine fuelled with biodiesel and its blends with diesel. *Int. J. Recent Technol. Eng.(IJRTE)*, *1*, pp.134-138.

[8] Potdukhe, S.P. and Deshmukh, M.M., 2013. Modeling and Energy Analysis of a Diesel and Biodiesel Fuelled Engine. *International Journal of Science and Research (IJSR),* 4(5):89-93

[9] Arun, B., 2014. A study on production, performance, and emission analysis of Tamanu oil-diesel blends in a CI engine. *International Journal of Science and Research*, *3*, p.2319Ā7064.

[10] Hariram, V. and Bharathwaaj, R., 2016. Application of zero-dimensional thermodynamic model for predicting combustion parameters of CI engine fuelled with biodiesel-diesel blends. *Alexandria Engineering Journal*, *55*(4), pp.3345-3354.

[11] Krieger, R. B. & Borman, G. L. 1966. "The Computation of Apparent Heat Release for Internal Combustion Engines," ASME paper 66-WA/DGP-4,

[12] Dinh, Thi Luong, Hoang,Vu Nguyen. 2014. Determination of C/H/O fractions and lower heating values for diesel-biodiesel blends derived from Vietnam. *International Journal of Renewable Energy and Environmental Engineering;* 2(3): pp. 224-231.

[13] Bafghi, A.A.T. andChegeni, F.K., 2014. An Experimental Study on the Equivalence Ratio of Biodiesel and Diesel Fuel Blends in Small Diesel Engine. *Bull. Env. Pharmacol. Life Sci*, *4*, pp.40-44.

Experimental Investigation of configuration types and array of external heat sources on downward spreading fires

Vinayak Malhotra[1], **Saumya Shekhar** [2], **Bhushan Thombare** [3], **T.Selvakumaran** [4]

[1,2,3,4] Department of Aerospace Engineering, SRM Institute of Science and Technology, Chennai, India
vinayakm@srmist.edu.in

Abstract. Scholars and scientists have been attempting to discover ways to control and lessen the reverberation of concurrent fires such as forest fires, building fires, and other various space fires but no convincing solutions have been concluded from their studies so far. The fundamental genesis of these types of fires concerns the unstable nature of the flames and considerable unpredictability associated with them. This led us to make an effort to study the etiquette of such flames – an experimental setup with rather ideal conditions was devised and an in-depth study was carried out. The present study predominately covers the study of fire propagation phenomena and the zeal of fire control in our daily life. This study includes the review of the varying regression rates and fire spread rates of flames as noticed in matchsticks when spaced in a linear orientation. The deportment instability of the flames will give us an insight into the heterogeneous fire propagation phenomenon and its control. Intuition into the heterogeneous fire propagation is expected for essential fire safety and, on its basis, an algorithm for the same is to be formulated. With this understanding along with the existing information, it might give us some possible solution to the reduction of such kinds of fires in buildings, forests, space propulsion systems and, large-scale fires in industries.

Keywords: Forward Heat Transfer, Array of External heat sources, Downward Spreading fires.

1. Introduction

Fire is undeniably one of the exceptional inventions by mankind and it has been a continuous source of human evolution. Efficient utilization of fire has resulted in a magnificent development in the fields of Engineering, Industrialization, the Practical world, Functional and operational systems in diverse domains. While the fire is a very common occurrence almost everyone takes to use its field of engineering, Industrialization, Practical world, Functional and operational systems in diverse domains, it also has the capacity to be out of control, presenting a critical situation to human lives and properties. The fire has also been the origin of the biggest disasters that occurred in the forms of large-scale fires in industries, forest fires, building fires, Aircraft and Rocket crashes, which lead to illimitable loss of mankind, nature, resources, and every year an immense amount of money is being invested on research to prevent fires. From the data given by the National Fire Prevention Association (NFPA), it was recorded that around 8.5 million acres of land are lost due to forest fires, around 10.7 billion dollars (USD) was estimated to be structured fire loss. The loss of residential, industrial, educational institutions was evaluated to be around 10,700 million dollars (USD). The report suggests that the rate of loss is rising at an alarming rate of 77% each year. The most recent forest fires in Amazon, Australia, and California lead to the loss of more than 1,676,473 hectares of land. The loss is estimated to be around more than USD 1.620 billion (USD) only for California fires have caused utmost damage and danger to human and animal life, property, and the environment. Apart from the forest, the fire propagation in aircraft like in collision of Saudi Arabian Airlines Flight 763 and Kazakhastan airlines B747 leads to the loss of 349 lives.

Figure. 1 Amazon rainforest.

Figure 2. Apollo 1 (1967).

Figure 3. Kazakhstan Airlines

The present work highlights elementary the in-depth knowledge of the fire propagation phenomenon. To know the physics of fire-related issues, it is required to know about the fire propagation events. The energy spends by the combustion of fossil fuels such as natural gas, oil, and coal is more than 80% of total energy. Combustion is defined as the rapid chemical fusion of a substance with oxygen that involves the production of heat and light. The main factors required to support combustion are (a) oxygen insufficient amount, b) fuel or combustible material, and (c) appreciate heat to reach the ignition temperature. The process of combustion is mainly classified into two types (1) The heterogeneous reaction of solid fuel with an oxidizer is called smoldering combustion, whereas (2) the homogenous reaction of gaseous fuel with the oxidizer which releases more heat is called flaming combustion. The present work emphasis on flaming combustion. In this type of combustion, the flame spread is observed which is the rate of flame movement across the fuel surface. The opposed flame spread rate is defined as the direction of airflow that is opposite to that of flame propagation. The Propagation of the reaction zone or combustion wave through the combustible mixture is Flame Propagation. When the transport of free radicals and heat has started a chemical reaction in an adjacent layer of the combustible mixture, the layer itself becomes the source of heat and radicals and is then efficient of starting reaction in the following layer. The temperature increases to the maximum temperature that flame can attain. For many decades, humankind has experienced plenty of accidents that have occurred due to fire. In essence, most of these accidents transformed as a result of fire (or flame) propagation. Researchers have attempted to quantify the fire spread rate for more than 60 years, and many mathematical models have been developed.

From the conventional work done by Fons [1] firstly by assuming that the fuel is preheated by convective, conductive, and radiative heat transfer from the flaming zone to ignition. Most work investigating flame spread through separate fuels has been performed using wooden sticks of various sizes where the flame spreads in the either horizontal or inclined configuration. Vertical wooden matchsticks were used by Vogel and Williams [2] to model horizontal discrete fuel flame propagation by determining necessary conditions for discrete fuel flame spread and by changing the spacing and length of matchsticks. A convection-controlled theory was started by using a constant flame temperature and stand-off distance, which had good agreement with experimental results. Miller [3] extended the study to arrays in the form of a rectangular lattice. The number of matchsticks per column was 8 and the number of matchsticks per row was varied from 1 to 9. The flame propagation rate appeared to increase asymptotically as the number of rows was increased. A solid array of paper strips separated by space in increasing amount was used by Emmons and Shen [4] for measuring fire-spread rates. Prahl and Tien [5] analyze horizontal fame spread over vertically-oriented matchsticks and paper strips by adding an opposed wind velocity. Correlations between fuel height, fuel spacing, wind velocity, and flame spread were explored by considering fundamental theory. From work, by C.C hwang [6] it was concluded that the rate of flame propagation along vertically orientated matchstick arrays increases as the angle of plate is increased (positive for uphill burning and negative for downhill burning). From the photographs of the flame zone taken by schlieren and thermocouples, it was concluded that the flame propagation is controlled by the width of the preheated region. Recently, Gollner et al. [7] explored flame spread through a single vertical column of matchsticks. Based on the observed phenomenon, buoyancy was expected to control fame spread between matchsticks, and a theory based on convective heat transfer was developed to predict fame spread and burning rates. Recently, numerical studies on flame spread over parallel thin fuel samples in a reduced gravity environment were reported for the concurrent flame spread in Shih [8] and opposed flow flame spread in Malhotra et al. [9]. In both these studies, the flame spread rate exhibited a non-monotonic variation with a decrease in separation distance. In the following work on composite fabric by Johnston et al. [10], On long thin composite fabric fuels made of 25% fiberglass and 75%, cotton upward flame spread experiments were conducted. The width of the fuel specimen was varied from 2.5 cm to 8 cm. Effect of fuel width on flame blow-off was reported. From the experiments done by Avinash [11], the flame propagation was investigated from degrees -90 to +90 for thin cellulose papers. It was found out that the maximum rate of propagation is at +90 degrees and as the number of external fuel cells increases fluctuations in flame spread rate are observed for upward flame propagation cases and are much larger for the case of 90°. These fluctuations are due to the formation and shedding of vortices, and also because of a sudden shift in the flame location as a result of cracking of the thin fuel. Results from several models based on energy balance were compared against the experimental results. Tiwari et. al., [12] investigated to gain physical institution into the acoustic thermal energy interaction and related implications on smoldering combustion. Experiments were carried on incense stick with a proper variation of fuel to sound source distance with a fixed frequency and variation of surface orientation. Under varying conditions, the effect of acoustics was evaluated with variation in fuel regression rates. The result showed that by varying source separation distance the forward heat transfer rate increases by Thermo acoustics interactions. From the experimental work done by Jiang [13], they studied the flame propagation for different spacing for wooden dowel arrays. It was found out for the smaller 0.75 and 0.875 cm spacings, flames would spread both vertically and horizontally simultaneously were as for the largest spacing, 1.5 cm, flames would spread only vertically along the central column. By considering these predictions, the mass-loss rate was also predicted. Tatnell et. al., [14] inspected the coupling and confinement of current in thermoacoustic phased arrays. The work showed that arrays of these sources generate sound unique to this mechanism. From the sound alone, the current flow was spatially resolved by varying the film geometry and electrical phase.

The specific objectives of the work are:
1) To study the effect of number of external heat sources in the form of distinct configurations on the pilot fuel.
2) To fundamentally understand the role of key controlling parameters.

2. Experimental Setup and Solution Methodology

Figure 4. Mild Steel plate with holes.

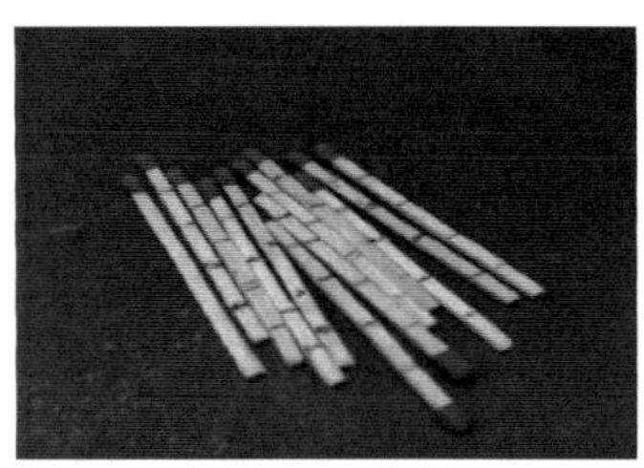

Figure 5. External Heat sources.

In a state to study the fire propagation phenomenon, a simple apparatus was upraised coupled with external heat sources. The apparatus consists of a) a mild steel plate, b) a rod to vary the orientation of the plate, c) Matchsticks as a fuel source, and d) a protractor (As shown in fig 4). The match sticks had a 0.5 cm separation region followed by three 1 cm markings (as shown in fig 5). A mild steel plate was used to fix the match sticks perpendicular to the base. The interspace distance between the match sticks was 0.5 cm. The study is about the fire propagation phenomena and study of fire for various regression rates and fire spread rate of flames.

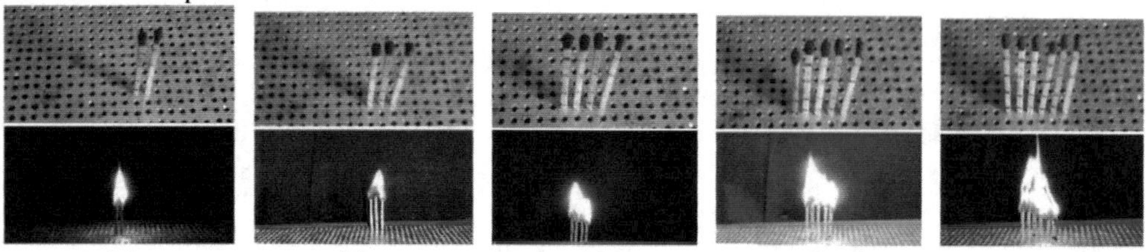

(a) (b) (c) (d) (e)

Figure 6. For Unilateral Configuration (a) n=1, (b) n=2, (c) n=3, (d) n=4, (e) n=5.

(a) (b) (c) (d) (e)

Figure 7. For Bilateral Configuration (a) n=1, (b) n=2, (c) n=3, (d) n=4, (e) n=5.

(a) (b) (c) (d) (e)

Figure 8. For Trilateral Configuration (a) n=1, (b) n=2, (c) n=3, (d) n=4, (e) n=5.

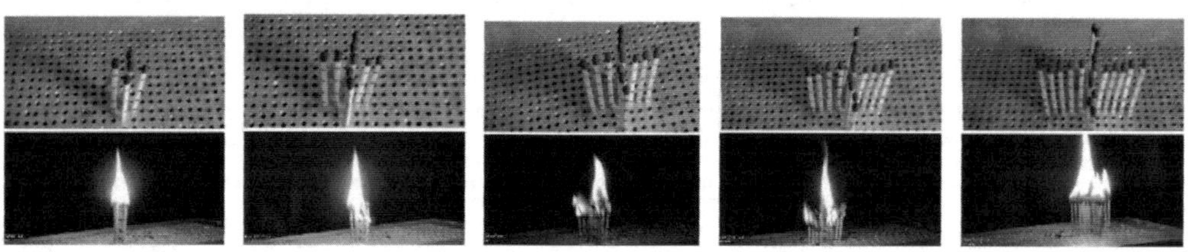

(a) (b) (c) (d) (e)

Figure 9. For Quad-lateral Configuration (a) n=1, (b) n=2, (c) n=3, (d) n=4, (e) n=5.

Forward Transfer Theory

From the Second Energy Equation we know that:

Energy Change = Energy produced – Energy lost

So,

$$\rho_s C_s V(dT/dt) = q_p - q_L \quad (1)$$

where, $$q_p = \Delta H_c V C_i A * e^{(E_a/RT)} \quad (2)$$

$$q_L = hA(T - T_a) \quad (3)$$

Hence, according to Classical Forward Heat Transfer Theory, the Flame spread rate (r) is given by:

$$r = \frac{\int q_{net}}{\rho_s \tau_s c_s (T_{surface} - T_{\infty})} \quad (4)$$

Measurement

The rate at which the surface burns is called the Flame spread rate which is linearly calculated as:

$$r = \frac{Distance\ burnt}{Time\ taken\ to\ burn\ that\ region} \quad (5)$$

The term q_{net} is the difference between the energy generated and the energy lost.
Where,

ρ_s : Density of the solid fuel
τ_s : Thickness of the solid fuel
c_s : Speed of sound in that medium
$T_{Surface}$: Temperature of the surface
T_∞ : Temperature of the surroundings
dT/dt : Change in temperature
q_p : Energy produced
q_L : Energy lost
A* : Pre-exponential factor
E_a : Activation energy
R : Ideal gas constant
T : Temperature in kelvin
H : Thermal Conductivity constant
A : Cross-sectional area of the material
T_a : Ambient temperature
q_{net}: Total Energy
c_s : Speed of sound in the material
T_∞: Temperature of the surroundings
r : Flame spread rate.

The experimentation was conducted at normal room temperature and readings were taken properly ensuring orderliness and conformity in every case. *It is important to note that every data presented here represents the repeatability and reproducibility of third order.*

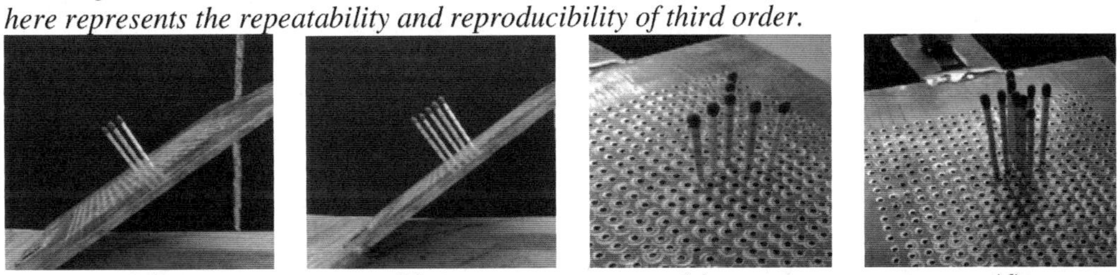

(a) (b) (c) (d)

Figure 10. Match sticks placed in 4 configurations (a) unilateral (b) bilateral (c) 'Y' (d) '+'.

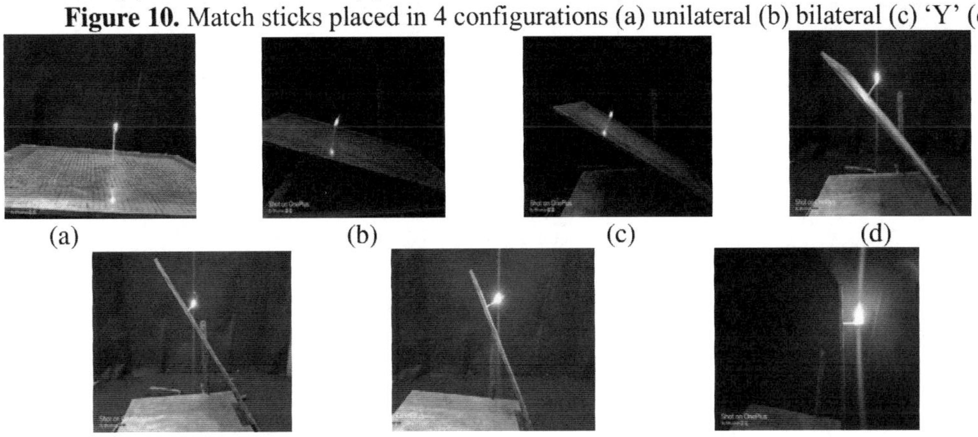

(a) (b) (c) (d)

(e) (f) (g)

Figure 11. Matchstick at different orientations (a)0^0 (b)15^0 (c)30^0 (d)45^0 (e)60^0 (f)75^0 (g) 90^0

The match sticks were fixed on the mild steel plate for four different configurations (Fig 10): (a) unilateral (b) bilateral (c) Y (Trilateral) and (d) + (Quad lateral). The burn rate was measured at various angles varying from 0 to 90 degrees with a constant interval of 15 (Fig 11). The main objective was to carry out sets of experiments of the mentioned configurations at various degrees. In each case, the flame spread rate of the pilot fuel was measured by noting down the time taken to burn 1cm of the match stick. Images and videos were taken in order to capture the special cases of the phenomenon.

3. Results and Discussion

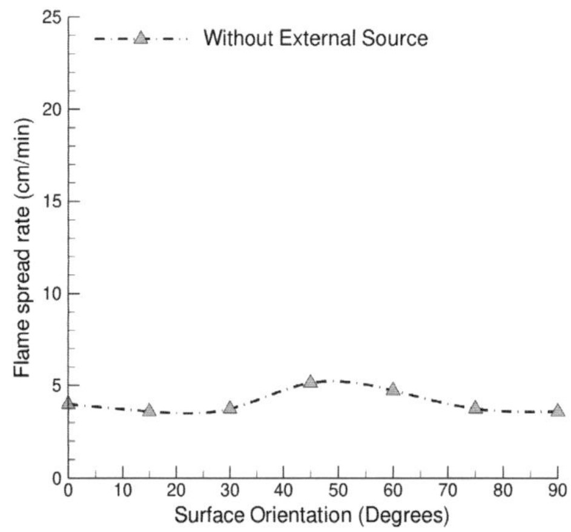

Figure12. Effect of Surface orientation on flame Spreading phenomenon.

The initial experimentation was performed to obtain the base case values of the flame spread rate. It is important to note that the flame spread rate of the pilot fuel was measured at all orientations. From the plot, it can be observed that the trend is non-monotonic with the maximum burn rate of 5.142 cm/min at 45 degrees. These values are taken as a reference for comparison of flame rate for different configurations coupled with external heat sources.

In accordance with traditional heat transfer theory over thin solid fuels, the propagating front spreads by heat feedback (forward heat transfer) from the burning to the unburnt solid fuel upstream. This results in an increase or decrease in regression rates. It is due to convective buoyant flow, localized velocity, and temperature fields are formed around the fuel surface. Additional energy and preheating are provided by High-temperature smoke which carries heat moving parallel to the fuel surface. High cumulative heat transfer from burnt to unburnt fuel can be attributed to high regression rates. The further results were divided into three cases.

1. Heat Sink Zone: The heat transfer in this regime drops due to the decrease in the localized temperature around the pilot fuel. The reason is not enough oxygen gets enter to fume the pilot fuel. Hence the regression rate for this case is less compared to the pilot fuel.
2. Neutralizing zone: In this case, the heat transfer is constant to that of the pilot fuel. Here the regression rate is the same as that of the base case of pilot fuel.
3. Heat source zone: The cumulative heat transfer is dominating in this region because of buoyancy that carries the heat from burnt to the unburnt fuel. Hence the regression rate for this case is greater than compared to the pilot fuel.

For a particular orientation, the ratio of the flame spread rate at that particular orientation for a given number of external sources to that of flame spread rate without any external source is called the MTS number. These numbers are divided into different regimes MTS<1 is the Heat sink zone, MTS>1 is the Heat source zone and MTS=1 is the neutralizing zone. For a particular orientation, a non-dimensional number was determined. It was done to validate the effectiveness of the configuration and to analyze the effect of configurations, orientations, and external sources to that of single-pilot fuel. The MTS number is defined as the ratio of flame spread rate of the external fuel for any given source (at any orientation) to the flame spread of pilot fuel without any external source (at that orientation).

$$\text{MTS number} = \frac{Spread\ rate\ in\ persence\ of\ external\ source\ (cm/min)}{Spread\ rate\ of\ pilot\ fuel\ without\ any\ external\ source\ (cm/min)}$$

Based on MTS number new regime is defined – **VBS regime**

1. **VBS-I:** Potential Heat sink Effect - Here the ratio of MTS number is less than 1. (MTS<1) In this effect energy from external sources is absorbed by pilot fuel.
2. **VBS-II:** Neutralizing Effect. Here the ratio of MTS number is equal to 1. (MTS=1). It does not take part in positive or negative heat sources.
3. **VBS-III**: Heat source Effect. Here the ratio of MTS number is greater than 1. (MTS>1). In this Effect, energy is supplied to external sources by pilot fuel.

Effect of number of external source variation on Pilot fuel spreading rate

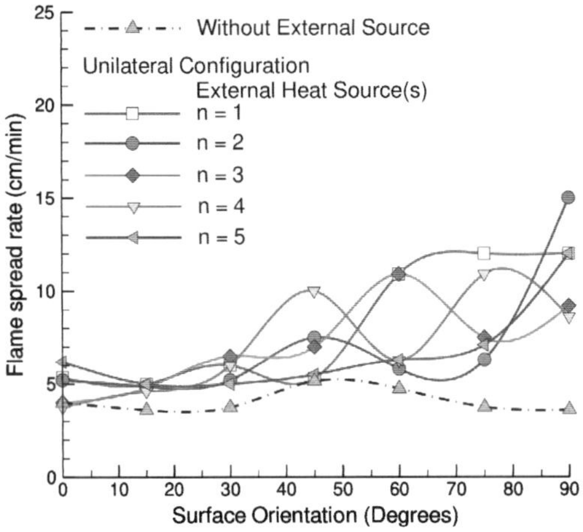

Figure 13. For Unilateral Configuration.

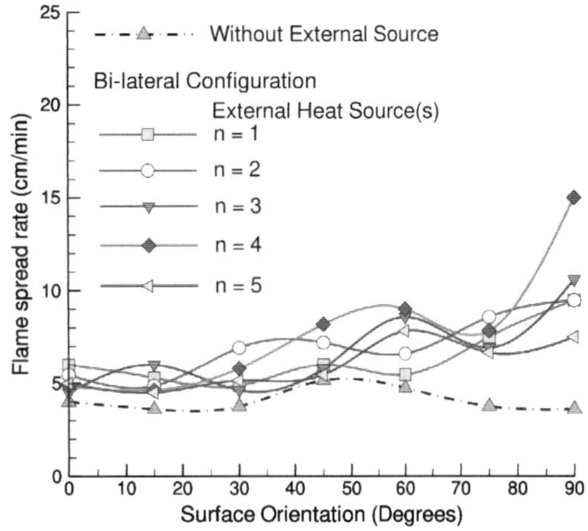

Figure 14. For Bilateral Configuration.

Figure 13 shows the effect of the regression rates of pilot fuel for unilateral configuration at different orientations. The external sources vary from n=1 to n=5. The plot is compared with the base case of pilot fuel without any external source. The flame spread rate is the same for 0^0 orientation and the presence of 3 external heat sources (n=3). Which comes under the Neutralizing zone. An interesting thing to note is at 0^0 orientation when n=4 the regression rate drops by **6%.** It comes under the heat sink zone. From the plot, it can be observed that at 90^0 orientation and when the Number of external heat sources(n) =2 the regression rate is maximum at the percentage of **76.019%.** In most cases, the regression rate comes under MTS>1. The reason is heat transfer is dominating in this region because of buoyancy that carries the heat from burnt to the unburnt fuel.

Table 1. MTS number variation for Unilateral Configuration.

Orientation (degree)	n=1	n=2	n=3	n=4	n=5
0	1.34	1.30	**1.00**	**0.94**	1.55
15	1.38	1.35	1.33	1.28	1.39
30	1.60	1.39	1.73	1.60	1.33
45	1.03	1.46	1.36	1.94	1.07
60	2.30	1.22	2.30	1.32	1.33
75	3.20	1.68	2.00	2.91	1.89
90	3.33	4.17	2.56	2.39	3.33

Figure 14 shows the effect of the flame spread rate of pilot fuel for Bilateral configuration at different orientations. The external sources vary from n=1 to n=5. The plot is compared with the base case of pilot fuel without any external source. From the plot, it can be observed that at 90^0 orientation and when the number of the external source(n) = 4, the regression rate is maximum with a rise of **76.019%.** All the values for flame spread rate for bilateral configuration come under the heat Source zone (MTS > 1). Heat transfer is dominating in this region because of buoyancy that carries the heat from burnt to the unburnt fuel. An interesting thing to note here is they were no Heat sink effect and neutralizing effect for this configuration.

Table 2. MTS number variation for Bilateral Configuration.

Orientation (degree)	n=1	n=2	n=3	n=4	n=5
0	1.50	1.36	1.13	1.22	1.25
15	1.47	1.35	1.67	1.28	1.25
30	3.75	1.85	1.23	1.55	1.37
45	1.17	1.40	1.13	1.59	1.06
60	1.15	1.39	1.81	1.90	1.65
75	2.00	2.29	1.85	2.09	1.78
90	2.63	2.63	2.94	4.17	2.08

Figure 15, shows the effect of the regression rates of pilot fuel for Trilateral (Y) configuration at different orientations. The external sources vary from n=1 to n=5. The plot is compared with that of the base case of pilot fuel without any external source. At 45^0 orientation coupled with 3 external sources(n=3) the flame spread rate is the same as that of the base case. Which makes it a Neutralizing effect. An interesting thing to note here is at 45^0 orientation and when n=2 the regression rate drops by **20%** which is the minimum for this configuration**.** It comes under the heat sink zone (MTS<1). The maximum rise occurs at 90^0 orientation and when n=3 the regression rate rises by **77.97%.** In most cases, the regression rate comes under the Heat source zone (MTS>1).

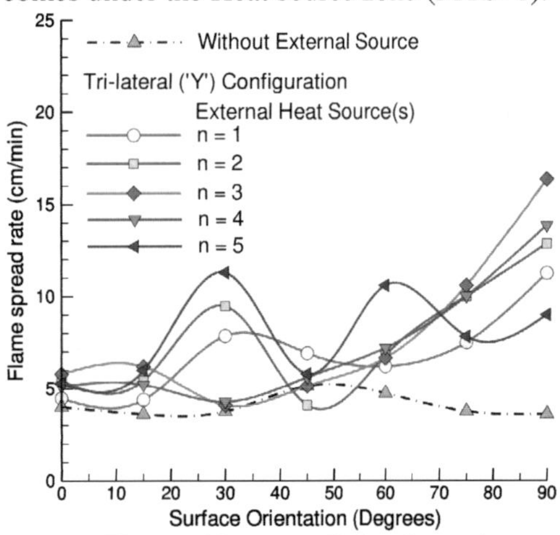

Figure 15. For "Y" Configuration.

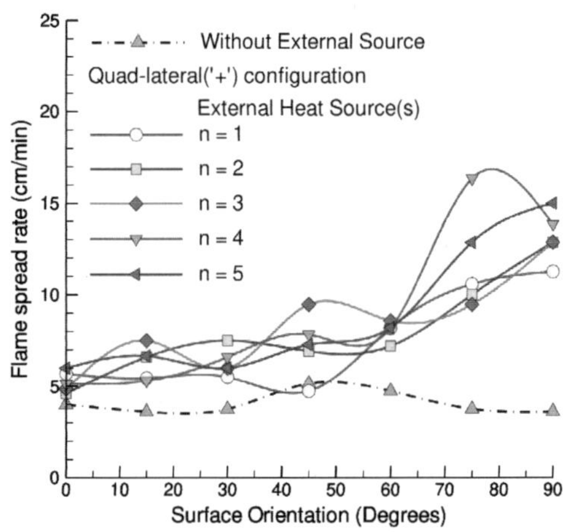

Figure 16. For "+" Configuration.

Table 3. MTS number variation for "Y" Configuration.

Orientation (degree)	n=1	n=2	n=3	n=4	n=5
0	1.12	1.36	1.45	1.29	1.32
15	1.22	1.51	1.72	1.45	1.67
30	2.10	2.53	1.09	1.14	3.01
45	1.35	**0.80**	**1.00**	1.09	1.13
60	1.31	1.44	1.41	1.52	2.23
75	2.00	2.67	2.82	2.67	2.09
90	3.13	3.57	4.54	3.84	2.50

Table 4. MTS number variation for "+" Configuration.

Orientation (degree)	n=1	n=2	n=3	n=4	n=5
0	1.43	1.15	1.22	1.29	1.50
15	1.51	1.83	2.08	1.47	1.85
30	1.47	2.00	1.60	1.76	1.60
45	**0.92**	1.35	1.84	1.52	1.42
60	1.73	1.52	1.81	1.73	1.73
75	2.82	2.67	2.53	4.36	3.43
90	3.13	3.57	3.57	3.84	4.17

Figure 16 shows the effect of the flame spread rate of pilot fuel for Quad-lateral "+" configuration at different orientations. The external sources vary from n=1 to n=5. The plot is compared with that of the base case of pilot fuel without any external source. From the plot, it can be observed that at 75^0 orientation and when the number of the external source is equal (n=4), the regression rate is maximum with a rise of **77.06%.** Most of the values for flame spread rate for "+" configuration comes under the heat source zone (MTS>1). An interesting thing to note here is at 45^0 orientation and when n=1 the flame spread rate is minimum with a drop of **8%.** Another point to be noted from this plot is that there is no Neutralizing effect occurred for this configuration.

Effect of configuration type on Pilot fuel spreading rate

Figure 17 shows the comparison of flame spread rate for various configurations at different orientations. For this, the number of external sources is kept constant (n=1) and the flame spread rate of pilot fuels for all configurations are compared. It is interesting to note that the highest flame spread rate is found for the Unilateral configuration for this case. The flame spread rate increases by **69.96%.** The reason is, forward heat transfer is dominating in this region because of buoyancy that carries the heat from burnt to the unburnt fuel. The flame spread rate at 45^0 orientation for Quad-lateral configuration is minimum with a drop of **8%**. It comes under the heat sink regime (MTS<1). For all the other cases the heat source effect is observed (MTS >1). Another point to be noted is that there is no Neutralizing zone for this case.

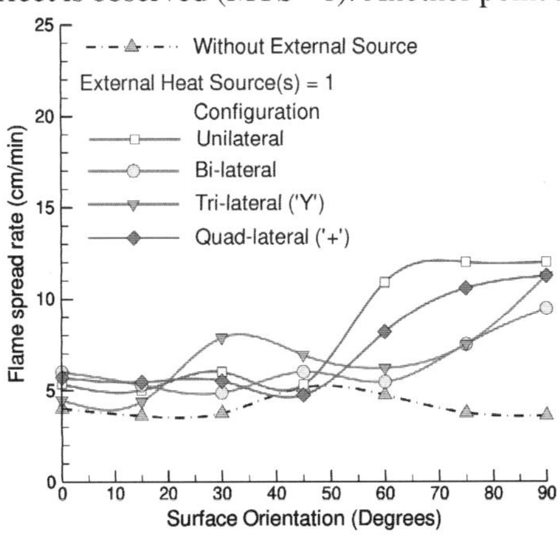

Figure 17. Number of external sources(n=1).

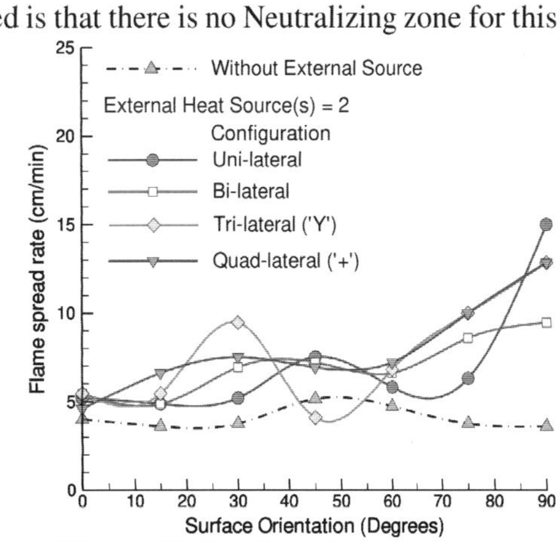

Figure 18. Number of external sources(n=2).

Similarly, in Figure 18, the number of external heat sources is increased to 2, and values of flame spread for pilot fuels were measured for different configurations at an interval of 15 degrees. The values were compared with the Regression rate of pilot fuel without any external source. An interesting thing to note here is that for Tri-Lateral configuration the flame rates drop by **20%** which is the minimum for this case**.** It comes under the heat sink zone (MTS<1). This is because of less oxygen entering the fume of pilot fuel. The maximum rise of about **76.019%** for unilateral configuration at 90^0 orientation is achieved. The reason is, forward heat transfer is dominating in this region that carries the heat from burnt to the unburnt fuel.

For Figure 19, the regression rate of pilot fuel for all different configurations is measured when the number of external sources is fixed to 3 (n=3). It was compared with that of the base case (without external source). Here the interesting thing to note is for Unilateral configuration at 0^0 orientation and for trilateral configuration at 45^0 orientation, the regression rates were found to be similar to that of the base case. It comes under the Neutralizing zone (MTS=1). The maximum rise in regression rate was found out at 90^0 orientation for Trilateral configuration. The maximum rise in flame spread rate was **77.97%.** Another point to be noted from this plot is that there is no heat sink zone (MTS<1). Most of the regime comes under the Heat source zone (MTS>1). The reason is, forward heat transfer is dominating in this region. Hence the regression rate for this case is greater than compared to the pilot fuel.

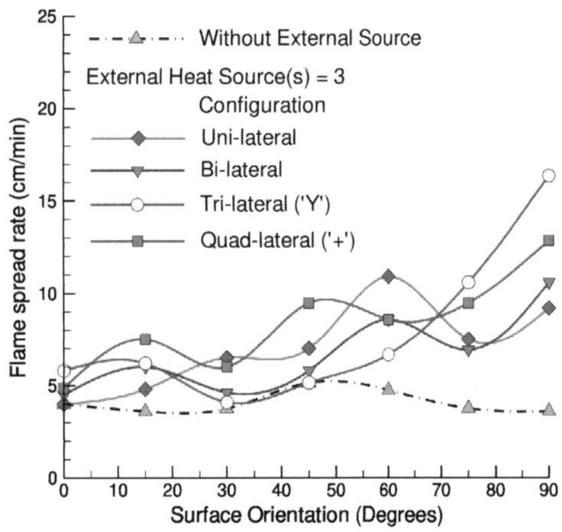

Figure 19. Number of external sources(n=3).

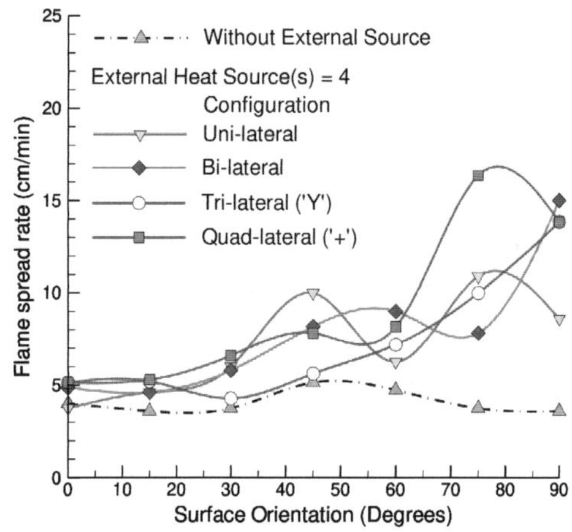

Figure 20. Number of external sources(n=4).

Figure 20 shows the comparison of flame spread rate for various configurations at different orientations. For this, the number of external sources is kept constant (n=4) and the flame spread rate of pilot fuels for all configurations are compared. It is interesting to note that the highest flame spread rate is found at 75^0 orientation for quad-lateral configuration. The flame spread rate rises by **77.06%.** The reason for this rise is connective heat transfer, which is dominating in this region. The flame spread rate at 0^0 orientation for unilateral configuration drops by **6%** which is the minimum for this case**.** It comes under the Heat sink regime (MTS<1). For all the other cases the Heat source effect is observed (MTS>1). Another point to highlight from this plot is that there is no Neutralizing effect (MTS=1) observed in this case.

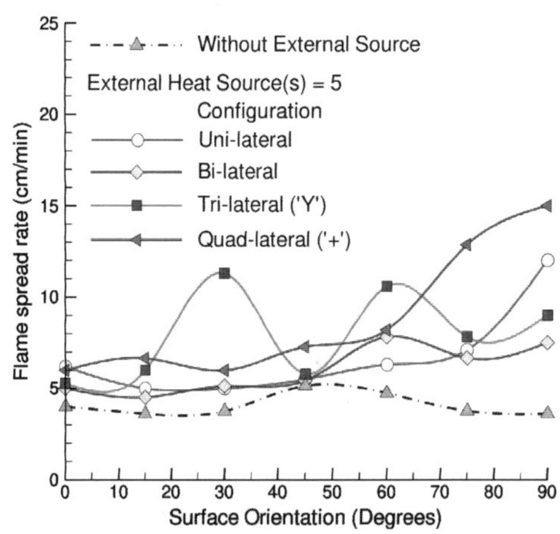

Figure 21. Number of external sources(n=5).

In Figure 21, the regression rate of pilot fuel for all different configurations is measured when the number of external sources is fixed to be five (n=5). It was compared with that of the base case (No External heat source). The maximum rise in regression rate was found out at 90^0 orientation for Quad-lateral configuration. The rise in flame spread rate was **76.01%.** Another point to be noted from this plot is that there is no heat sink zone (MTS<1) and no neutralizing zone (MTS=1). All of the regression rates come under the Heat source zone (MTS>1). Hence the regression rate for this case is greater than compared to the pilot fuel.

Effect of configuration type on external heat source spreading rate

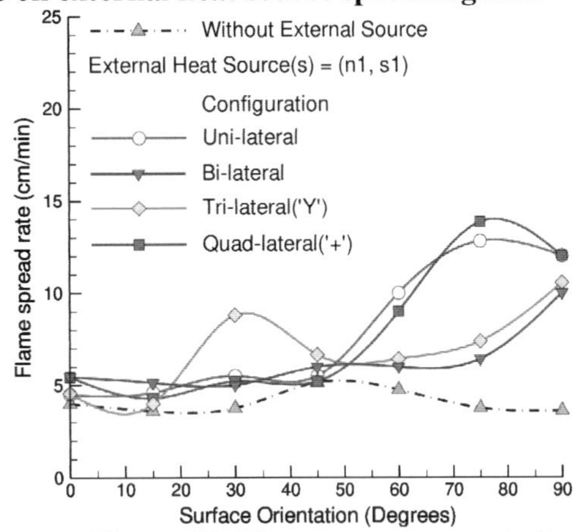

Figure 22. Flame spread rate at 'n1'.

In Figure 22, the regression rate of 1st external source (n1) for all the configurations at different orientations is compared with the pilot fuel without any external source. The maximum rise in regression rate was found out at 75^0 orientation for Quad-lateral configuration with a rise of **72.89%.** The minimum raise is about **0.99%** at 45^0 orientation for Quad-lateral configuration. For all the configurations the MTS regime is in the Heat source regime (MTS>1), due to the fact connective heat transfer is dominating in this region, which carries the heat from burnt to the unburnt fuel. Hence the regression rate for this case is greater than compared to the pilot fuel.

Table 5. MTS Number variation of 'n1' for different configurations.

Orientation (degree)	Unilateral configuration	Bilateral configuration	Y configuration	+ configuration
0	1.13	1.36	1.14	1.36
15	1.28	1.43	1.11	1.19
30	1.47	1.33	2.35	1.39
45	1.07	1.17	1.30	1.01
60	2.11	1.27	1.36	1.90
75	3.41	1.71	1.97	3.69
90	3.33	2.78	2.93	3.33

Effect of Energy interaction on spreading rate

In Figure 23, the flame spread rate of unilateral configuration for the first external source (n1) is compared with that of pilot fuel without an external source and with the external source. The maximum rise is at 75^0 orientation with a rise of **70.61%.** The minimum raise is about **6.54%** at 45^0 orientation. Most of the values for flame spread rate for unilateral configuration comes under the Heat Source zone (MTS>1). This is due to forward heat transfer. When compared to the pilot fuel the MTS number rises by **3.63%** for 45^0 orientation and maximum drop is obtained at 0^0 orientation of about **13.87%.** An interesting point to be noted here is for 60^0 and 90^0 orientations the flame spread rate is constant for both pilot and first external source(n1). It comes under the Neutralizing zone (MTS=1).

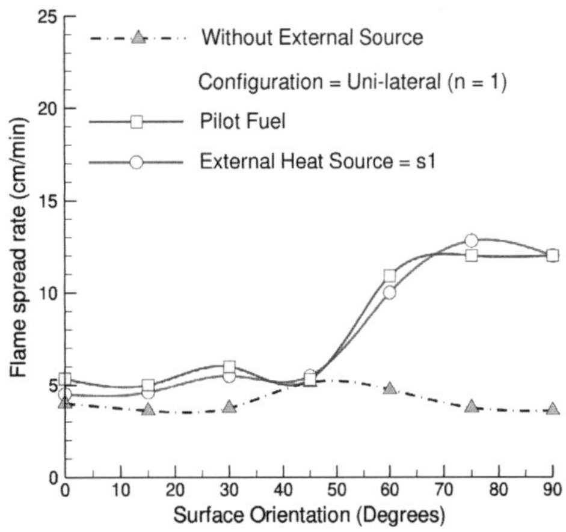

Figure 23. For Unilateral Configuration.

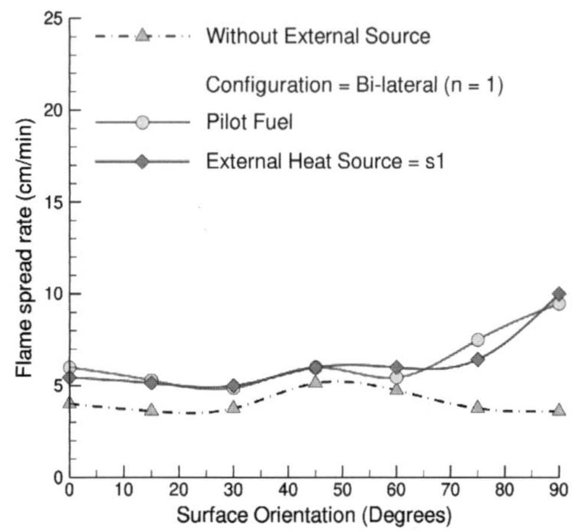

Figure 24. For Bilateral Configuration.

In Figure 24, the regression rate for Bilateral configuration, for the first external source (n1) is compared with that of pilot fuel without external source and with the external source. The maximum rise is at 90^0 orientation with a rise of **64.02%.** The minimum raise is about **14.52%** at 45^0 orientation. A point to be noted is there is no heat sink effect and no neutralizing effect observed for this case. When compared to the pilot fuel the MTS number rises by **2.05%** for 90^0 and the maximum drop is obtained at 30^0 of about **40.52%.** An interesting point to be noted here is for 45^0 orientation the flame spread rate is constant for both pilot and first external source. Here Neutralizing effect is observed.

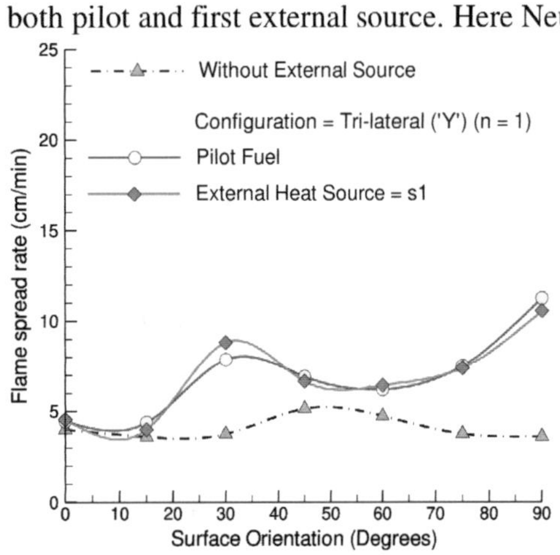

Figure 25. For 'Y' Configuration.

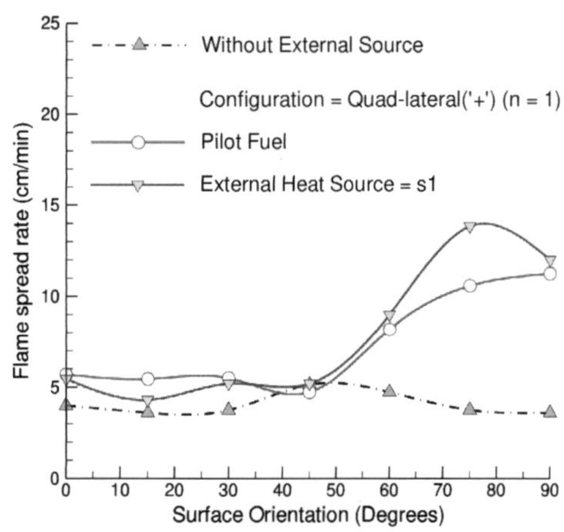

Figure 26. For '+' Configuration.

In Figure 25, the flame spread rate for the trilateral configuration for the first external source (n1) is compared with that of pilot fuel without an external source and with the external source. The maximum rise is at 90^0 orientation with a rise of **65.87%.** The minimum raise is about **9.90%** at 15^0 orientation. All the values for flame spread rate for trilateral configuration come under Heat Source zone (MTS>1). When compared to the pilot fuel of the configuration the MTS number rises by **8.87%** for 15^0 and the maximum drop is obtained at 45^0 orientation of about **6.55%.** Another point to be noted from this plot is that there is no neutralizing zone.

In Figure 26, the regression rate for Quad-lateral configuration for the first external source (n1) is compared with that of pilot fuel without an external source and with the external source. The maximum rise is at 75^0 orientation with a rise of **72.89%.** The minimum raise is **0.99%** at 45^0 orientation. All the values for flame spread rate for Quad-lateral configuration come under the Heat source zone (MTS>1). This is due to the connective heat transfer, which is dominating in this region that carries the heat to move from burnt to unburnt fuel. A point to be noted is there is no heat sink zone (MTS<1) and no Neutralizing zone (MTS=1) in this case. When compared to the pilot fuel the MTS number rises by **8.93%** for 45^0 and the maximum drop is obtained at 15^0 of about **17.81%.**

4. Acknowledgement

Research work done by us could not be feasible without the considerable work done by the former scientists and researchers. Furthermore, we acknowledge the hard work and intense efforts made by firefighters all over the world for the everlasting service they have been delivering us for the past many decades.

5. Conclusion

An experimental investigation of the characteristics of spread rate for matchsticks arranged in different configuration types (Uni, Bi, Y, and +) and an array of external heat sources were studied on downward spreading fires. The influences of angular orientation and spread rate variation on pilot fuel for various external sources are investigated systematically. Angular orientation was varied from 0 to 90° in the interval of 15. The number of external sources is varied from n=1 to 5 and the effect of flame rate on pilot fuel for each configuration was studied. Also burn rate for the first external source was studied in detail for all four configurations. Based on the results it can be concluded that the external heat source affects forward heat transfer from burnt to unburnt pilot fuel.

1. For Unilateral configuration the plots vary non- monotonically with almost every case (MTS>1). At zero degree for n=3 the MTS number is equal to 1 and n=4 the MTS number is less than 1.
2. For bilateral configuration, flame rates vary non-monotonically and are always in the Heat Source zone for pilot fuel and the first external source.
3. For Tri and quad lateral configuration, the flame rates vary non-monotonically for both pilot fuel and external heat source. For trilateral at 45 degrees for n=2, the MTS number is less than 1, that is it is in the heat sink zone whereas at n=3 the MTS number is equal to 1.
4. For quadrilateral at n=1, the MTS number is less than 1. For quadrilateral, there is no Neutralizing zone. External heat source results in increasing flame rate owing to enhanced heat transfer to unburnt fuel. Forward heat transfer with a parallel heat source is mostly dominated by radiation from external heat source and conduction from pilot fuel. The governing physics of phenomena include the alternation of the reaction zone.
5. **Applications of work**: The work carries a wide range of applications including engineering viz., combustion and propulsion, defense systems validation, testing and upgradation like missile systems, industrial with power generation systems, practical, functional, scientific applications.

6. References

[1] Fons, W.L., Journal of Agricultural Research, "Analysis of Fire Spread in Light Forest Fuels" , 72, pp. 93-121, (1946).
[2] Vogel, M., and Williams, F.A. 1970. "Flame propagation along matchstick arrays". Combust. Sci. Technol., l(6), 429–436.
[3] Miller, R. A. (1970). "Flame propagation in two-dimensional matchstick arrays". Unpublished Report, The Guggenheim Laboratories, Princeton University.
[4] Emmons, H., and Shen, T. 1971. "Fire spread in paper arrays". Proc. Combust. Inst., 13(1), 917–926
[5] Prahl, J.M., and Tien, J.S. 1973. "Preliminary investigations of forced convection on flame propagation along paper and matchstick arrays". Combust. Sci. Technol., 7(6), 271–282
[6] C. C. HWANG & YUSHENG XIE (1984) "Flame Propagation Along Matchstick Arrays On,

Inclined Base Boards", Combustion Science and Technology, 42:1-2, 1-12.
[7] M.J Gollner, Y. Xie, M. Lee, et al. Combust. Sci. Techol. 184 (5) (2012) 585-607
[8] H.Y.Shih, "Flame spread and interactions in an array of thin solids in low speed concurrent flows", Combust. Theory Model. 13(3) (2009) 443-459
[9] V. Malhotra, C. kumar, A. kumar, "Opposed flow flame spread over an array of thin solid fuel sheets in a microgravity environment", Combust. Theory Model. 17 (5) (2013) 835-857
[10] M.C. Johnston, J.S Tien, D.E Muff, X. Zhao, S.L Olson, P.V Ferkul, "Self induced buoyant blow off in upward flame on thin solid fuels", Fire Saf. J. 71 (2015) 23-27
[11] Govindaraju Avinash , Amit Kumar, Vasudevan Raghavan. "Experimental analysis of diffusion flame spread along thin parallel solid fuel surfaces in a natural convective environment" 0010-2180/© 2015 The Combustion Institute.
[12] Tiwari, P., Ramanan, V., and Malhotra, V., 'An Experimental Insight into Thermoacoustic Smoldering'. Journal of Space Exploration, volume 6, issue 1, 2017
[13] Lin Jiang , Zhao Zhao , Wei Tang , Colin Miller, Jin-Hua Sun , Michael J. Gollner. "Flame spread and burning rates through vertical arrays of wooden dowels". 1540-7489 © 2018 The Combustion Institute
[14] Tatnell, D. M., Heath, M. S., Hepplestone, S. P., Hibbins, A. P., Hornett, S. M., Horsley, S. A. R., and Horsell, D. W., 'Coupling and confinement of current in thermoacoustic phased arrays'. Science Advances, Applied Physics, 6, 2020.

Finite element simulation of residual stresses in welded steel butt joints and their experimental verification and mitigation

V Eswara Kumar [1], Dr. D V N J Jaggannadha Rao [2] and Dr.Y Seetharama Rao [3]

[1,2,3]DEPARTMENT OF MECHANICAL ENGINEERING, GAYATRI VIDYA PARISHAD COLLEGE OF ENGINEERING (AUTONOMOUS), JNTUK UNIVERSITY

Abstract. The localized, intense, and moving heat source existing in welded fabrication results in residual stresses and distortion. Residual stresses remain inside the material even after the external forces or thermal gradients are removed after welding. Residual stresses decrease fatigue life of the welded joint leading to premature failures and associated distortion cause problems in fitment during assembly. Estimation of residual stresses before actual fabrication is essential for elimination or mitigation after fabrication. In this work, finite element thermo-mechanical analyses are carried out to estimate the levels of residual stresses of thin welded similar butt joints of 1018 steel and SS 304 steel. The numerically obtained results of the joints are compared with experimental results of identical joint geometry, welding conditions, and material properties existing in the literature. The comparison of numerically obtained thermal histories with those from experiments yielded good agreements. Longitudinal residual stresses are found to be dominant and comparison of these longitudinal stresses with experimental data also found to be in good agreements. Further, the levels of residual stresses are computed with three different preheat temperatures (600°C, 700°C and 800°C) to study the effect of preheating temperature in the mitigation of residual stresses. It has been found that the peak value of longitudinal stress decreased considerably with preheat temperature. Increasing preheat temperature has a marginal reduction in the peak value of longitudinal stress.

Keywords: TIG welding, FEM, residual stress, mitigation, preheat.

1. INTRODUCTION

Residual stresses in welding appear due to improper fit between various parts, various phases, or various regions within the part that originating out of non-uniform thermal strain, and strains originating from the solidification and solid-state phase transformations [1]. Residual stresses can be classified into two categories one is macro stresses and the other is micro stresses [2]. Micro stresses occur within a grain of the component but macro stresses occur along with the component. Macro stresses are three types tensile, compressive, and shear stresses. Compressive stresses are good for welded components and cause stress relaxation, whereas tensile stresses also called longitudinal stresses harm the welded joint. Residual stresses decrease fatigue life of the joint leading to premature failure and associated distortion causes problems in fitment during assembly. The prediction of residual stresses and distortion before actual fabrication of the joint is necessary for the safe design of the joint. Experimental methods like strain gauging, X-ray diffraction, and neutron diffraction can be used for the measurement of residual stresses. However, these are prohibitively expensive and provide limited information. In this context, a numerical method like the finite element method (FEM) offers a cost-effective alternative. So, residual stresses and distortion can be computed within the whole domain are the problem which can be validated by limited experimentation.

Residual stresses in welding are a well-researched topic in literature. Many authors carried out experimental measurements using hole drilling, strain gauging, X-ray diffraction, and neutron diffraction. Two dimensional, axisymmetric, and three-dimensional finite element analyses are also carried out to assess the levels of residual stresses in welding problems. Mitigation methods are proposed to reduce the levels of residual stresses. A detailed review on welding residual stress distributions can be found in the publication of Syahida et al. [3]. A brief summary of the research work carried out recently in the welding of plates is given as follows.

Vijay et al. studied the variation of temperature in TIG welded SS 304 plate butt joint of 3 mm thickness using ANSYS workbench and found that comparison of temperature distributions predicted by numerical analyses are in good agreements with thermocouple measurements [4]. Siddique and Khwaja developed mathematical relationships using Taguchi design to predict temperature distribution and effect of process parameters like welding speed and input power of TIG welding of 304 stainless steel [5]. Kartal et al. studied the comparisons of the residual stress profiles generated by two different welding techniques (multi-pass and single pass welding) in 80 mm thick ferritic steel welds [6]. Bajpei et al. developed a finite element model using ANSYS software for the minimization of the residual stresses and distortions in thin butt-welded aluminium alloy 5052-H32 plates during gas metal arc welding process by using three different heat sink models employing thick backing plate and cooling fluids [7].Almeida et al. proposed a finite element model for the simulations of a TIG butt-welding process applied to thin plates made of an AISI 316L austenitic stainless steel [8]. Subsequent measurement of residual stresses using hole drilling method yielded good agreements with numerical predictions. Hashemzadeh et al. used regression methods to develop relationships to estimate the distortion and residual stresses in thin butt-welded plates using analytical and finite element methods [9]. Vekatkumar et al. studied the effect of welding heat input on temperature and residual stresses induced by single pass V-butt joint gas tungsten arc welding using a 3D Finite Element (FE) numerical model of 304 stainless steel [10]. Tianci et al. proposed a novel heat source model combining Gaussian surface and uniform volume heat source for the simulation of temperature distribution and residual stresses of a S355 steel T welding. The comparison between measured results by the hole drilling and X-ray diffraction (XRD) and simulation indicates that the combined heat source is sufficient to allow for an overall prediction of the entire welding process of T joints [11]. Luca et al. used the Finite Element (FE) method to numerically analyse the thermo-mechanical behaviour and residual stresses in dissimilar welded T-joints [12]. Balaram et al. carried out experimental and numerical investigation to analyze the residual stress distribution in the dissimilar TIG weldments of AISI 304 and Monel 400 and found that the residual stresses developed in weldments are within the yield limits [13]. A coupled thermo-metallurgical-mechanical numerical model is used for the simulation of residual stresses on bead on plate TIG welding of S355J2 steel by Zhang et al. This model is validated by measurement of residual stresses by X-ray diffraction and relaxation methods with good success by the authors [14]. Suman and Biswas investigated residual stresses in a square butt submerge arc welded Creep Strength Enhanced Ferritic (CSEF) steel plate experimentally and numerically. The mitigation of residual stresses using preheating is also discussed in this paper [15]. However, there is enough scope for research in this topic as it is associated with nonlinearities like phase change, temperature dependence on material properties, path dependence of stress-strain response, plasticity, moving distributed heat source, and filler metal deposition.

In this work, 3-dimensional thermo- mechanical simulation using ANSYS FE program is used to predict the residual stress distribution developed in the butt joints welded by the TIG welding technique. The similar joints are of AISI 1018 steel and SS 304 steel. The thermal histories at various locations of these joints are measured by thermocouples and residual stresses are measured by neutron diffraction as reported in the literature [16]. This experimental data is compared with present data obtained by finite element simulation of joints of identical geometries and welding conditions as reported in the previous literature cited above. The comparison of numerical thermal histories with those obtained experimentally yielded good agreements. Longitudinal residual stresses are found to be dominant. Comparison of numerically obtained longitudinal residual stresses with those obtained experimentally also yielded good agreements. Further the levels of residual stresses are computed with three different preheat temperatures (600°C,700°C, and 800°C) to study the effect of preheating temperature in the mitigation of residual stresses. It has been found that the peak value of longitudinal stress decreased considerably with preheat temperature. Increasing preheat temperature has a marginal reduction in the peak value of longitudinal stress.

2. FINITE ELEMENT SIMULATION OF TIG WELDING

FE simulation of welding is a challenging numerical problem involving localized, intense moving heat source, phase changes from solid to liquid and vice versa, the temperature dependence of material properties, and plasticity. Sequential thermo-mechanical analyses are to be carried out to evaluate residual stress distributions.

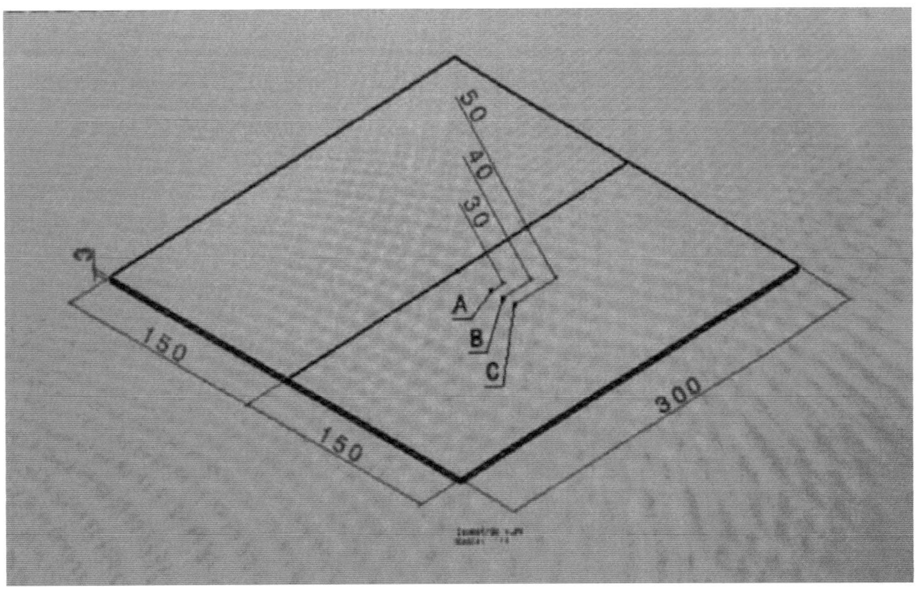

Figure 1. Geometry of the joint along with locations of thermocouples

The geometry of the butt joint and welding parameters used in this study are given in figure 1 and table 1 respectively. Dimensions of the two plates that are butt welded are 300 mm x 150 mm x 3 mm. The locations of placement of thermocouples are shown by letters A, B, and C in figure1 and these are at 30 mm, 40 mm, and 50 mm transverse from the weld centreline top and middle of the joint.

Table 1. Welding parameters

Voltage (V)	Current (A)	Bead height (mm)	Bead width (mm)	Efficiency	Speed (mm/s)
150	18	3	6	0.4	4

A numerical 3-D model has been created in ANSYS using mechanical APDL 19.2 for the TIG welding simulation process. The moving heat source in the welding processes has been modelled in many ways such as point heat source, volumetric heat source, Gaussian distribution model, and double ellipsoidal model in the literature. In the present work double ellipsoidal heat flux density distribution has been used because of its capability to model 3-D situations. The double ellipsoidal heat flux density distribution (figure 2) is modelled by the following equations [17]

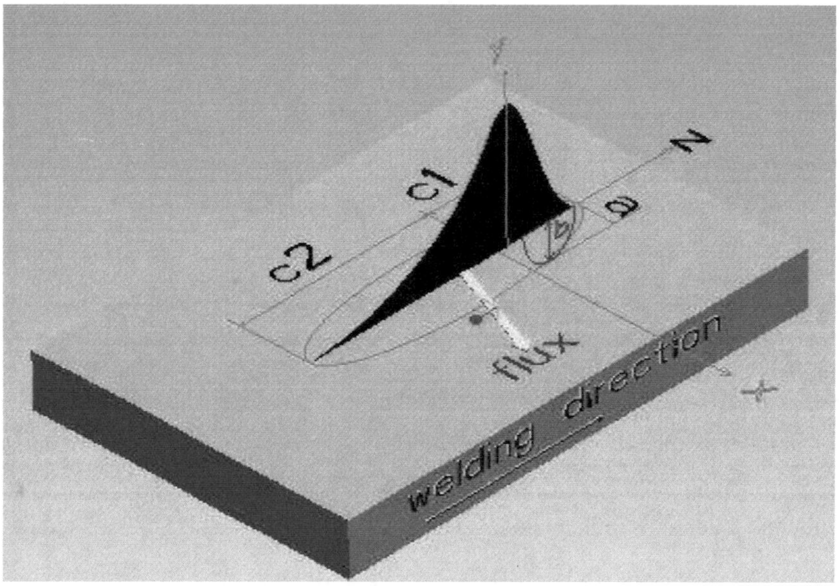

Figure 2. Double ellipsoidal heat flux density distribution.

$$Q = \eta VI \tag{1}$$

$$q_f(x,y,z,t) = \frac{6\sqrt{3}f_f Q}{\pi^{3/2}\, abc1} exp\left(-3\frac{x^2}{a^2}\right) exp\left(-3\frac{y^2}{b^2}\right) exp\left(-3\frac{(z-vt)^2}{c1^2}\right) \tag{2a}$$

$$q_r(x,y,z,t) = \frac{6\sqrt{3}f_r Q}{\pi^{3/2}\, abc2} exp\left(-3\frac{x^2}{a^2}\right) exp\left(-3\frac{y^2}{b^2}\right) exp\left(-3\frac{(z-vt)^2}{c2^2}\right) \tag{2b}$$

In the equation (1) Q is the total heat input, V is voltage, I is current and η is arc efficiency. The factors f_f and f_r in the above equations denote the fractional ratios of total heat input (Q) in both forward and backward welding directions. Here, v is the speed of welding, $a, b, c1$ and $c2$ are the parameters of heat source. In many cases these parameters are usually found to be correlated with the geometry of the bead. The benefit of applying these equations is to directly calculate the given flux density from the position of an individual node. In the current analysis, slightly changed parameters of double ellipsoidal model are used in which $c1 = c2 = a$, $f_f = 1/3$, and $f_r = 2/3$. It was recommended that the values of a and b to be (1.3) x (half of bead width) and (0.8) x (depth of penetration) respectively [18].

The geometry of the butt joint currently under consideration is symmetric along the weld centreline. The geometry is meshed with an 8-noded thermal element, SOLID70 available in the ANSYS element library to compute thermal solutions; corresponding 8-noded structural element, SOLID185 also available in the ANSYS element library is used to compute structural solutions. The geometrical and meshed models of the symmetric half of the butt joint are shown in figure 3 and figure 4 respectively.

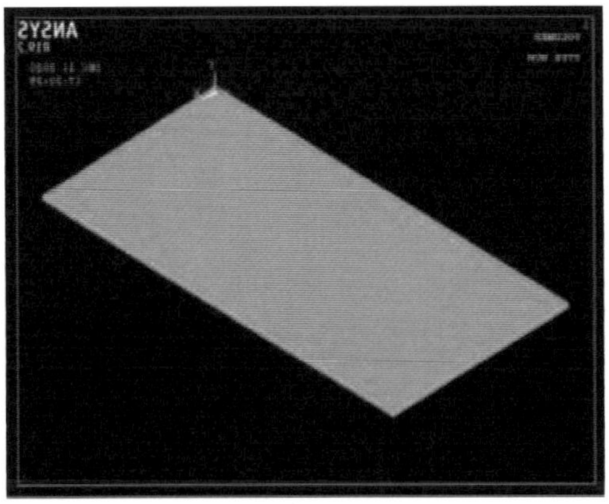

Figure 3. 3-D model of the symmetric half of the butt joint

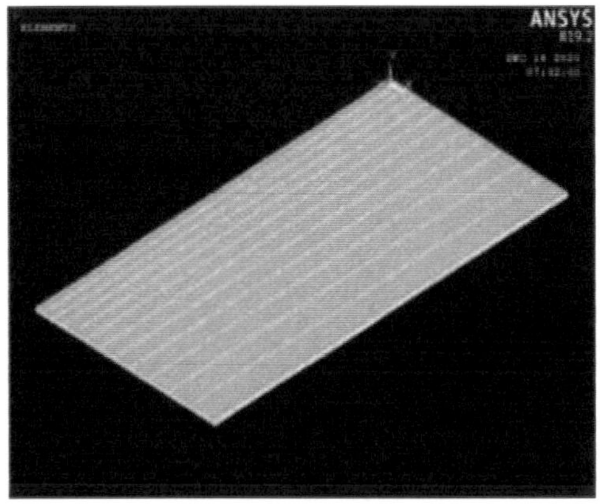

Figure 4. Meshed model of the symmetric half of the butt joint

In the sequential thermo-mechanical analyses first, the energy equation is solved with appropriate boundary conditions to get a thermal solution. This thermal solution is given as input to the succeeding structural analysis to get the stress solution. The welding and subsequent cooling process is divided into 200-time steps in which 150 steps are of the same time step during actual arc burning period and the remaining 50-time steps are for the cooling period of the joint. The time steps in the cooling period are in the geometric progression with a common ratio 1.1. The generation of thermal solutions and succeeding structural solutions are carried out sequentially for all 200-time steps. The butt joints analysed in this work are of two materials namely 1018 steel and SS 304 steel. The thermo-mechanical properties of these two materials are given in table 2 and table 3 respectively. The other properties density, latent heat, and melting temperatures of 1018 steel and SS 304 steel are given in table 4 and table 5 respectively are taken from the web.

Table 2. Thermo-mechanical properties of AISI 1018 steel [17]

Temperature (°C)	Thermal conductivity (W/m K)	Specific heat (J/kg K)	Enthalpy (10^9J/cum)	Poisson's ratio	Yield stress (MPa)	Stress At strain =1 (MPa)	Young's Modulus (GPa)	Coefficient of Thermal Expansion (10^{-6}/°C)
0	51.90	450	1.00	0.2786	290	314	200	10.0
100	51.10	499.20	2.00	0.3095	260	349	200	11.0
300	46.10	565.50	2.65	0.3310	200	440	200	12.0
450	41.05	630.50	3.80	0.3380	150	460	150	13.0
550	37.50	705.50	4.10	0.3575	120	410	110	14.0
600	35.60	773.30	4.55	0.3738	110	330	88	14.0
720	30.64	1080.4	5.00	0.4238	9.8	-	20	14.0
800	26.00	931.00	5.23	0.4738	9.8	-	20	15.0
1450	29.45	437.93	9.00	0.4990	9.8	-	2.0	15.0
1510	29.70	400.00	11.00	0.4990	9.8	-	0.2	15.0
1580	29.70	735.25	13.25	0.4990	0.009	-	0.00002	15.0

5000	42.20	400.00	28.40	0.4990	0.009	-	0.00002	15.5

Table 3. Thermo-mechanical properties of SS 304 [17]

Temperature (°C)	Thermal conductivity (W/m K)	Specific heat (J/kg K)	Enthalpy (10^9J/cum)	Young's modulus (GPa)	Thermal Expansion (10^{-6}/°C)	Poisson's ratio	Yield stress (MPa)	Tangent modulus (MPa)
20	15.0	442	1.036	200	19e-6	0.278	230	2800
200	17.5	515	1.73	185	19e-6	0.288	184	2590
400	20	563	2.59	170	19e-6	0.298	132	2380
600	22.5	581	3.51	153	19e-6	0.313	105	2142
800	25.5	609	4.46	135	19e-6	0.327	77	1890
1000	28.3	631	5.45	96	19e-6	0.342	50	9.6
1200	31.1	654	6.48	50	19e-6	0.350	10	5.0
1340	31.1	669	7.22	10	19e-6	0.351	10	1.0
1390	66.2	675	9.57	10	19e-6	0.353	10	1.0
2000	66.2	675	12.86	10	19e-6	0.357	10	1.0

Table 4. Other properties of 1018 steel

Property	Value
Density	7800 Kg/m^3
Melting temperature	1539 °C
Latent heat of solidification	260 kJ/kg

Table 5. Other properties of SS 304

Property	Value
Density	8000 Kg/m^3
Liquidus temperature	1390 °C
Solidus temperature	1340 °C
Latent heat of solidification	260 kJ/kg

A macro in ANSYS APDL is written to apply the moving double ellipsoidal heat flux density distribution with welding speed along the weld centreline. The applied heat flux density distribution when the torch is at the middle of the joint is shown in figure 5. Symmetry boundary conditions are applied at the symmetric plane of the joint for both thermal and structural analyses. Natural convection is applied on all boundary surfaces of the problem domain except symmetric surface as cooling load, the heat transfer coefficient is taken as 15 W/m^{2}K and the ambient temperature is taken as 30°C (figure 6). All nodes on a line at the bottom of symmetric surface are arrested in all three degrees of freedom to prevent rigid body motion in the structural solution. Kinematic hardening is assumed to represent plastic behaviour.

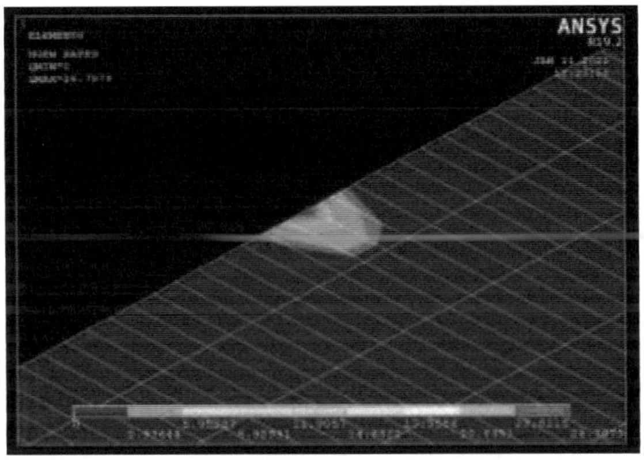

Figure 5. Heat flux density distribution of double ellipsoidal when the welding torch at the center of the joint

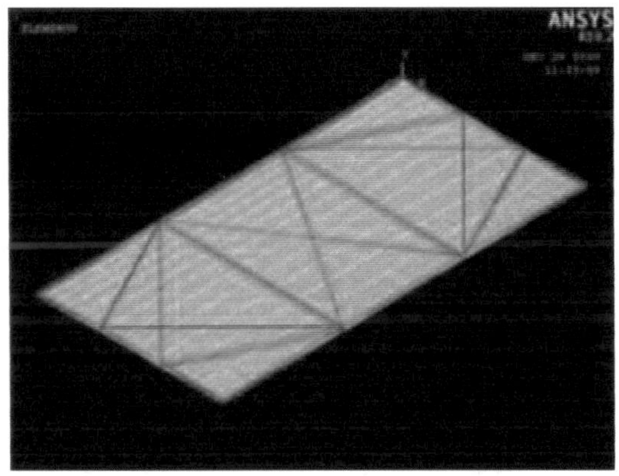

Figure 6. Convective boundary conditions (h = 15 W/m^2 K)

3. RESULTS AND DISCUSSION

Sequential thermo-mechanical analyses have been carried out to compute temperature distribution and levels of residual stresses of a welding problem of an experimental work reported in the literature. The butt joints are of two materials namely 1018 steel and SS 304 steel. Temperature distributions and residual stresses are computed for these joints separately.

The accuracy of a finite element solution is assessed by performing a convergence study. It is well-established fact that the solution tends to converge to the exact solution as the number of elements are increased so that element size becomes smaller and smaller. Three meshes (table 6) of increasing mesh density are tested for convergence of peak temperature at a particular location of the 1018 steel butt joint. Thermal histories at the distances of 30mm, transverse to the weld centreline top surface and middle section of the joint extracted from the solutions of three meshes and plotted in figure 7. Finally, it is found that all the thermal histories are concurrent. Hence, it can be concluded that convergence has been achieved.

Table 6. Meshes for convergence study

Mesh no	No of nodes	No of elements
Mesh 1	7852	5400
Mesh 2	13741	10800
Mesh 3	19630	16200

IOP Conf. Series: Materials Science and Engineering 1168 (2021) 012020 doi:10.1088/1757-899X/1168/1/012020

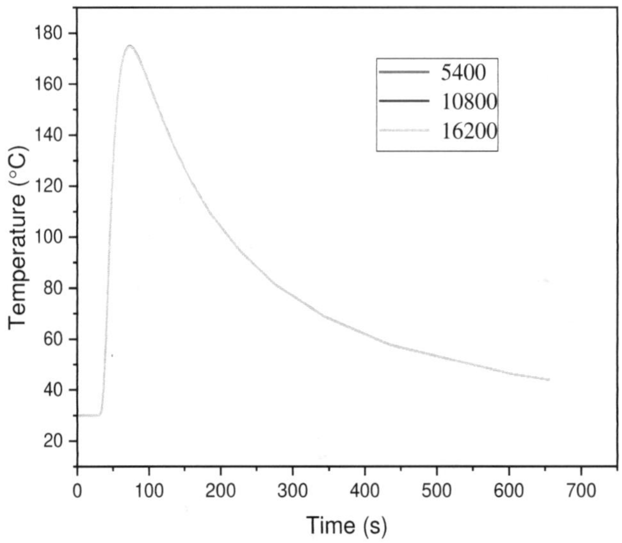

Figure 7. Convergence study: Thermal histories of three meshes at 30 mm distance from the weld centreline

Temperature distributions of 1018 steel butt joint after 25 s, 50 s, 75 s, 656 s from the start of welding are displayed in figure 8, figure 9, figure 10 and figure 11. The position of the torch and formation of weld pool (red color region) can be clearly seen from these figures. Thermal histories of 1018 steel at distances 30mm, 40mm, and 50mm, transverse to the weld centreline top and middle section of the joint are plotted in figure 12. From this figure, it is evident that the peak value of the temperature decreases with an increase in the distance from the weld centreline.

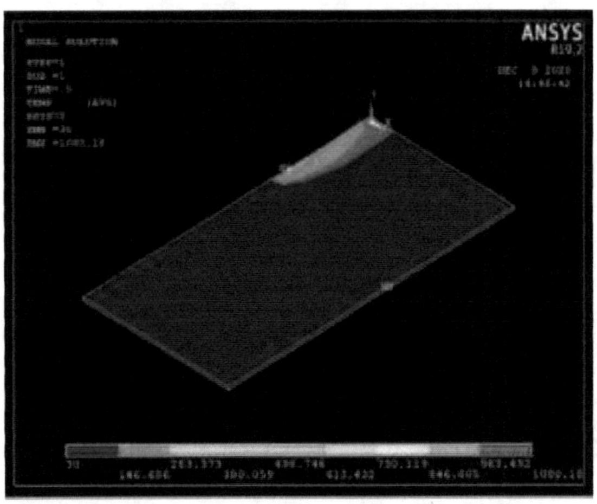

Figure 8. Temperature distribution after 25 s from the start of welding of 1018 steel

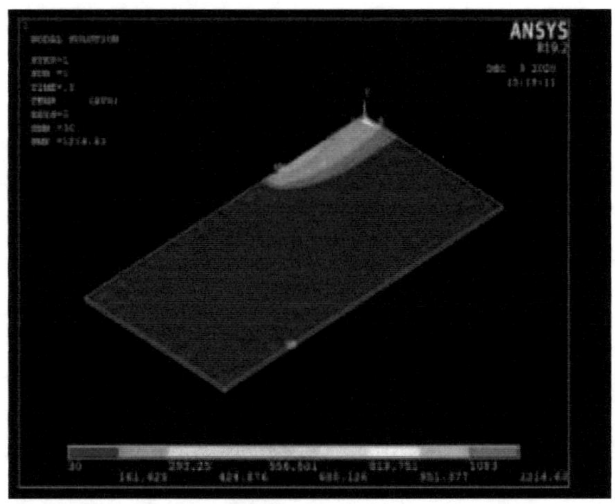

Figure 9. Temperature distribution after 50 s from the start of welding of 1018 steel

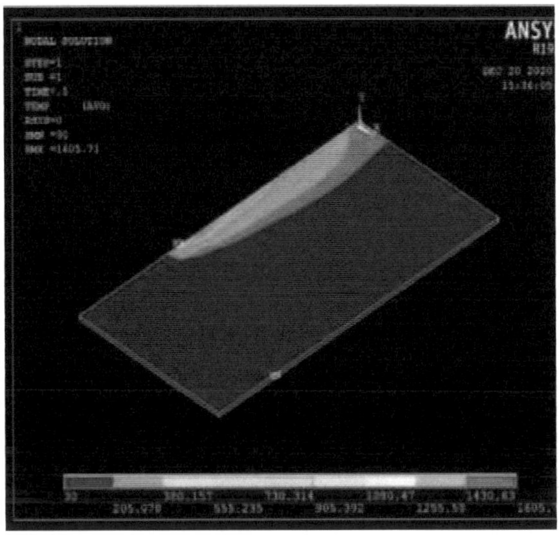

Figure 10. Temperature distribution after 75 s from the start of welding of 1018 steel

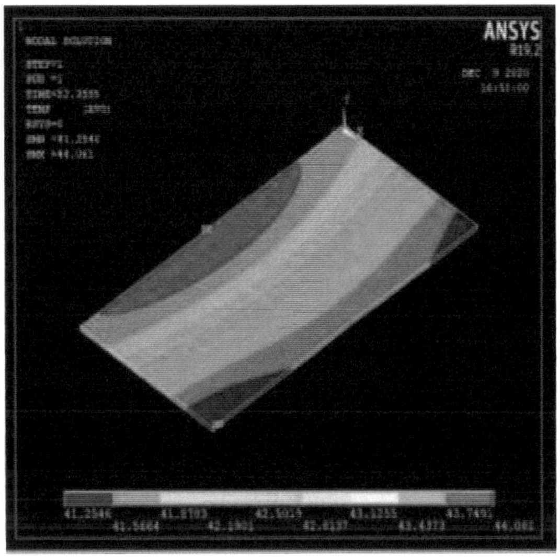

Figure 11. Temperature distribution after 656 s from the start of welding of 1018 steel

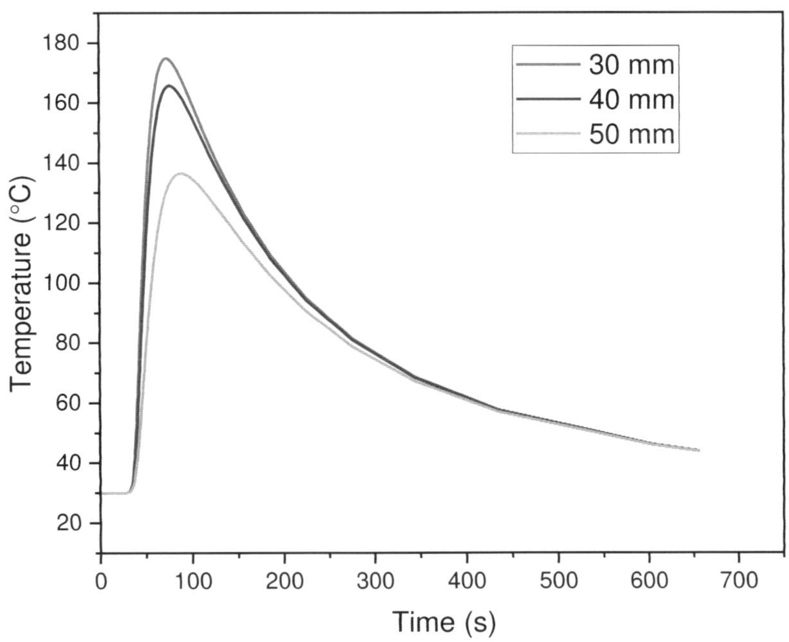

Figure 12. Thermal histories at various locations from the weld centerline top and middle of the joint (mild steel)

3.1. Comparison of experimental thermal histories with numerical thermal histories

The comparison of present numerically obtained thermal histories with those experimentally obtained in the literature of 1018 steel butt joint at 30 mm, 40, mm and 50 mm from weld centreline at the top and middle of the joint are shown in figure 13, figure 14, and figure 15 respectively. From these figures, it can be concluded that the agreements between corresponding thermal histories obtained numerically and experimentally are good. A similar observation can be made from figure 16 which represents the comparison of numerically obtained thermal history with that obtained experimentally in the case of SS 304 steel but joint also.

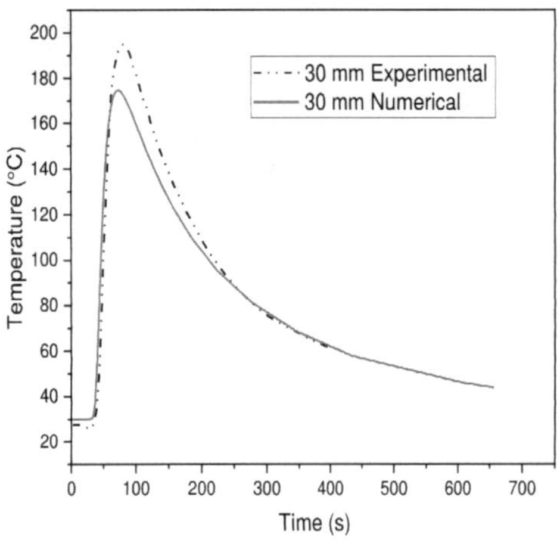

Figure 13. Comparison of numerical and test warm warm thermal histories of 1018 steel butt joint at a 30 mm from the weld centerline

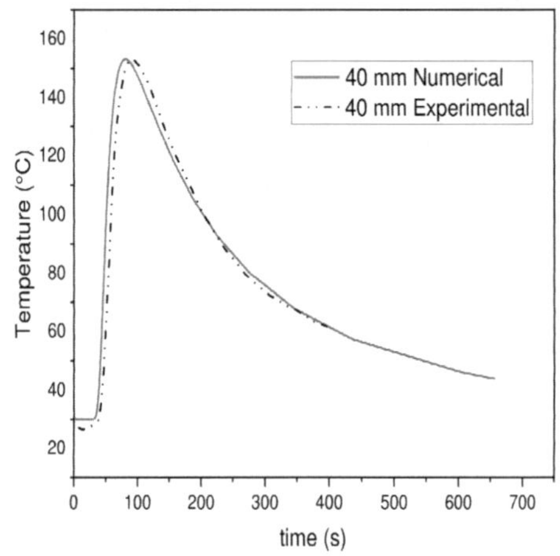

Figure 14. Comparison of numerical and test warm histories of 1018 steel butt joint at a 40 mm from the weld centerline

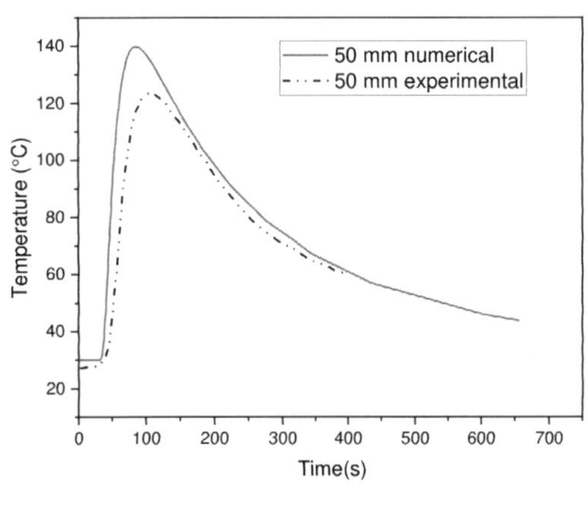

Figure 15. Comparison of numerical and test warm thermal histories of 1018 steel butt joint at a point 50 mm from the weld centerline top and middle of the joint

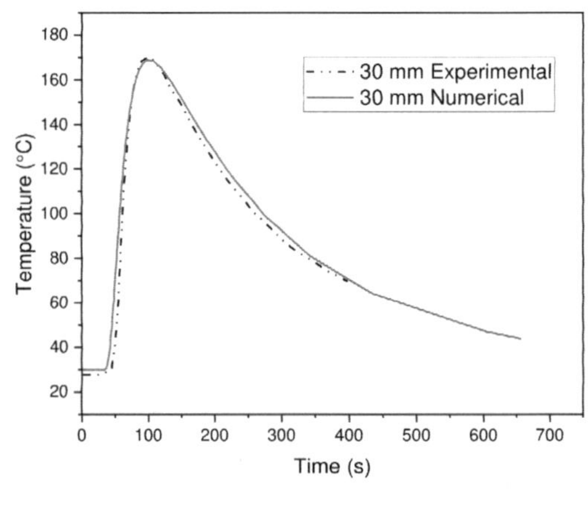

Figure 16. Comparison of numerical and test warm thermal histories of 304 stainless steel butt joint at a point 30 mm from the weld centerline top and middle of the joint

The peak temperature obtained in thermal histories of AISI 1018 steel is greater than the thermal histories obtained in SS 304 steel. The peak temperatures obtained at 30 mm, 40 mm, and 50 mm locations away from weld centerline are 174°C, 165°C and 136°C. Numerically obtained thermal histories are in good agreement with experimentally obtained thermal histories.

3.2. Comparisons of residual stress distributions

Numerically obtained transverse, normal, longitudinal and von-Mises residual stress distributions of 1018 steel are shown in figure 17, figure 18, figure 19, and figure 20 respectively. The transverse, normal, longitudinal and Von-Mises residual stress distributions on a line transverse to weld centreline middle and top of the joint are shown in figure 21. From this figure, it can be observed that normal stresses are nearly zero as they represent stresses normal to open surface. Longitudinal stresses are dominant from the rest of the stresses.

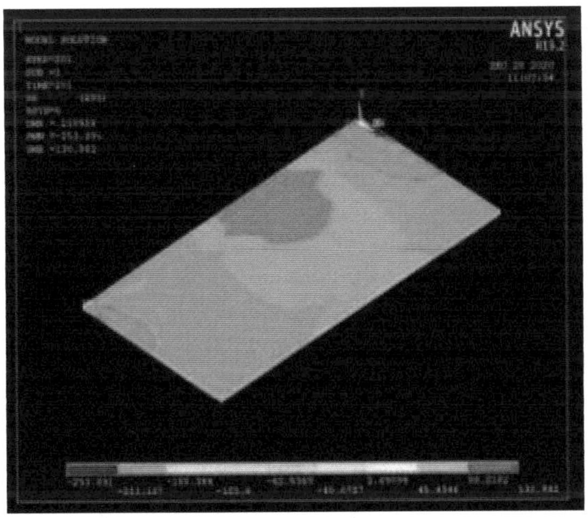

Figure 17. Transverse residual stress distribution of 1018 steel

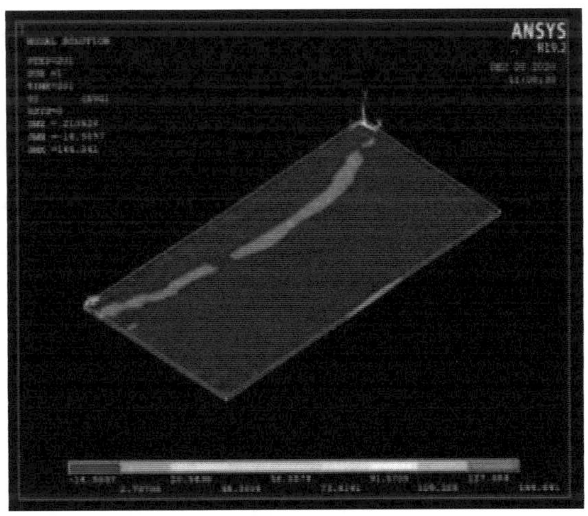

Figure 18. Normal residual stress distribution of 1018 steel

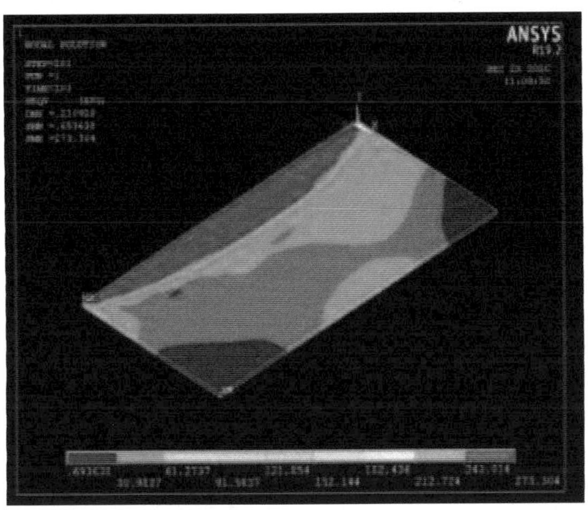

Figure 19. longitudinal residual stress distribution of 1018 steel

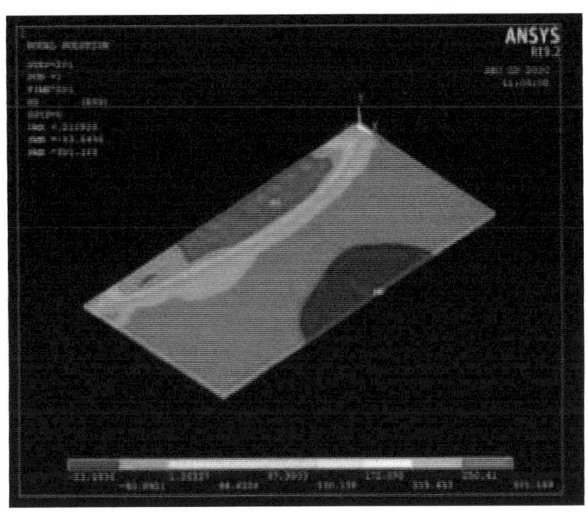

Figure 20. Von-mises residual stress distribution of 1018 steel

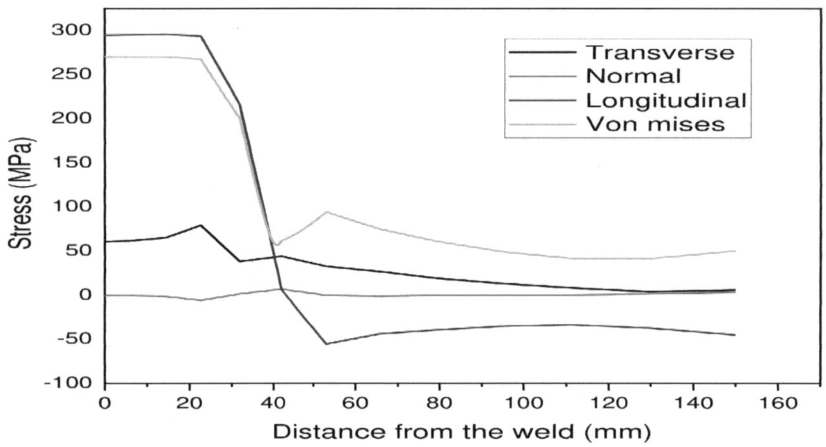

Figure 21. Component residual stress distributions on a line transverse to weld centreline top and middle of the joint of 1018 steel

The comparison of longitudinal residual stresses obtained numerically with those obtained experimentally using neutron diffraction which are existing in the literature on a line transverse to weld centreline middle and top of the 1018 steel butt joint in figure 22. From the figure, the comparison between numerical and experimental results has been good. However, the numerical simulation could not able to capture the stresses near the weld centreline. This may be attributed to the omission of weld metal deposition in the FE simulation. A similar observation (figure 23) can be made from the comparison of numerically obtained and experimentally obtained longitudinal residual stress distributions on the same line mentioned above given in the case of SS 304 steel butt joint. The peak value of numerically obtained longitudinal residual stress of 1018 steel under present welding conditions is 289 MPa whereas the same for SS 304 steel is 237 MPa. This can be attributed to the lower value of virgin yield stress of SS 304 steel when compared with that of 1018 steel.

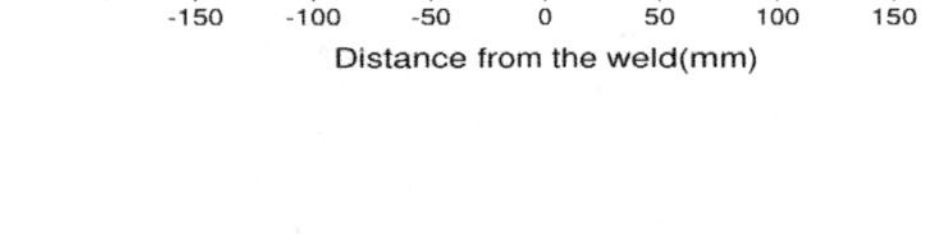

Figure 22. Comparison numerical and experimental longitudinal residual stress distributions on a line transverse to weld centreline top and middle of 1018 steel

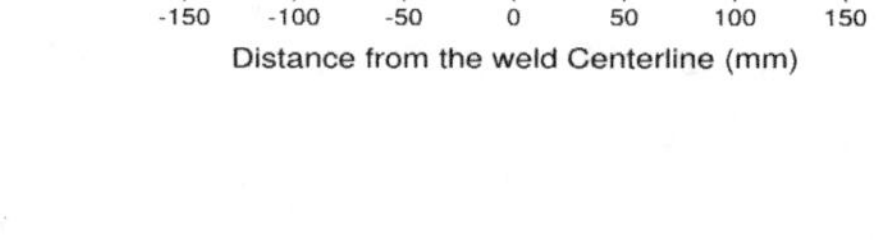

Figure 23. Comparison numerical and experimental longitudinal residual stress distributions on a line transverse to weld centreline top and middle of SS 304 steel

3.3. Effect of preheat temperatures on longitudinal stress distribution

Preheating the plates to be joined before actual welding is a method to reduce residual stresses. Analyses are carried out with three preheat temperatures 600 °C, 700 °C and 800 °C of SS 304 steel joint under identical welding parameters of the case with no preheating. The comparisons of longitudinal residual stress distributions with these preheat temperatures and that obtained with no preheating are illustrated in figure 24. It may be observed from this figure that the peak value of residual stress decreased considerably with preheating and there is a marginal decrease in peak value of residual stress with an increase in preheat temperature also as reported in table 7. This is because of the fact that preheating the joint before welding reduces temperature gradients consequently reduces residual stresses.

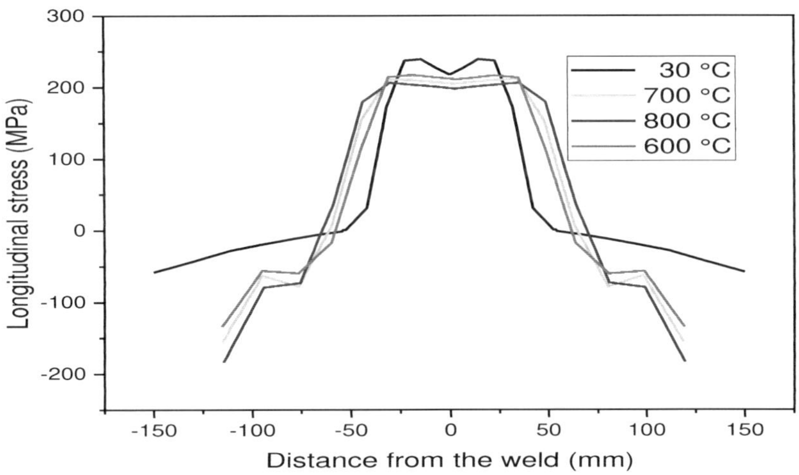

Figure 24. Effect of preheat temperature on longitudinal residual stresses

Table 7. Peak values of longitudinal residual stress with preheating temperatures of SS 304 steel

Preheat temperature (°C)	Peak value of longitudinal residual stress (MPa)
30 (without preheat)	237
600	216
700	213
800	205

4. CONCLUSIONS

Sequential thermo-mechanical analyses are carried out to compute temperature distribution and residual stress distribution of a welding problem whose experimental results are existing in the literature using the finite element method. Following observations are made from this work:

- Comparison of experimental thermal histories with numerical thermal histories yielded a good agreement.

- Longitudinal residual stresses are dominant in comparison with other component stresses.
- The peak value of numerically obtained longitudinal residual stress of 1018 steel (289 MPa) is more than that of SS 304 steel (237 MPa) under identical welding conditions. This may be due to the fact that virgin yield stress of 1018 steel (290 MPa) is more than that of SS 304 steel (230 MPa)
- Comparison of experimentally obtained longitudinal residual stresses are in good agreement with corresponding numerical results. However, FE simulation could not able to capture stresses near the weld centreline. This may be due to the omission of weld bead within the FE simulation.
- The peak value of longitudinal residual stress is decreased with preheating.
- There is a marginal decrease in longitudinal residual stress with an increase in preheat temperature.

5. REFERENCES

[1] Pramod Kumar, Rajesh Kumar , Abdul Arif and M Veerababu, " Investigation of numerical modelling of TIG welding of austenitic stainless steel (304L)," *Material Today:Proceedings,* vol.18, pp.2-5, 2019.

[2] Varma Prasad V M, Joy Varghese V M, Suresh M R and Siva Kumar D, "3D simulation of residual stress developed during TIG welding of stainless-steel pipes," *Procedia Technology*, vol.24, pp.364-371, 2016

[3] Nurul Syahida Mohd Nasir, Mohammad Khairul Azhar Abdul Razab, Sarizam Mamat and Muhammad Iqbal Ahmad, " Review on welding residual stress," *ARPN Journal of Engineering and Sciences,* vol.11, pp.6166-6175, 2016.

[4] Vijay Gohel , Jatin Makwana and Riteshkumar Ranjan, " Thermo-Mechanical analysis in TIG welding of SS 304," *International Journal of Engineering Development and Research,* vol.4, pp.1704-1710, 2016.

[5]Mohd Anees Siddiqui and K.M.Moeed, " Effect of process parameters on temperature distribution and weld pool geometry during TIG welding of 304 stainless steel," *International Journal for Scientific Research & Development*, vol.4, pp.162-168, 2016.

[6] Kartal M E, Kong Y H, Korsunsky A M, Cooks A C F and Bouchard J P, "The influence of welding procedure and plate geometry on residual stresses in thick components", *International Journal of Solids and Structures*", vol. 80, pp:420-429, 2016

[7] Bajpei T, Chelladurai H and Ansari M Z, "Mitigation of residual stresses and distortions in thin aluminium alloy GMAW plates using different heat sink models", *Journal of Manufacturing Processes*, vol. 22, pp:199-210, 2016

[8] Almaeida D F, Martins R F and Cardoso J B, "Numerical simulation of residual stresses induced by TIG butt-welding of thin plates made of AISI 316L stainless steel", *Procedia Structural Integrity*, vol. 5, pp:633-639,2017

[9] Hashemzadeh M, Garbtov Y and Soares C G, "Analytically based equations for distortion and residual stress estimations of thin butt-welded plates", *Engineering Structures*, vol. 137, 115-124, 2017

[10] Venkatkumar et al., "Finite Element Analysis of Heat Input Effect on Temperature, Residual Stresses and Distortion in Butt Welded Plates", *Materialstoday: Proceedings*, vol. 5(2), pp:8328-8337, 2018

[11] Tianci L, Zhang L, Chang C and Wei L,"A Uniform-Gaussian distributed heat source model for analysis of residual stress field of S355 steel T welding", *Advances in Engineering Software*, vol. 126, pp:1-8, 2018

[12] Luca A D, Greco A, Mazza P and Caputo F, "Numerical investigation on the residual stresses in welded T-joints made of dissimilar materials", *Procedia Structural Integrity*, vol. 204, pp:800-809,2019

[13] Balaram Y, Babu B S, Vardhan T V, Ramana G V and Chakradhar G B P, "Residual stress analysis of dissimilar tungsten inert gas weldments of AISI 304 and Monel 400 by numerical simulation and experimentation", *Materials today:Proceedings,* vol. 19(20, pp;478-483, 2019

[14] Zhang K, Dong Wand Lu S, "Finite element and experiment analysis of welding residual stress in S355J2 steel considering the bainite transformation", *Journal of Manufacturing Processes*, vol. 62, pp:80-89, 2020

[15] Tsai C L, Han M S and Jung G H, " Investigating the Bifurcation Phenomenon in Welding, " *Welding Journal*,pp. 151s-160s, 2006.

[16] H.Eisazadeh, E.A.Payzant, P.A.Cornwell, J.R.Bunn and D.K.Aidun, "Exploring the Cooling Process for Residual Stress Reduction in Dissimilar Welds," *welding journal,*vol.97, pp.315-325, 2018.

[17] Jagannadha Rao D.V.N.J, Residual Stresses in Welding: Some Experimental and Numerical Studies ,PhD thesis, IIT Kharagpur,2015.

Application of thermal spraying process in advancement of welding Technology

Syed Qaisar Husain[1] , Divyansu Sharma[2], Vikash Singh[3], Gagan Varshney[1], Girendra Bhati[1], Sudesh Singh[4] and Avinash Ravi Raja*[1] ,
[1]Greater Noida Institute of Technology, Greater Noida, India
[2]National Institute of Technology Jalandhar, India
[3]Accurate Institute of Management and Technology, Greater Noida, India
[4]Sharda University, Greater Noida, India
E-mail: meesam.amu123@gmail.com; divyanshu.nitj@gmail.com; singh.vikash628@gmail.com; gagan.me@gnit.net.; girendrabhatime@gmail.com; sudesh.sengar@gmail.com, avinash129@gmail.com*

Abstract: In the contemporary world, with the incubation of advanced technologies and tremendous outbursts of research works, simplification of large cumbersome tasks with assets to saving time and money is perhaps one of the most crucial aspects in the current scenario. This paper focuses primarily on advancing welding technology using thermal spraying as the major tool for improving the cost, labor and time associated with some of the common welding processes. The analysis has been carried out on mild steel plates having the thickness of 3 mm & 4 mm thick. This has indeed provided multi-faceted benefits such as weldability of similar and dissimilar metals keeping in mind the quality of welded joint, cost of operation and minimizing time for achieving optimum results etc. The study has also compared different welding processes to that of thermal spraying for achieving the same results based on well-defined parameters, as discussed in further sections. Ultimately, now with thermal spraying a plethora of processes can be accompanied such as coating, cladding etc, in addition to welding technology as well.

Keywords: Thermal Spraying; Oxy-acetylene Welding; BronzoChrome-10185; SEM; Microstructure

1. Introduction:

The primary factors influencing the credentials of any welding technology depends entirely on cost, time and labor associated with the particular processes. Although in some cases this may seem to be a herculean task to be accomplished, keeping the feasibility of the process in mind. However, with the incorporation of thermal spraying technology, welding of a variable material type has been made easier compared with the past processes such as Gas Metal Arc Welding, Shield Metal Arc Welding etc.[1-3]. Thermal spraying is a group of processes in which finely divided metallic and nonmetallic materials are deposited in a molten or semi molten state on a prepared substrate [4,5]. Among the various types of thermal spraying processes, major emphasis has been laid on the usability of flame spray process in our experimental research and analysis. The temperature of the flame spray varies around till 3200^{0}C with a low particle velocity of around 110 m/s. The coating thickness varies from 5 mm- 10mm, keeping the spraying distance constant in the process.

For successful completion of the research project, a plethora of papers were comprehended in a detailed manner. Herman et al. tried to analyze the strength and variations of coating thicknesses of various thermal spraying cladded plates and sheets [6]. Similarly, Planche et al. tried to innovate the thermal spraying incorporating the plasma, flame and HVOF process [7].

2. Methodology and Experimentation

The major advantage of using Thermal Spraying for welding processes is that the application of heat input incurred on base metal can be optimized using thermal spraying. In more simple words, being low heat input process compared to other processes, resulting in low HAZ (Heat Affected Zone) ,this will inhibit grain coarsening leading to a higher strength of welded material [8,9].

The foremost step incubated in the process was majorly related to investigating the weld ability of the plates in this regard. However, as coating was not our aim here, hence we tried to optimize the spraying distance with major modifications for incurring a well defined weld bead. The powder used for thermal spraying was mainly 'bronzo chrom 10185', a eutalloy of N-B-Si, suitable for metal-to-metal friction and corrosion resistance [10]. The above shows the concentration of Ni-B-Si-Al-Fe present on reaction with low Carbon steel on thermal spraying.

First, the bead profile using mild steel plates (4mm thick) was majorly obtained for 3 different samples with different spraying distances for optimizing our research work.

Table 1. Representing the sample type along with spraying distances

SAMPLE TYPE	SPRAYING DISTANCE
Sample 1	3-4 mm
Sample 2	1-2 mm
Sample 3	50 mm

By optimizing the spraying distance, the quality of weld bead was improved. Moreover, it was observed that at large spraying distances from weld bead, such as 50 mm or so, the thickness as well as width of weld bead was significantly increased. Figure 1 represent the microstructure of 3 different samples obtained on 100 X magnification.

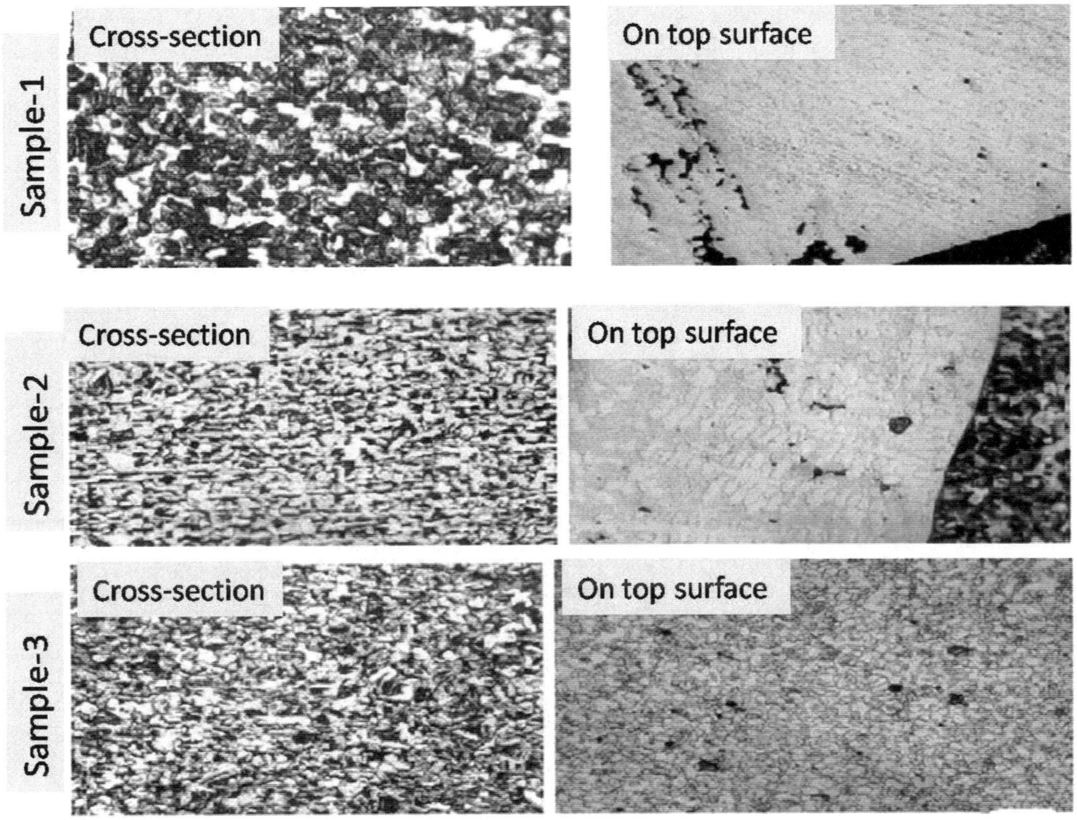

Figure 1. Microstructure of Sample-1, Sample-2, and Sample-3

For microstructure, etching was mainly done using NITAL (2% HNO_3 and 98% C_2H_5OH). Nital develops ferrite grain boundaries in low carbon steels(here mild steel) and thus produces contrast between pearlite and cementite or ferrite network ultimately helping us to further clearly observe grain boundaries under transformation.

Since the weld beads obtained with different samples helped us to examine the feasibility of the process. Hence, we accelerated our approach with obtaining welded joints using Mild steel and stainless steel plates. The mild steel plate used was low Carbon Steel 3mm thick plate and with stainless steel majorly 316-L was used for research analysis. Our major objectives included welding of two mild steel plates followed by two stainless steel plates and finally we had dissimilar plates for investigating the feasibility of the process.

Etching of mild steel in all the cases incurred was carried about using Nital as the major etchant and with stainless steel, predominantly two etchants that is, Carpenter and Picral was used for obtaining the microstructure.

Designing of hopper for thermal spraying

A metallic hopper suitable for regulating the powder flow, improving the durability and powder-output etc, were some prominent reasons for designing of hopper as shown in figure 2. Since low heat input is incurred in Thermal Spraying, there is no possibility of melting of metallic hopper. The designing of metallic hopper was majorly done using the following:

1. Kitchen Storage steel jar (cylindrical with 3.42 cm radius)
2. Copper ring attached to the periphery and then welded to find support inside the torch.
3. Storage jar cap for optimizing, storing and mixing of different powders responsible forwelding (if available)

Figure 2. Designed Hopper attached with thermal spraying.

3. Results and Discussions

Weldability of two mild steel plates was determined using thermal spraying as our broad objective (refer figure 3). 4 mm mild steel plate has been welded through thermal spraying by keeping the distance between the welded plates was kept to be 4 mm and spraying distance of around 10 mm for best optimization.

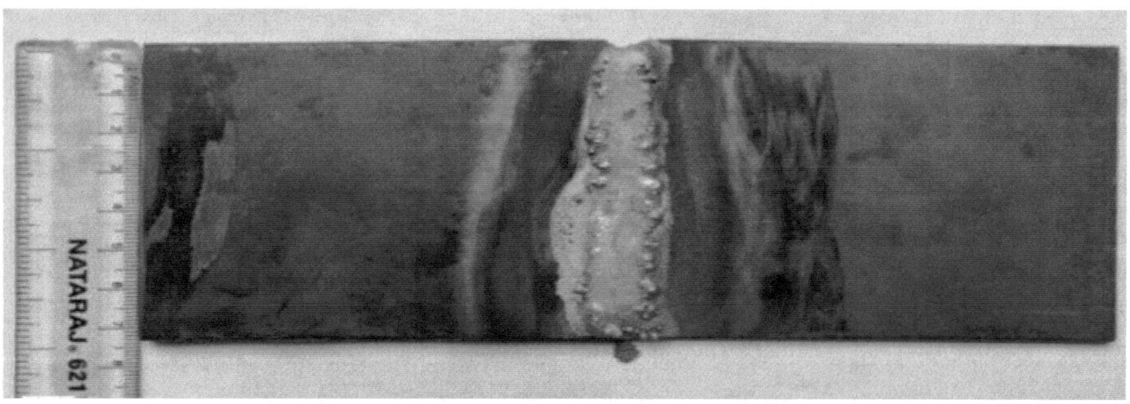

Figure 3. Welding of 2 mild steel plates with a dominated weld zone

For obtaining microstructure, mainly Nital was preferred, and 3 major surfaces were analyzed using Optical Microscope. This included the Weld joint-reinforcement, root enforcement and interface, HAZ (Heat affected Zone) and finally the base metal. The following figures depict the microstructure obtained for varied surfaces of the same specimen.

Microstructure Analysis

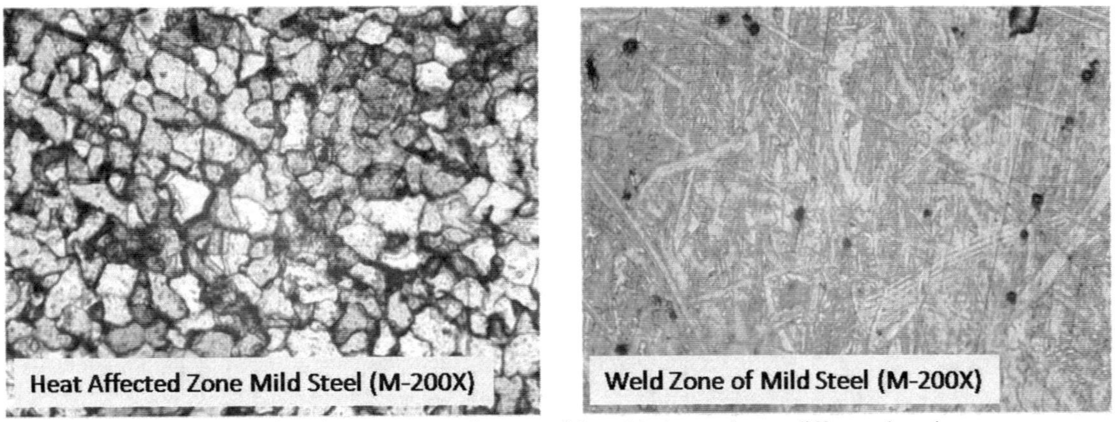

Figure 4. Microstructure of successful welded samples at different locations.

As evident from the microstructure, a lot of porosity has been found in the welded portions especially the interface and welded zone of mild steel plates. As evident from the figure 4, this may be due to several reasons such as high gas flow to cause O_2 or H_2 molecules to easily interact with the weld pool. Powder moisture or loose fittings in the gas hose from gas cylinders to the power house welding torch, gun etc. Figure 5 represent the Interface Zone of the welded sample.

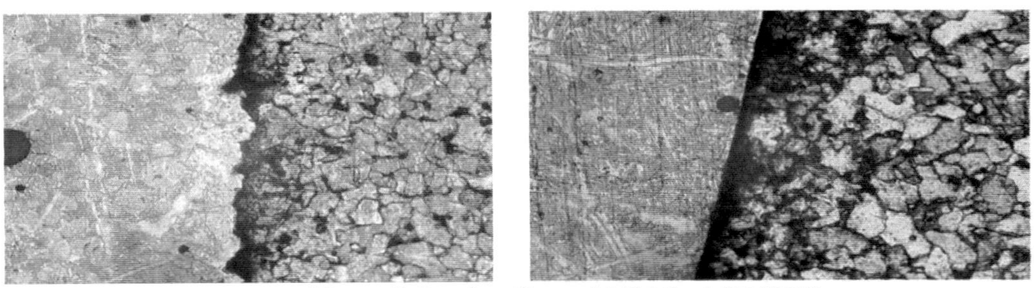

Figure 5. Interface Zone of Mild Steel (M-200X)

4. Conclusion

Welding using thermal spraying is entirely a new concept embedded with the newness and feasibility of the process compared to many other processes at stake. In simple words, as clear from Results and Discussions section, a significant amount of research strategies can be impregnated using thermal spraying , adding to the advantage of welding dissimilar plates is also very important in this regard. However, we must realize that there are many unturned objectives in the coming future which mainly includes : use of trolley or torch-fixer cart for keeping the spraying distance fixed, determining and advancing the compatibility of a powder with similar characteristics as to Bronzo Chrom (10185) , optimizing the flame/ gas velocity for getting well defined weld bead etc . These are some of the future endeavors that can be realized by scrutinizing more efforts with detailed research analysis in this context. Undoubtedly, a lot has been achieved especially combating to the very innovative idea of Welding – using Thermal Spraying.

5. References

[1] Gupta SK, Raja AR, Vashista M, Yusufzai MZ. Effect of Gas Metal Arc Welding on Magnetic Response of Ferritic Stainless Steel. Arabian Journal for Science and Engineering. 2020 Feb;45(2):1293-303.

[2] Gupta SK, Raja AR, Vashista M, Yusufzai MZ. Effect of heat input on microstructure and mechanical properties in gas metal arc welding of ferritic stainless steel. Materials Research Express. 2018 Dec 12;6(3):036516.

[3] Selvi S, Sankaran SP, Srivatsavan R. Comparative study of hardfacing of valve seat ring using MMAW process. Journal of materials Processing technology. 2008 Oct 16;207(1-3):356-62.

[4] Davis JR, editor. Handbook of thermal spray technology. ASM international; 2004.

[5] Fauchais P, Vardelle A, Dussoubs B. Quo vadis thermal spraying?. Journal of thermal spray technology. 2001 Mar;10(1):44-66.

[6] Herman H, Sampath S. Thermal spray coatings. InMetallurgical and ceramic protective coatings 1996 (pp. 261-289). Springer, Dordrecht.

[7] Planche MP, Liao H, Normand B, Coddet C. Relationships between NiCrBSi particle characteristics and corresponding coating properties using different thermal spraying processes. Surface and Coatings Technology. 2005 Dec 21;200(7):2465-73.

[8] Fang CF, Meng XH, Hu QX, Wang FJ, He RE, Wang HS, Yu GU, Ming MA. TANDEM and GMAW twin wire welding of Q690 steel used in hydraulic support. Journal of iron and steel research, International. 2012 May 1;19(5):79-85.

[9] Raja AR, Yusufzai MK, Vashista M. Micro-magnetic analysis of friction stir welded steel plates. The International Journal of Advanced Manufacturing Technology. 2018 Jul;97(5):2051-9.

[10] Redwood M. Gloves and Glove-making. Bloomsbury Publishing; 2016 Feb 25.

IOP Conf. Series: Materials Science and Engineering 1168 (2021) 012022 doi:10.1088/1757-899X/1168/1/012022

Optimization of FDM 3D printing process parameters using Taguchi technique

Kuldeep Sharma, Kapil Kumar, Rishi Kumar Singh and M S Rawat

Greater Noida Institute of Technology, Greater Noida, UP, India

Abstract. Fused deposition modelling (FDM) is a fast growing and low-cost 3D printing technology in order to comply most prominent demands of today's industries in terms of capability to fabricate complex parts along with high flexibility in design. The dimensional accuracy, is an urgent need of final parts printed by FDM process, that is primarily influenced by the process parameters. Optimizing the process parameters which significantly influence the dimensional accuracy is the primary goal of this study in order to achieve the ultimate final part quality. This experimental study investigates the effect of different process parameters viz. layer height, raster angle, nozzle temperature and surrounding pressure on thickness of the final part for Poly Lactic Acid (PLA) filament. Experiments, based on Taguchi's L9 orthogonal array, were performed and subsequently experimental data have been analysed by ANOVA. It has been observed that the layer height is the most significant factor in order to achieve the dimensional accuracy.
Key words- FDM, Taguchi Method, ANOVA

1. Introduction

3D (3 Dimensional) printing or Additive manufacturing (AM) has gained great popularity over the past few years due to its ability to produce complex objects with ease, available sizes, flexibility of usable materials, easy handling and wide range of applications such as engineering industry, medical sciences, food industry, construction, aeronautics, textile industry, automotive industry and so on [1]. There are various methods of Additive Manufacturing such as stereolithography, syringe extrusion, selective layer sintering, fused deposition modelling(FDM)/fused filament fabrication(FFF) being used over the field of its applications as per the requirements of industry but Fused Deposition Modelling (FDM) has become the most widely employed rapid prototyping technique among other methods [2]. FDM uses a temperature controlled head to extrude semi liquid thermoplastic through a nozzle of fixed orifice in layer by layer formation, shown in figure 1[3], the movement of printing head is controlled by a computer aided manufacturing (CAM) software[4].

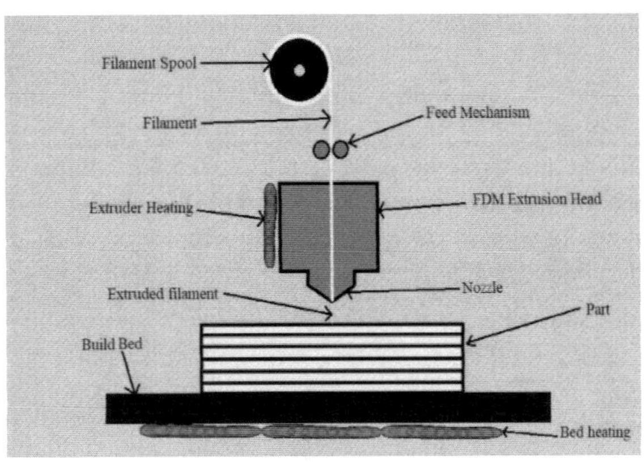

Figure 1 FDM Process Schematic

Researchers are continuously working towards improving different characteristics of FDM produced parts by tweaking with different process parameters and stating a range of optimum settings for a FDM machine and material at which the strength or production time or production cost or any other aspect is optimized. Commonly optimized process parameters are layer height, line width, nozzle temperature, bed temperature, print speed, raster angle etc [5–8]. Most researchers follow conventional technique for experimentation i.e., varying only one parameter while keeping others constant at one time. This conventional approach of experimentation is very time consuming and demands enormous number of resources as number of experiments grows exponentially as more optimizable parameters are to be taken into consideration. Moreover, interaction terms of various process parameters are not entertained in this study. Taguchi method is a powerful tool to screen significant process parameters from many while performing a significantly lesser number of experiments[5]. Taguchi's L9 OA is adopted here to study the effect of each processing parameter on dimensional accuracy of parts produced by FDM, and then to determine optimum process parameters.

2. Literature Review

Parametric optimization of any manufacturing process is a prime necessity of the researchers for producing quality final product in terms of dimensional accuracy and mechanical properties. This section provides an extensive review of many researchers regarding optimization of process parameters of FDM process. Srivastava *et al.*[6] proposed a hybrid approach to optimize model material volume and build time in fused deposition modelling. RSM based experimentation was conducted on FDM Maxum modeler for ABS workpiece and a conical primitive of STL size X=20mm, Y=20mm and Z=69.99mm was taken. Established optimal condition by grey analysis approach are as follows: an air gap of 0.0254mm, contour width of 0.654mm, orientation of 0.0° and raster angle of 0°. Prashantha and Roger [2] used fused deposition modelling based three dimensional (3D) printing system for the fabrication of conductive polymer nanocomposites. Printing was done at 210°C with a nozzle of diameter 0.4 mm. Bed temperature at the time of printing was 60°C and distance between bed surface and nozzle was fixed at 0.2mm. Specimens printed with two surface layer shells, layer height is fixed at 0.2mm and infill is set to100% linear infill pattern with a stacking alternating between 45° and -45°. Panda *et al.*[7] used a recently developed technique such as bacteria foraging was adopted as its superior performance over other random search techniques as there are only few parameters needed to be adjusted. The control factors considered in this study are thickness, build orientation, raster angle, raster width and air gap. Functional relationship between strength and process parameters is developed in this work using surface methodology for prediction process. Hmeidat *et al.*[8] used an epoxy resin, Epon 826 in their study, a Bisphenol A diglycidyl ether (DGEBPA) resin with 178-186 weight per epoxide with a density of 1.162 g/cc. These 3D printed nanocomposites had a strength in range of 80 to 143 MPa, which is significantly higher than any previously reported values for 3D printed thermoset composites, including short fiber-reinforced materials. These materials have excellent layer-to-layer bonding because of the crosslinking process which occurs after the component has been printed. Papon and Haque[9] produced CNF/PLA nanocomposite filaments using a laboratory mixing extruder. The matrix material used was polylactic acid-biopolymer (PLA) granulates of 3mm size with a glass transition temperature of 55°C and melting temperature of 175°C. The reinforcing material they used was CNF with a maximum length of 400nm and average diameter of 50nm. It was observed that geometry of nozzle affects the bead spreading architecture. Improved ultimate strength and modulus of elasticity was seen in CNF/PLA nanocomposites with 0.5% CNF as compared to neat PLA. Bead orientation direction and orientation of CNF was seen to be affected by extruder orifice and printer nozzle. Cataldi *et al.*[10] used a single screw extruder for producing nanocomposites consisting of polyvinyl alcohol and cellulose nanocrystals. Thermoplastic polyvinyl alcohol was used as Matrix for the composite in this work. Stiffness of filament and 3D printed specimens was increased after the introduction of CNC. The toughness of 3D printed material was increased due to good edition between CNC and PVOH. Lower densification was achieved upon introduction of CNC. Alafaghani *et al.*[7] examined the effect of infill percentage, building direction, infill pattern, printer speed, extrusion temperature and layer height independently on mechanical properties and dimensional accuracy. Extrusion temperature, layer height and building direction affected dimensional accuracy more than infill pattern, infill percentage and printing speed. B. Coppola *et al.*[8] studied the effects of printing temperature on PLA/clay nanocomposites in 3D printing. Semi-crystalline polylactic acid (PLA) was used as matrix in this study. They used FDM technique Tu 3D print dog bone specimens at three different temperatures: 185, 200 and 215°C to mechanically test in tensile mode. A higher elastic modulus was exhibited by nanocomposite 3D printed samples as compared with PLA specimens and it increases with increase in printing temperature. They concluded that printing temperature strongly affects properties of final object. Ceretti *et al.*[9] produced multi-layered PCL scaffolds and analysed the influence of process parameters and extrusion technology on the deposited material. The three configurations used were: 0.6mm x 0.6mm, 0.8mm x 0.8mm and 1mm x 1mm and path heights were 0.3mm, 0.4mm and 0.5mm for each configuration respectively, while the area of the whole grid was 10mm x 10mm. Human foreskin fibroblasts cells were seeded at a cellular density of 40000 cells/cm to verify the effectiveness of produced scaffolds. They concluded that extrusion head don't have significant effects on resulting geometry of the samples and for the extrusion of poly-caprolactone, grain extrusion head is preferable. Bartolomeo Coppola *et al.*[10] used a layered silicate-reinforced polylactic acid (PLA) for manufacturing samples with additive manufacturing. During the melt-compounding process, the temperature profile used was: 160°C-180°C-180°C-180°C-180°C-180°C-180°C-170°C (from hopper to die). The single screw extruder had a die of 3mm, operating temperature profile: 180°C-180°C-160°C (from hopper to die) and had a screw speed of 10rpm. At the time of printing of specimens, bed temperature was fixed at 50°C, layer height was fixed at 0.2mm, faster angle was fixed at +/- 45° and nozzle diameter were fixed at 0.35mm. Printing temperature had a direct effect on specimen transparency, the transparency increases with increase in printing temperature. They concluded that, printing temperature should be chosen considering polymer architecture and nanocomposite morphology. Balchan and Drickamer [11] studied the effects of pressure from 60 to over 400 kbar on resistance of iodine and selenium. A rapid drop in resistance of selenium was seen between 60 and 128 kbar then a discontinuous drop. When pressure if further increased, its behaviour is apparently metallic. Iodine shows rapid drop in resistance between 60 and 255 kbar, further increasing pressure shows slower drop in resistance. Sobczak and Asthana[12] reviewed the effect of pressure on solidification structure formation in metal castings. They analysed effects of pressure on nucleation growth, phase diagrams, interfacial energy and diffusion coefficient. Refinement of structure was seen to be promoted due to high pressure which may be caused by reduced critical radius and reduced rate of new phase nucleation and decreased interfacial free energy. Russell J. Hemley[13] studied the effect of high pressure on molecules such as halogens, nitrogen and group V, oxygen and chalcogens, CO and CO_2, Simple hydrocarbons, H_2O, related hydrogen bonded systems, simple molecular mixtures, clusters and complex molecules. Resit Unal and Dean[14] used Taguchi approach in their study to design and optimize quality and cost. They presented and overview of Taguchi method with steps involved. They found out Taguchi method is a powerful tool which can simultaneously improve quality and cost. They concluded, Taguchi method emphasizes on pushing quality back to design stage, resulting in designing a product or process which is robust to cause quality problems. Mohan Pandey *et. al.* [15] presented and classified different slicing algorithms used in rapid prototyping or layered manufacturing process into two categories, one which consider build edge as rectangular and other which consider build edge as a slope. Ahn *et al.*[16] studied the surface roughness in parts produced by fused deposition modelling. They presented a new approach to formulate surface roughness. Effects of surface angle, layer thickness, cross-sectional shape of filament and overlap interval on surface roughness were showed. They concluded surface roughness expression can predict roughness of FDM parts. Boschetto and Bottini [17] formulated a method to predict dimensional accuracy and surface roughness of FDM parts as a function of process parameters such as layer thickness and deposition angle. They investigated 60 different deposition angles between 0-180°. The used ABS for their research. They performed a case study which evidenced that it is possible to predict dimensional deviations and geometrical tolerances. They concluded that dimensional deviations are greatly influenced by deposition angle. Bellini *et al.*[18] studied new developments in field of fused deposition modelling of ceramics. They developed a new extrusion system which uses granules instead of filament. Core finding of this study is overcoming the limitations presented by material being in filament form by using granular form. Pandey *et al.*[19] presented a semi empirical model for evaluation of surface roughness of a FDM part. One of the major problems in FDM part is surface roughness caused by staircase effect. Hot cutter machining (HCM) is used to address this problem. To find significance index of process variables ANOVA is used. They concluded that HCM can achieve a surface finish of the order 0.3 μm with 87% confidence level.

The process parameters taken into consideration for optimization for a FDM process are generally the parameters which could be controlled by the FDM machine itself. In other words, only the parameters bound to the FDM machine are to be optimized while external factors like surrounding temperature, pressure, humidity remains unchanged. These external factors can greatly affect the quality and characteristics of parts produces by FDM. In this study, surrounding pressure is treated as a process parameter and its value is varied to analyse the effect of surrounding pressure as well as other process parameters on strength of FDM manufactured parts. Effect of pressure on properties and structure of material is extensively studied by Nobel laureate physicist P.W. Bridgman[23]. His work showed importance of pressure for studying changes in structure and properties of matter. It was suggested by J.D. Bernal in 1928 that all matter should become metallic at sufficiently high pressure due to electron dislocation caused by forced overlap of electrons[15,24].

3. Experimental Methodology

3.1 Selection of process parameters

Based on literature, it was analysed that the process parameters such as layer height, raster angle and nozzle temperature greatly affect the dimensional accuracy and mechanical properties of printed parts. In this study, a new process parameter, surrounding pressure is used with all other process parameters for evaluating its effect on dimensional accuracy in terms of thickness, pressure is not taken as a process parameter in any study before as per the literature review, so surrounding pressure is a novel parameter in this study. Three levels of parameters are considered for sample preparation because Taguchi l9 method is being used for process parameter selection in this study. The levels of these parameters were selected with the help of machine manual and performing the preliminary test on a 3D printer. On the basis of preliminary experimentation, the ranges and subsequently the levels of the printing parameters were chosen as shown in Table 1. The parameters which are kept fixed throughout experimentation are shown in Table 2.

Table 1 Setting of levels for control factors

Control Factors	symbols	units	Levels		
			I	II	III
Layer Height	A	mm	0.1	0.15	0.2
Raster Angle	B	degree	0	45	90
Nozzle Temperature	C	°C	200	210	220
Surrounding Pressure	D	psi	0	50	100

Table-2: List of fixed parameters

S.No.	Parameter	Fixed Value
1	Bed Temperature	60°C
2	Nozzle Diameter	0.4mm
3	Infill Percentage	100%
4	Printing Speed	3600 mm/min
5	Solid Layers	3
6	Extrusion Width	0.4mm

3.2 Materials

The specimens used in this study are printed using white 1.75±0.03 mm diameter Polylactic Acid (PLA) filament. The filament was purchased from amazon.in and sold by 3D bazaar. PLA is a thermoplastic polyester with chemical formula $(C_3H_4O_2)_n$. It the most widely used filament material in FDM printers.

3.3 Experimental Setup

Inferno i3 FDM 3D printer (Figure 2) is used in this research for producing specimens. This printer is capable of producing specimen up to a maximum size of 220*220*250mm by using polylactic acid (PLA), Acrylonitrile butadiene styrene (ABS), Polyvinyl Alcohol (PVA), High-density polyethylene (HDPE) and similar materials as this machine is equipped with heated print bed. This machine is running on marlin firmware and can print over USB while connected to a computer or via SD card which have G-code file for the part. To control and vary pressure around FDM printer, an air compressor and an air-tight enclosure is used, shown in Figure 2. The compressor employs a pressure switch which could be set to a desired value of pressure and a pressure difference, when the desired pressure is reached. This pressure switch automatically detects the pressure difference and turns off the compressor if there is any pressure drop and turns the compressor on, allowing the setup to maintain constant pressure. The printer is kept inside the enclosure then the lid of the container is sealed with the help of fasteners.

3.4 Sample Preparation

The samples used in this study are modelled based on ISO 527-2 (Model 5A) which is standard used for determination of tensile properties of reinforced and non-reinforced plastics [25]. Figure 3 indicates the dimensions and shape used to create CAD model. All test specimens are printed using Polylactic Acid (PLA) on Inferno i3 FDM printer. To evaluate the influence of processing parameters over the dimensional accuracy and repeatability of samples, the samples were measured and compared with CAD model. Thickness of printed parts of all 9 samples was measured using a vernier calliper (least count of 0.01 mm) at 3 different points along its axis and average of 3 was taken as a result (Figure 4).

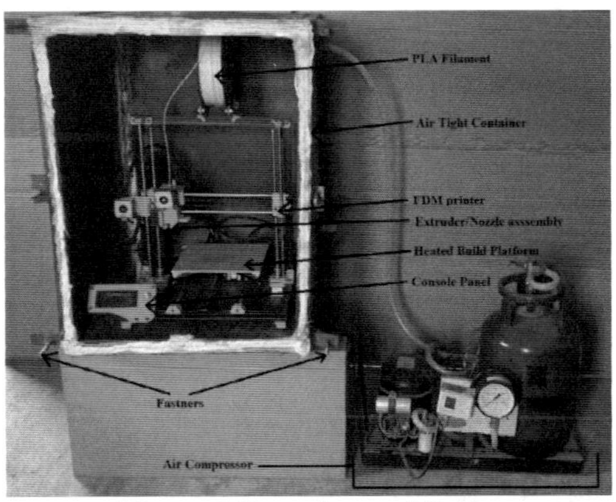

Figure 2 Experimental Setup

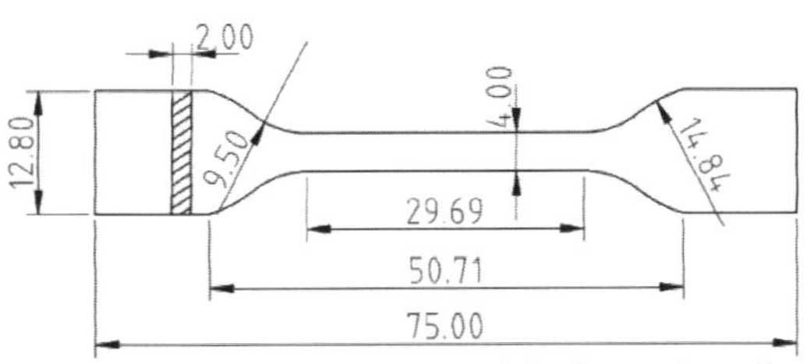

Figure 3 Specimen CAD model, dimensions in mm

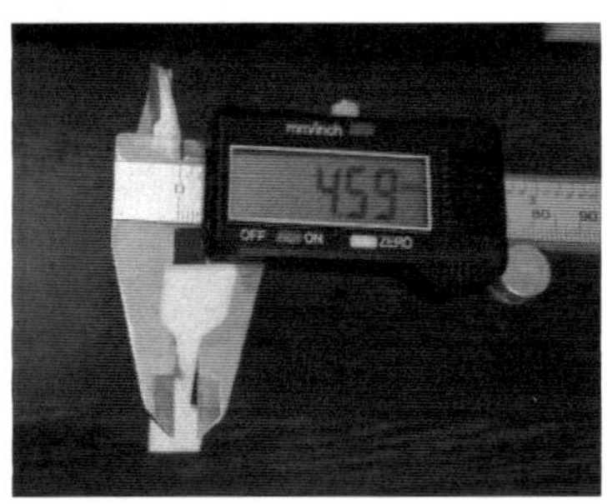

Figure-4: Measurement of dimensions

3.5 Experimental design based on Taguchi Method

In order to evaluate the effect of printing parameters of FDM process in terms of dimensional accuracy of width of parts, a Taguchi method is used here to optimize the process. The Taguchi method has become a powerful tool for the systematic application of design and analysis of experiments for the purpose of designing and improving the product quality[26].

According to Taguchi methodology, the characteristic that a target value represents best performance in terms of dimensional accuracy, is called Nominal is best type of problem. The S/N ratio can be calculated as shown below:

$$S/_N = -10 \log_{10} \frac{((y_1-m)^2+(y_2-m)^2+\cdots)}{n} \quad (1)$$

Where, m = Target value

n = Number of repetitions

Table 3 Response values using L9 orthogonal array

Experiment Number	**A**	**B**	**C**	**D**	**Thickness (mm)**				
					Trial			Average value	S/N ratio (db)
					1	2	3		
1	1	1	1	1	1.99	2	1.98	1.99	37.78
2	1	2	2	2	2.01	2.01	2.04	2.02	32.22
3	1	3	3	3	1.97	2.01	2.01	2.00	34.36
4	2	1	2	3	2.02	2.05	2.02	2.03	29.59
5	2	2	3	1	2.01	2.02	2.07	2.03	27.45
6	2	3	1	2	2.03	2.01	1.98	2.01	33.31
7	3	1	3	2	2.1	2.09	2.14	2.11	19.01
8	3	2	1	3	2.09	2.06	2.15	2.10	19.43
9	3	3	2	1	2.07	2.15	2.13	2.12	18.31

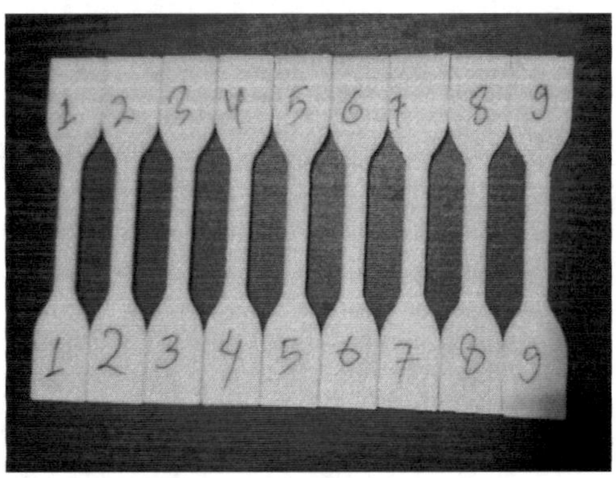

Figure 5 Printed parts by FDM process

L9 orthogonal array chosen for the experimentation in which 9 rows corresponding to the number of experiments, with 4 columns at three levels as shown in Table 3. The experiments were performed for each combination of rows as per selected L9 orthogonal array and the results are indicated in Table 3. Figure 5 shows the all 9 printed parts under printing parameters shown in Table 3. Figures 7, 8, 9 show the optical micrographs of the surface of the FDM printed parts under different printing conditions. Dimensional accuracy plays an important role in the assembly of the final product.

4. Result and Discussion

In the present study, the first step in the analysis is to review the test results for each experiment conducted by the Taguchi's L9 OA. The S/N ratios are calculated for each experiment by applying Nominal the Best characteristics, as shown in Table 3. The plot of average values of S/N ratio for each factor at different levels are shown in Figure 6. In S/N analysis, the greatest value of S/N represents a more desirable condition for a particular control factor. Figure 6 suggests that the factors at levels A1, B1, C1, D2, are the best levels that give the target value of the thickness.

Effects of layer height on dimensional accuracy are clearly visible in this study. It is seen that as the layer height is increased, dimensional accuracy is decreased. When a layer is printed on top of previous layer, a step is formed and this is called stepping effect. At lower layer height, this stepping effect is minimized an it results in a smoother and even surface finish. Lower layer height is also known as higher resolution for 3D printed parts because better surface finish and so have greater dimensional accuracy. In this study, 0.1mm layer height is optimum layer height because it shows minimum deviation and highest accuracy. The same results have been observed by [18] in his study. It is observed that as the nozzle temperature is increased, the dimensional accuracy is decreased within the temperature range taken in this study, which might be a result of increasing fluidity of extruded PLA with increase in extrusion temperature. The extruded material, which is in semi liquid form, tends to flow in any direction it is free to flow, but it gets a very small fraction of time to do so because it begins to cool as soon as it is out of nozzle. When extrusion temperature is increased, the extruded material takes more time to cool down and thus have a little longer to flow, which causes dimensional inaccuracy. In this study, 200°C is optimum nozzle temperature. Higher dimensional accuracy is seen in parts which are printed at 0° raster angle in this study which might be because when printing at 0° raster angle, all the layers are printed parallel to x-axis and subsequent layers are also parallel which results in higher accuracy.

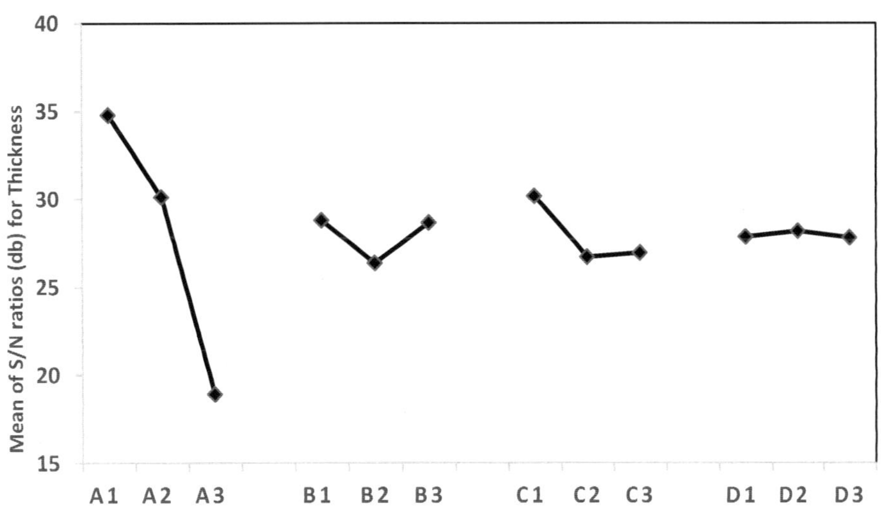

Figure 6 Effect of control factors on thickness

Analysis of Variance (ANOVA) is executed on S/N data and indicated in Table 4. Table 4 describes that the percent contribution of factor D (surrounding pressure) is very small (0.06 %) in comparison to factors A, B and C. Hence factor D can be treated as an error. Table 5 indicate pooled ANOVA at 95%

confidence interval for S/N data. Table 5, the percent contribution of parameters affecting the target value of thickness, i.e., layer height (92.15%), raster angle (0.53%) and nozzle temperature (1.14%), indicates the influence of layer height is more than the other process parameters. In Table 5, the error percentage contribution is less than 15%. Therefore, it is verified that no significant process parameters are omitted during the experiment, therefore, no opportunity for the further improvement.

Table 4 ANOVA for S/N ratio

Factor	DOF	Sum of Square (SS)	Variance (V)	Percent Contribution (%P)
A	2	399.11	199.55	92.15
B	2	11.16	5.58	2.58
C	2	22.57	11.28	5.21
D	2	0.26	0.13	0.06
Error	0	0	indeterminate	
Total	8	251.45		

Table 5 Pooled ANOVA for S/N ratio

Factor	DOF	Sum of Square (SS)	Variance (V)	F-ratio	Pure Sum (S´)	Percent Contribution (%P)
A	2	399.11	199.55	1511.36	398.35	92.09
B	2	11.16	5.58	42.26	2.32	0.53
C	2	22.57	11.28	85.48	4.95	1.14
Error	2	0.26	0.13			**6.23**
Total	8	251.45				100

$F_{0.05}(2,2) = 19.0$

4.1 Estimation of Optimum Quality Characteristics

After analysis, the selected significant process parameters are A1, B1 and C1. The optimum value of width can be evaluated as:

$$\mu = \bar{T} + (\overline{A_1} - \bar{T}) + (\overline{B_1} - \bar{T}) + (\overline{C_1} - \bar{T}) \quad (2)$$

$$\bar{T} = 27.94$$

$$\overline{A_1} = 34.79$$

$$\overline{B_1} = 28.79$$

$$\overline{C_1} = 30.17$$

$$\mu = 37.74 \text{ mm}$$

$$F_{0.05}(2,2) = 19.0 \quad \text{(Tabulated)}$$

$$\text{Confidence Interval} = \text{C.I.} = \pm\sqrt{F_{\propto}(2,2)\frac{V_e}{N_e}}$$

F (2,2) = F-Value from F-Table for minimum DOF (2) and error DOF (2)

V_e= Error Variance = 0.13

N_e= Effective Number of Replications = $\frac{Total\ Number\ of\ Results}{Dof\ of\ Mean + Dof\ of\ all\ Factors}$

$$N_e = \frac{9}{1+6} = 1.28$$

$$\text{Confidence Interval} = C.I. = \pm 1.92$$

$$37.74 - 1.92 < \mu < 37.74 + 1.92$$

For 95% C.I., the predicted optimum thickness in decibel is

$$35.82 < \mu < 39.66$$

For 95% C.I., the predicted optimum thickness in mm is

$$2.01 < thickness < 1.97$$

5. Confirmation Experiments

The confirmation step is the final step to authenticate the results obtained from the analysis phase. The expected value of thickness at optimum condition is equal to 37.74 db (= 2.00 mm) which is already calculated in Section 4. Three confirmation experiments are conducted for significant parameters at their optimal levels obtained in the analysis phase. The average value of thickness at the optimal setting of process parameters of the FDM process was observed to be 36.68 db. The result of the confirmation experiment is shown in Table 6. The value of width lies within the range calculated in the analysis phase. An error of 1.58 % is observed. Therefore, the optimal setting of process parameters, predicted in the analysis phase, can be recommended. Optical micrograph of confirmation experiment is shown in Figure 9.

Table 6 Result of confirmation experiments

	Optimal printing parameters	
	Prediction	Experimental
Levels	A1B1C1D2	A1B1C1D2
Thickness (db)	37.74	37.14

6. Surface morphology

The samples were observed under an optical microscope (Nikon YTV55) to understand and study their microscopic structure, voids and irregularities. The samples were observed in such a manner that different layers could be seen. When samples were observed under microscope, the irregularities in extruded PLA were clearly visible. The parts printed at 0.1mm layer height have less irregularities and as the layer height is increased irregularities gradually increased, shown in Figures 7, 8 and 9.

Figure 7 Optical micrograph of part printed by experiment 1 (A= 0.1 mm, B= 0°, C= 200° C, D= 0 psi)

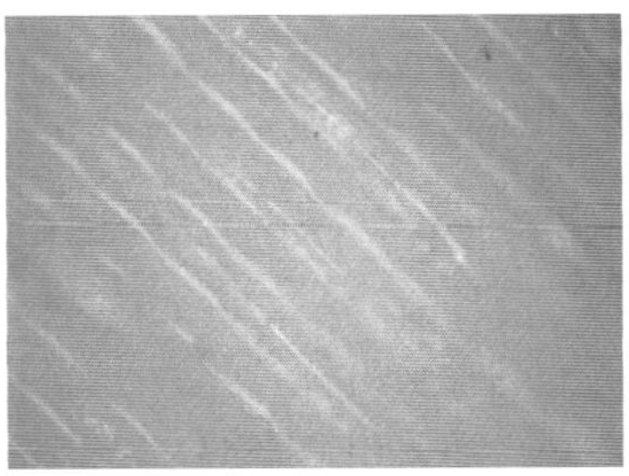

Figure 8 Optical micrograph of part printed by experiment 9 (A= 0.2 mm, B= 90°, C= 220° C, D= 0 psi)

Figure 9 Optical micrograph of part printed by confirmation experiment (A= 0. mm, B= 0°, C= 200° C, D= 0 psi)

The irregularities obtained in the parts at higher layer height might be caused due to large compressive force (due to weight of a layer) experienced by extruded material at thick layer height and it caused the extruded semi liquid material to squeeze. As the layer height is decreased, the value of compressive force is reduced and material got more space to settle which resulted in less squeezing of material and hence less irregularities.

7. Conclusions

1. The lower value of layer height (0.1mm), lower value of raster angle (0°) and lower value of nozzle temperature (200° C) are significant printing parameters to achieve the target value (2 mm) of thickness.

2. Statistically, influence of layer height is more than any other parameters to maintain dimensional accuracy.

3. The value of part's thickness (37.14 db) obtained by confirmation experiment is very close to the predicted value (37.74 db) at optimal setting of process parameters.

4. The predicted optimal range of thickness at 95% confidence interval is

$$2.01\ mm < thickness < 1.97\ mm$$

5. Effects of surrounding pressure on dimensional accuracy are not clearly visible in the range taken in this study (0-100psi). Therefore, it can be said that the parameter surrounding pressure is ineffective within this range on dimensional accuracy of FDM parts. Further increasing pressure may have some effects and there is scope to study this parameter further.

References

[1] Espalin D, Muse DW, MacDonald E, Wicker RB. 3D Printing multifunctionality: Structures with electronics. Int J Adv Manuf Technol. 2014;72(5–8):963–78.

[2] Prashantha K, Roger F. Multifunctional properties of 3D printed poly(lactic acid)/graphene nanocomposites by fused deposition modeling. J Macromol Sci Part A Pure Appl Chem. 2017 Jan 2;54(1):24–9.

[3] Alafaghani A, Qattawi A, Alrawi B, Guzman A. Experimental Optimization of Fused Deposition Modelling Processing Parameters: A Design-for-Manufacturing Approach. Procedia Manuf. 2017;10:791–803.

[4] Zein I, Hutmacher DW, Tan KC, Teoh SH. Fused deposition modeling of novel scaffold architectures for tissue engineering applications. Biomaterials. 2002;23(4):1169–85.

[5] Dean EB, Unal R. Taguchi Approach to Design Optimization for Quality and Cost: An Overview. Annu Conf Int Soc Parametr Anal. 1991;1–10.

[6] Srivastava M, Maheshwari S, Kundra TK, Rathee S. Estimation of the effect of process parameters on build time and model material volume for fdm process optimization by response surface methodology and grey relational analysis. In: Advances in 3D Printing and Additive Manufacturing Technologies. Springer Singapore; 2016. p. 29–38.

[7] Panda SK, Padhee S, Sood AK, Mahapatra SS. Optimization of Fused Deposition Modelling (FDM) Process Parameters Using Bacterial Foraging Technique. Intell Inf Manag. 2009;01(02):89–97.

[8] Hmeidat NS, Kemp JW, Compton BG. High-strength epoxy nanocomposites for 3D printing. Compos Sci Technol. 2018 May 26;160:9–20.

[9] Papon EA, Haque A. Tensile properties, void contents, dispersion and fracture behaviour of 3D printed carbon nanofiber reinforced composites. J Reinf Plast Compos. 2018 Mar 1;37(6):381–95.

[10] Cataldi A, Rigotti D, Nguyen VDH, Pegoretti A. Polyvinyl alcohol reinforced with crystalline nanocellulose for 3D printing application. Mater Today Commun. 2018 Jun 1;15:236–44.

[11] Coppola B, Cappetti N, Di Maio L, Scarfato P, Incarnato L. Influence of 3D printing parameters on the properties of PLA/clay nanocomposites. In: AIP Conference Proceedings. American Institute of Physics Inc.; 2018.

[12] Ceretti E, Ginestra P, Neto PI, Fiorentino A, Da Silva JVL. Multi-layered Scaffolds Production via Fused Deposition Modeling (FDM) Using an Open Source 3D Printer: Process Parameters Optimization for Dimensional Accuracy and Design Reproducibility. In: Procedia CIRP. Elsevier B.V.; 2017. p. 13–8.

[13] Coppola B, Cappetti N, Maio L Di, Scarfato P, Incarnato L. 3D printing of PLA/clay nanocomposites: Influence of printing temperature on printed samples properties. Materials (Basel). 2018 Oct 11;11(10).

[14] Balchan AS, Drickamer HG. Effect of pressure on the resistance of iodine and selenium. J Chem Phys. 1961;34(6):1948–9.

[15] Sobczak JJ, Drenchev L, Asthana R. Effect of pressure on solidification of metallic materials. Int J Cast Met Res. 2012;25(1):1–14.

[16] Hemley RJ. Effects of high pressure on molecules. 2000;c:763–800.

[17] Unal, Resit; Dean EB. Taguchi approach to design optimization... Annu Conf Int Soc Parametr Anal. 1991;1–10.

[18] Mohan Pandey P, Venkata Reddy N, Dhande SG. Slicing procedures in layered manufacturing: A review. Rapid Prototyp J. 2003;9(5):274–88.

[19] Ahn D, Kweon JH, Kwon S, Song J, Lee S. Representation of surface roughness in fused deposition modeling. J Mater Process Technol. 2009;209(15–16):5593–600.

[20] Boschetto A, Bottini L. Accuracy prediction in fused deposition modeling. Int J Adv Manuf Technol. 2014;73(5–8):913–28.

[21] Bellini A, Shor L, Guceri SI. New developments in fused deposition modeling of ceramics. Rapid Prototyp J. 2005;11(4):214–20.

[22] Pandey PM, Reddy NV, Dhande SG. Improvement of surface finish by staircase machining in fused deposition modeling. J Mater Process Technol. 2003;132(1–3):323–31.

[23] Bridgman PW. The physics of high pressures. London: G. Bell & Sons, Ltd.; 1931.

[24] Bernal JD. The problem of metallic state. 1929;

[25] ISO 527-2. International Standard. 2019.

[26] Kumar K, Agarwal S. Multi-objective parametric optimization on machining with wire electric discharge machining. Int J Adv Manuf Technol. 2012;62(5–8):617–33.

Estimation of temperature during TIG welding of titanium

Avinash Ravi Raja*[1], Anuj Dixit[1], Ayush Jain[2], Syed Qaisar Husain[1],Gagan Varshney[1], Sudesh Singh[3]

[1]Greater Noida Institute of Technology, Greater Noida,India
[2]Indian Institute of Technology (BHU), Varanasi, India
[3]Sharda University, Greater Noida, India
Email:avinash129@gmail.com*; anuj.me@gniot.net.in; ayush.jain.mec17@itbhu.ac.in; meesam.amu123@gmail.com; gagan.me@gnit.net.; sudesh.sengar@gmail.com

Abstract: In this experimentation work, the welding of titanium (commercially pure of grade 2) has been performed. The technique used to perform the welding is Tungsten Inert Gas Welding. Several experimentations have been performed to get the preferred welding. On which further studies have been done. Temperature has been measured during the welding process to get the estimation.

Keywords: Gas tungsten arc welding; TIG Welding; HAZ; microstructure.

1. Introduction

TIG welding process is an arc welding process which uses a non consumable tungsten electrode to form an electric arc between work piece and the tungsten electrode [1,2]. Tungsten has a high melting temperature of about 3422°C; this extremely high melting point of tungsten helps in the formation of the arc without causing the electrode to melt [3,4]. Usage of TIG welding is mainly observed in aircraft industry [5]. TIG welding especially improved the welding of hard to weld metals like aluminium and magnesium [6-8].

Titanium and its alloys have proven themselves to be very useful and effective in a lot of different applications. They have great mechanical and material properties at elevated temperatures. It can form oxides and cause the weld to be brittle which is not desirable and makes its welding a difficult task. Corrosion resistance of the metal and high strength-to-density ration of titanium can be considered its most useful properties [9]. Because of these properties it is vastly popular in aerospace and marine engineering, military and industrial processes etc [10,11].

Generally, in regards with the alloying elements and microstructure of the titanium alloy, the alloys are classified into four major groups:

1. α Alloys
2. Near α Alloys
3. $\alpha + \beta$ Alloys
4. 4. $Metastable \beta Alloys$

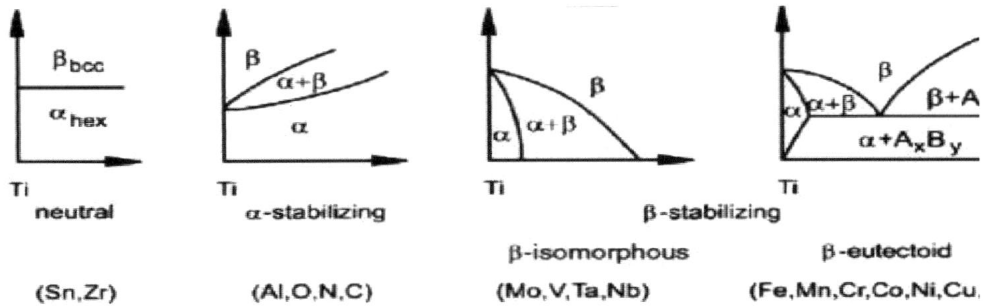

Figure 1. Effects of alloying elements on phase diagramof Ti Alloys

Table 1. Various processes for Welding Ti-Alloys

Processes	**Approx. plate thickness**
Tungsten inert gas welding (TIG) Welding	Up to 5mm
Plasma arc welding (PAW)	5-15 mm
Electron beam welding	Up to 50mm
Laser beam welding	Up to 15mm
Gas metal arc welding	3-12 mm
Friction stir welding	Up to 6mm
Resistance spot welding	Up to 4mm

2. Experimentation

Material Used: Commercially pure titanium of grade 2 was used in this experiment. Grade 2 of titanium can be welded by a variety of welding processes with specific measures. Tensile strength of grade 2 titanium lies between 482 MPa to 620 MPa. The composition of commercially pure grade 2 Titanium is given in table 2. Table 1 represents the approx weld plate thickness with respect to the different welding technique.

Table 2. Chemical composition of Grade 2 Titanium

Element	**Content**
Titanium,Ti	≥98.9
Iron, Fe	≤0.30
Oxygen,O	≤0.25
Carbon,C	≤0.080
Nitrogen,N	≤0.030
Hydrogen,H	≤0.015

The experiments were done with Lincoln Electric Aspect 300. An optimized setting that produced acceptable weld was found by experimenting. The measurements were then conducted on that setting. The parameters used are shown in the table 3. Before the start of the experiment, the speed of the movable tractor was calibrated to get a required welding speed.

Table 3. Optimized parameters used for experimenting

Parameter	Value
Arc Current	115 A
Arc Voltage	11.5 V
Welding Speed	180 mm/min
Gas Flow Rate	20 lit/min
Electrode Diameter	2.4 mm
Electrode Tip Angle	45°

Figure 2. TIG Welding Setup

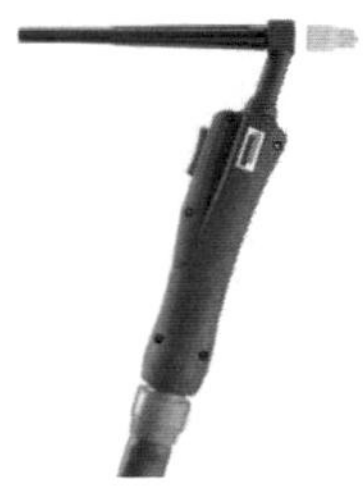

Figure 3. Torch used in the experimentation

Beads were made on plates of 3 mm thickness of commercially pure titanium of grade 2. After the pilot experimentation an optimal setting of parameters was found which was used to base the conclusion. Argon was used as shielding gas to provide the contamination from the atmospheric environment.

3. Result and Discussion

The temperature was measured using autogenously TIG welded bead on plate. The temperature cycle observed during the welding process is shown in fig.4. From the figure it can be observed that heating cycle of the process is very fast when compared to the cooling cycle.

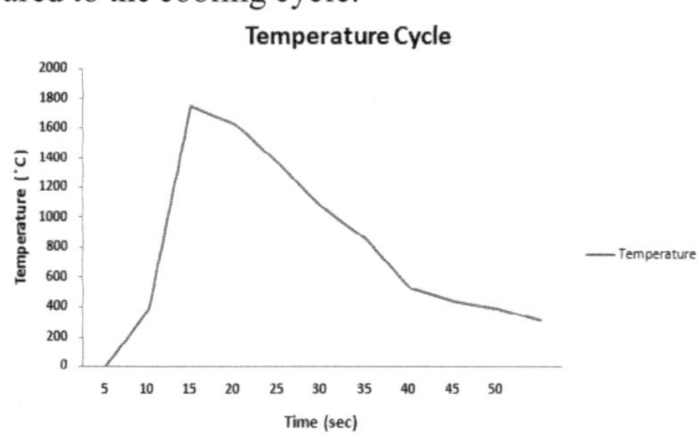

Figure 4. Temperature distribution during the process

The microstructure of HAZ and base of the welded plate observed can be seen in figures below. It was seen that the temperature greatly affects the physical characteristics of the material during welding. Low thermal conductivity of the titanium reduces the cooling rate after welding resulting in the formation of the observed microstructure.

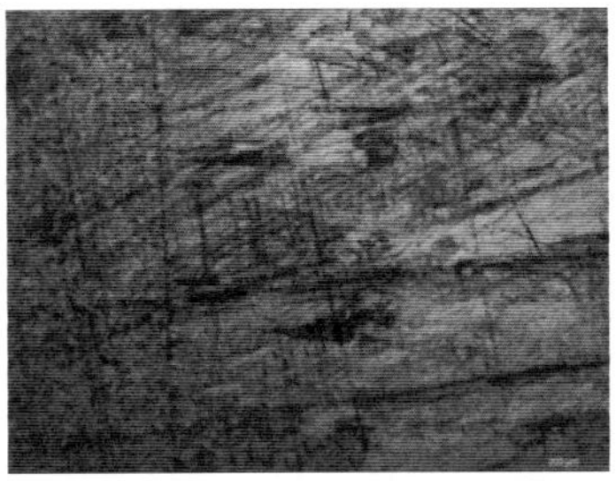

Figure 5. Observed microstructure of HAZ

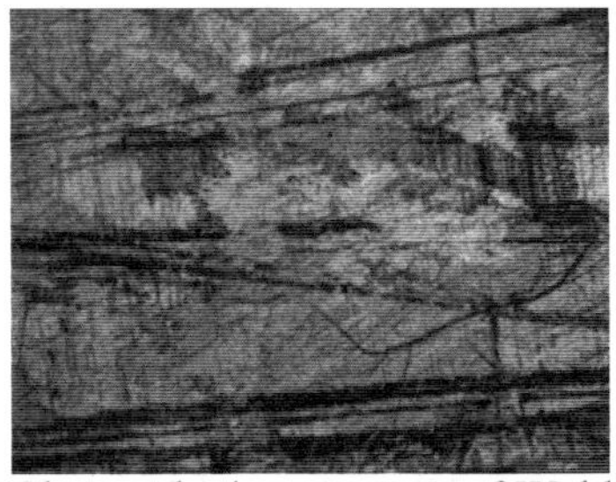

Figure 6. Observed microstructure of Weld

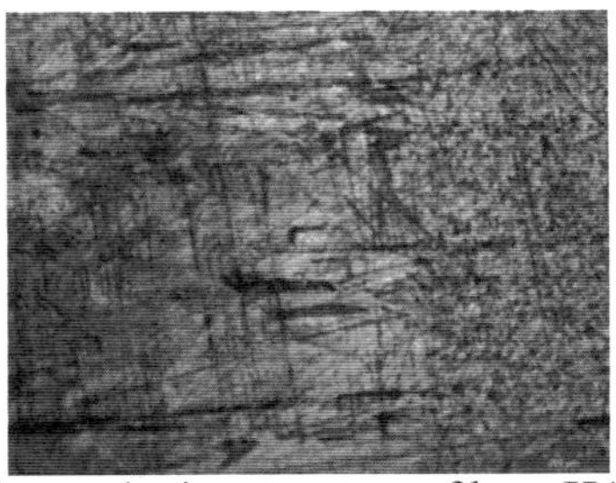

Figure 7. Observed microstructure of base HAZ interface

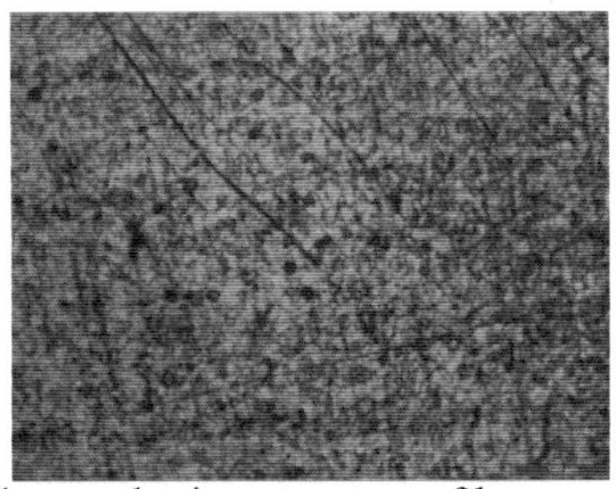

Figure 8. Observed microstructure of base

4. Conclusion

- Suitable welding current range to get acceptable weld for TIG welding of titanium is 110-120 A. And suitable welding voltage range is 10-12 V.
- Mechanical and physical properties of titanium are greatly affected by the temperature involvement during the welding process.
- Temperature distribution during welding process greatly affects important welding attributes like HAZ, microstructure of weld, etc.

5. Reference

[1] Yunlian Q, Ju D, Quan H, Liying Z. Electron beam welding, laser beam welding and gas tungsten arc welding of titanium sheet. Materials Science and Engineering: A. 2000 Mar 15;280(1):177-81.
[2] Modenesi PJ, Apolinario ER, Pereira IM. TIG welding with single-component fluxes. Journal of materials processing technology. 2000 Mar 1;99(1-3):260-5.
[3] Wu CS, Ushio M, Tanaka M. Analysis of the TIG welding arc behavior. Computational Materials Science. 1997 Jan 1;7(3):308-14.
[4] Strigul N, Koutsospyros A, Arienti P, Christodoulatos C, Dermatas D, Braida W. Effects of tungsten on environmental systems. Chemosphere. 2005 Oct 1;61(2):248-58.
[5] Mendez PF, Eagar TW. Welding processes for aeronautics. Advanced materials and processes. 2001 May 1;159(5):39-43.

[6] Faraji AH, Goodarzi M, Seyedein SH, Barbieri G, Maletta C. Numerical modeling of heat transfer and fluid flow in hybrid laser–TIG welding of aluminum alloy AA6082. The International Journal of Advanced Manufacturing Technology. 2015 Apr 1;77(9-12):2067-82.

[7] Huang Y, Fan D, Fan Q. Study of mechanism of activating flux increasing weld penetration of AC A-TIG welding for aluminum alloy. Frontiers of Mechanical Engineering in China. 2007 Oct;2(4):442-7.

[8] Marya M. Theoretical and experimental assessment of chloride effects in the A-TIG welding of magnesium. Welding in the World. 2002 Jul;46(7):7-21.

[9] Schiff N, Grosgogeat B, Lissac M, Dalard F. Influence of fluoride content and pH on the corrosion resistance of titanium and its alloys. Biomaterials. 2002 May 1;23(9):1995-2002.

[10] Gorynin IV. Titanium alloys for marine application. Materials Science and Engineering: A. 1999 May 15;263(2):112-6.

Modelling and Analysis of SUV Chassis

Manikandan H[1], Cyril S[2]

[1,2] Assistant Professor, Department of Mechanical Engineering, CMR Institute of Technology, Bengaluru 560037, Karnataka, India.

Abstract. Truck chassis is like a skeleton of the vehicle. The chassis undergoes various stresses and deformation, when the vehicle is in motion. The present work aims at deriving the static and dynamic characteristics of the chassis, by modelling the chassis and carrying out meshing and then subjecting it to static and dynamic analysis as well as modal analysis. Various materials were chosen for doing the Finite Element Analysis. Modal analysis is done by attuning the selective material properties and the predicted natural frequencies and mode shapes were validated against experimental results. The final results were incorporated in the model to optimise the chassis with respect to strength, vibration and weight.

Keywords: Automotive chassis, Static characteristics, Natural frequency, Modal Analysis.

1. Introduction

A chassis possess an inner vehicle frame that supports an artificial object in its construction and also provide overall protection internal parts. The chassis acts as a frame and engine, along with various components like wheels, axles, brakes, suspensions etc. are mounted on it. Thus, a chassis receives the reactive forces of the wheels during acceleration of the vehicle. Also, at the time of braking, it absorbs aerodynamic forces and shocks developed. In short, the chassis should be designed rigorously for maximum payload and performance capabilities. Most commonly, a ladder chassis frame is considered as it satisfies the above design characteristics and also provides torsional flexibility at certain regions where it is indispensable. Generally, analytical chassis design involves various materials and cross sections to arrive at optimal design. This is because, while static loading conditions, the frame is subjected to bending loads due to the self-weight of the members. In dynamic analysis, not only the forces induced by the engine components are considered but also external factors like the surface profiles of the roads, which directly induces some excitation forces to the chassis while the vehicle is in motion is also been given as input. The mode shapes at different natural frequencies plays an important role in identifying the critical mounting points of the engine, suspension and braking systems as well.
V. Veloso, and H.S. Magalhaes [1] investigated a stress distribution caused by 35 KN input load applied in the chassis' centre of the stringer. The maximum Von Mises stress value was 442 Mpa. This value is greater than the UTS of the material. Yuen Ren and Yongchang Yu [2] investigated force on the frame during its operation, mainly refering to bending moment, bending strength and deformation of the frame.. Abd Rahman et.al [3] investigated stress analysis on a heavy-duty truck chassis using FEM. Results had shown the occurance of critical point of stress at opening of chassis due to its contact with the bolt. Thus, it is important to reduce to stress at specific location. Teo Han Fui et.al [4] have studied a 4.5 ton truck chassis containing two C section side rails and five numbers of cross members against road roughness and excitations. Vibration of the components fixed to the chassis due to excitation and road roughness were studied. Chassis responses were examined by stress distribution and displacement. The suitable locations of engine and suspension systems were determined by the

IOP Conf. Series: Materials Science and Engineering 1168 (2021) 012024 doi:10.1088/1757-899X/1168/1/012024

results of mode shape. Analysis results reveal that the major disturbance of the chassis is caused due to road excitation. Yongjie Lu et.al [5], have investigated tyre dynamic load by numerical simulation and experimental field test. MSC. ADAMS software was used to model the virtual prototype of heavy vehicle. Shock absorber, tire, suspension and road profiles were modelled to simulate the real time conditions. Experimental test was carried out in a four way lane road by using truck manufactured by Dongfeng Motor Company model conforms DFL1250A9. The GVW of the truck is 24.9 tons. Seven piezoelectric accelerometers were placed on the driver seat, front axle head, intermediate axle head and rear axle head. Effects of speed, mass, road roughness and tyre stiffness on dynamic load were studied in detail. The study revealed that increase in speed may have considerable effect on tire and road damage. The proposed virtual model has efficient and realistic simulation and the new methodology unveiled to study the interaction between road and vehicle.

2. Methodology

Figure 1 indicated the methodology adopted in this work. It starts from the CAD model of the chassis. The present work used CATIA V5 for generating the model of the SUV chassis.The dimensions adopted are as shown in Figure 2. Then Meshing is done using Hypermesh Software. Generally CAD file is served in '.igs' format. And the job should be carefully checked for geometric features like Free edges, Scar lines, Duplicate surfaces, Small fillets, Small holes, Beads and Intersection of parts (Assembly). Also during meshing, the critical areas have to be addressed properly. Critical areas are high stress locations. Dense meshing and structured mesh is recommended in these regions. Those areas away from the critical areas are general areas. Also, there are different types of meshes, viz., Quad, Trip and Mixed mesh. Better mesh pattern is generally produced in mixed mesh. The present work adopted the mixed mesh. The meshed model of SUV Chassis is as shown in figure 4. Also during quality check, different quality parameters like skew, aspect ratio, included angle, Jacobian, stretch etc. are the measured for assessing any sort of deviation of the given element from the ideal shape. After meshing, a quality check is done on the mesh generated and then the initial conditions and boundary conditions are defined. Then, FEA analysis is done using Ansys software. The chassis is subjected to static analysis, dynamic analysis and finally modal analysis.

The following specifications indicate the design parameters used in modelling of chassis.

Overall length: - 3937 mm
Overall width: - 1150 mm
Overall height: - 215mm
Capacity of car =weight per passenger x no. Of passengers
= 80kg x 8
= 640kg Capacity of car with 1.25%=800kg

Total load acting on car chassis
= Gross vehicle weight + weight of passenger
= 800+800
=1600kg
=15696N

Chassis has two members which are longitudinal and hence load will be shared by these two members. Therefore, Load acting on each member,

$$= \frac{15696}{2}$$

$$= 7848 \text{ N}$$

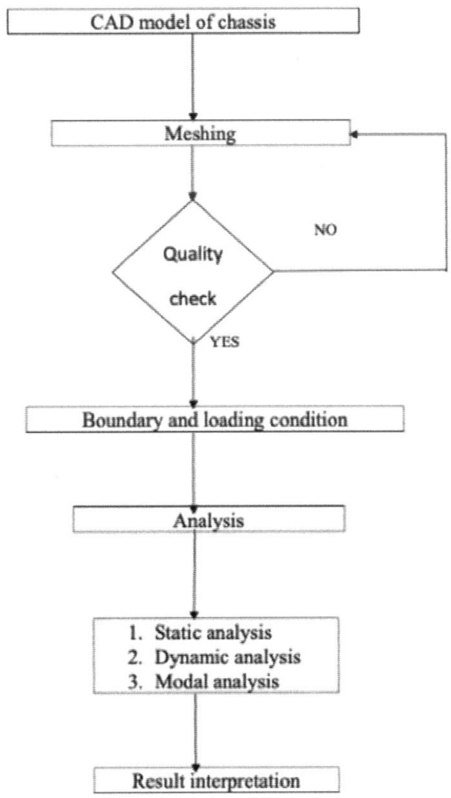

Figure 1 Flowchart of methodology used

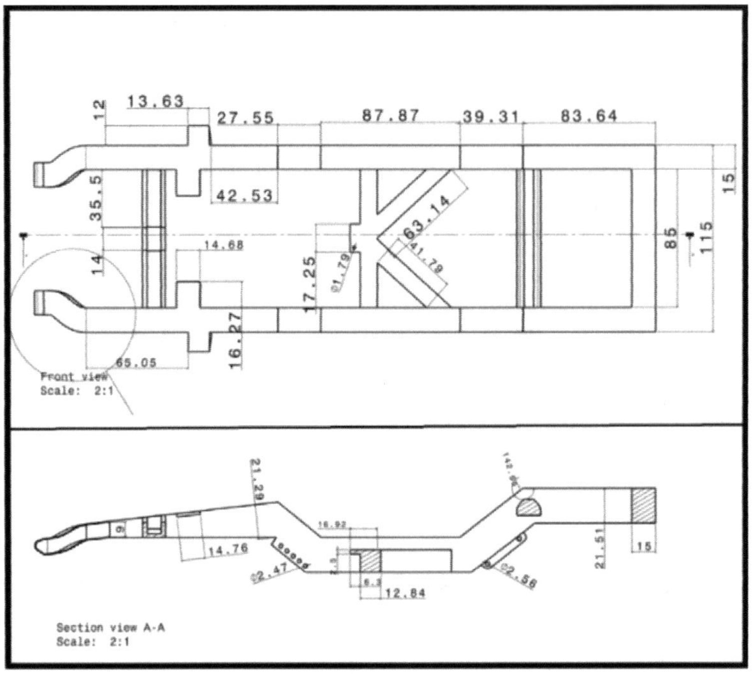

Figure 2 CAD Model of Chassis showing dimensions adopted

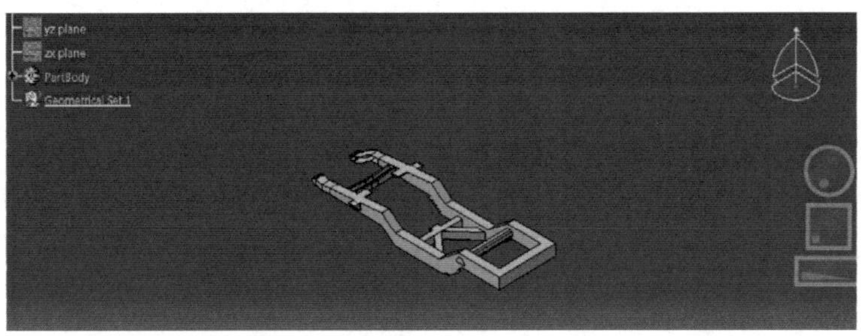

Figure 3 3D model of SUV chassis using CATIA V5

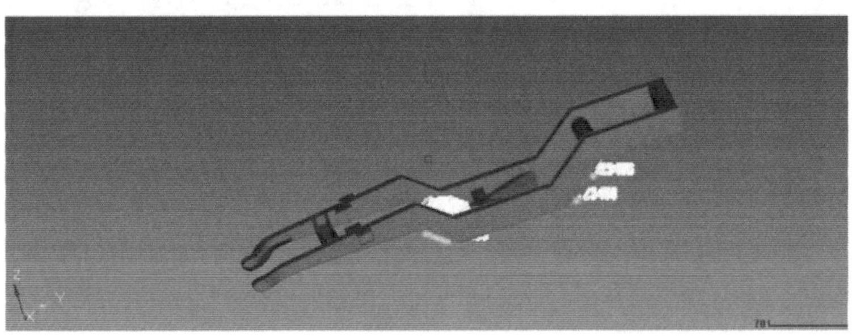

Figure 4 Meshed Model with Boundary Condition

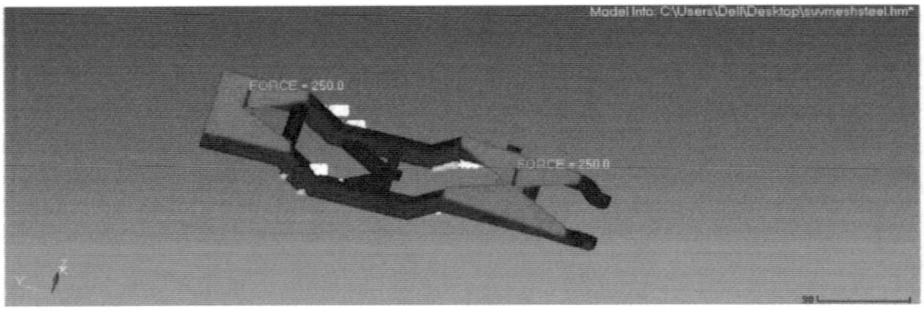

Figure 5 Meshed chassis with boundary conditions and loading condition

3. Results and Discussion

In Hyper Mesh software, boundary conditions are usually stored in load collectors. The meshed model with boundary conditions is as shown in figure 4. In loading condition, uniform load is applied on the chassis. Meshed Chassis with a Boundary and loading condition is as shown in figure 5. The deflection and stress pattern in the chassis model is obtained by carrying out static analysis. The results of the model are shown in tables 1,2 and 3 respectively. The design stress for the Q235 steel chassis material is 250Mpa. Also, the deflection produced and the stress induced for a load of 1150 kN. The different cross sections chosen here are rectangle, square and tubular. The different materials chosen are Aluminium alloy, Q235 steel and Cast iron. These materials are chosen based on the literatures available. During static condition the chassis frame is subjected to bending loads due to the self weight. Cross sections are decided based on bending equation. Then the same cross sections are analyzed by using FEM as well as hyper mesh. Also, a linear system model is chosen to perform dynamic analysis, whose inputs are loads or forces and the output include displacement response or acceleration response.

Table 1 Static analysis results on material Q235 Steel

Load (kN)	Maximum Displacement (mm)	Maximum Stress (N/mm^2)
100	0.01149	10.54
150	0.01723	15.81
160	0.01838	16.86
200	0.02298	21.07
220	0.02528	23.18
240	0.02758	25.29
260	0.02987	27.39
300	0.03447	31.61
400	0.04596	42.14
500	0.05745	52.68
600	0.06894	63.22
700	0.08043	73.75

800	0.09192	84.29
900	0.1034	94.83
1000	0.1149	105.4
1200	0.1379	126.4
1400	0.1609	147.5
1600	0.1838	168.6
1800	0.2068	189.7
2000	0.2298	210.7
2040	0.2344	214.9
2100	0.2413	221.3
2160	0.2482	227.6
2200	0.2582	231.8
2220	0.2551	233.9
2240	0.2574	236.0
2260	0.2597	238.1
2280	0.2620	240.2
2300	0.2643	242.3

Table 2 Static analysis on material Aluminium Alloy

Load (kN)	Maximum Displacement (mm)	Maximum Stress (N/mm^2)
200	0.063	25.28
400	0.126	50.56
600	0.189	75.83
800	0.252	101.11
1000	0.315	126.40
1100	0.3465	139.00
1120	0.3528	141.60
1140	0.3591	144.10
1180	0.3717	149.10
1220	0.3843	154.20
1260	0.3969	159.30
1300	0.4090	164.30
1340	0.4221	169.40

Table 3 Static Analysis on material Cast Iron

Load (kN)	Maximum Displacement (mm)	Maximum Stress (N/mm^2)
100	0.063	25.28
200	0.126	50.56
400	0.189	75.83
500	0.252	101.11
600	0.315	126.40
640	0.3465	139.00
680	0.3528	141.60

Modal analysis is performed once the chassis finite element model is created and meshing in free-free state is done with constraints. The results for the first 25 frequency modes is calculated and it is revealed that road simulations are the most important problematic for truck chassis. This analysis made use of RADIOSS Solver. 3 modes, which are related to the chassis displacement in x, y and z directions and 3 modes are related to chassis rotation about x, y and z axes. Above results shows related natural frequencies and mode shapes for chassis with maximum displacement in y direction in each mode, first mode having a frequency of about 451 Hz and 6th mode having frequency of 1329 Hz. Found natural frequencies from modal analysis of truck chassis, are used for determining the suitable situations for truck parts in working conditions.

Considering a safe load of 100 KN applied on the chassis maximum displacement and maximum stress values and maximum load up to which chassis can withstand are all given below in Table 4.
The dynamic analysis on the chassis of material Q235 steel under load of 250 KN and moment of 85 N m, in this loading condition chassis have a maximum displacement of 0.099 mm and with maximum stress value of 59.64 N/mm^2.

4 Conclusion

The outcome of the paper is that the maximum load that the given chassis into consideration can withstand during static and dynamic conditions have been found out. This varies for various chassis structures for different vehicles . We have adopted a ladder type chassis here. The Modelling was performed on CATIA V5 software. And for meshing the chassis, Hypermesh is used. Also, various quality checks were performed while importing the igs file into hypermesh. Finally, static analysis, dynamic analysis and modal analysis is performed in ANSYS software and the results are quite convincing. For validation, different materials like Q235 steel, Cast iron and Aluminium alloy were used. Also, the modal analysis in unconstrained state is also performed on the chassis for the above-mentioned materials and the mode shapes are obtained. Considering a safe load of 100 KN applied on the chassis, maximum displacement and maximum stress values and maximum load up to which chassis can withstand are found out, which is listed in table 4. The dynamic analysis on the chassis of material Q235 steel under load of 250 KN and moment of 85 N m, in this loading condition chassis have a maximum displacement of 0.099 mm and with maximum stress value of 59.64 N/mm^2.

References

1. V Veloso, HS Magalhaes, GI Bicolho, ES Palma. Failure investigation and stress analysis of a longitudinal stringer of an automobile chassis. Engineering failure analysis 16(2009)1696-1702.
2. Yuan Ren, Yongchang Yu, Binbin Zhao, Chuanhui Fan, He Li. Finite element analysis and optimal design for the frame of SX360 dump trucks. procedia engineering 174(2017)638-647.
3. Abd Rahman, Roslan, Mohd Nasir Tamin, and Ojo Kurdi. "Stress analysis of heavy duty truck chassis as a preliminary data for its fatigue life prediction using FEM." *Jurnal Mekanikal* 26.2 (2008).
4. Fui, Teo Han, and Roslan Abd Rahman. "Statics and dynamics structural analysis of a 4.5 ton truck chassis." Jurnal Mekanikal 24.2 (2007).
5. Lu, Yongjie, et al. "Numerical and experimental investigation on stochastic dynamic load of a heavy duty vehicle." Applied Mathematical Modelling 34.10 (2010): 2698-2710.

IOP Conf. Series: Materials Science and Engineering 1168 (2021) 012025 doi:10.1088/1757-899X/1168/1/012025

Substrate material selection and design optimization of patch antenna

Anil Kumar Dubey
Department of Electronics and Communication Engineering
Greater Noida Institute of Technology, Greater Noida, India
anilkumardubey10@gmail.com

Abstract- *In this paper parametric analysis of different substrate materials used for microstrip patch antenna are analyzed based on its physical properties, also parametric calculation performed for selection of substrate working at operating frequency of 5.8 GHz. Proposed microstrip patch antenna was optimized by varying its design structure from rectangular patch to E-shaped structure. Method of moment based electromagnetic simulator IE3D was used for parametric calculation and optimization of proposed patch antenna design.*

1. Introduction

Patch antenna consists of conducting patch on a ground plane separated by dielectric substrate and it radiates electromagnetic signals at high frequencies. This type of miniaturized design of patch antenna was first conceived in 1953 by G. A. Deschamp in USA [1]. Later on in 1955 H. Gutton and D. Baissinot patented an aerial antenna of similar design working in UHF band in France [2]. In 1975 J. Q. Howell elaborated design of rectangular shaped patch antenna and circular shaped patch antenna [3]. The patch antenna can be used as an transmitting and receiving device [4]. ISM Band frequency 5.8 GHz is chosen for designing of patch antenna as radio can support upto 1300 Mbps wireless speed and in 2.4 GHz it is between 450 Mbps to 600 Mbps[9]. Different substrate materials are considered for designing of patch antenna having dielectric constant lies between 2 to 12 i.e., $2 < \epsilon_r < 12$ [6]. Patch antenna can be designed in the shape of rectangle, circle, triangle etc. Further its geometry can be modified using analyzing its radiation properties and its broadband applications [5]. Different parametric results like return loss, radiation pattern, gain, VSWR etc. are analyzed on simulating its design on electromagnetic simulator IE3D.

2. Design flow of Patch Antenna

As operating frequency selected for patch antenna design is 5.8 GHz. Rectangular shaped patch antenna dimensions are calculated using the antenna design equation [7]. Parameters for designing of rectangular patch antenna are-

Resonating frequency (f_r) = 5.8 GHz (ISM Band)
Width of Rectangular patch antenna (W) = 29 mm
Length of Rectangular patch antenna (L) = 18.35 mm
Height of dielectric Substrate (h) = 1.580 mm

Feed point position on rectangular patch is calculated and simulated. Final proposed design of patch with feeding point is shown in figure1.

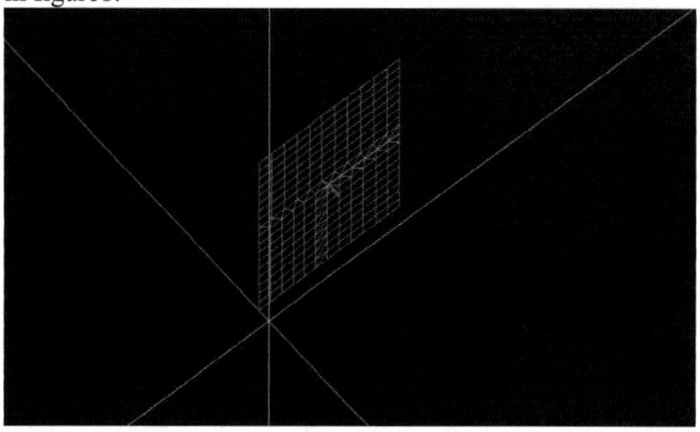

Figure1. Structural view of rectangular patch antenna

On designing and analyzing the patch antenna using different substrate material, following three substrate materials have better performance on the basis of return loss, VSWR, gain and radiation pattern etc [8].

1. Substrate 1: RT Duroid
   Dielectric constant ϵ_r of RT Duroid = 2.2, loss tangent δ = 0.001, water absorption = 0.02%, tensile strength = 450 MPa.

2. Substrate 2: RO4003
   Dielectric constant ϵ_r of RO4003 = 3.5, loss tangent δ =0.002, water absorption = 0.06%, tensile strength = 141 MPa.

3. Substrate 3: FR4 Glass Epoxy
   Dielectric constant ϵ_r of FR4 = 4.4, loss tangent δ = 0.001, water absorption < 0.25%, tensile strength < 310 MPa.

Parametric comparison of above chosen substrate materials after simulating different substrate material based antenna on electromagnetic simulator IE3D are compared and shown in Table 1 below. They are compared on the basis of return loss, VSWR and Gain.

Table 1: Comparison of various substrates

Parameters / Substrate Materials	Return Loss (dB)	VSWR	Gain (dB)
RT Duroid	-24.04	1.26	3.42
RO4003	-22.26	1.68	5.20
FR4	-20.64	1.28	1.89

IOP Conf. Series: Materials Science and Engineering 1168 (2021) 012025 doi:10.1088/1757-899X/1168/1/012025

On the basis of above parametric comparison RT Duroid is selected for further design of rectangular patch antenna as its return loss is -24.04 dB that will be good radiating and can be used for designing of antenna.

3. Proposed patch antenna design

Parametric outputs of proposed designed rectangular patch antenna after simulating it on IE3D Electromagnetic Simulator.

3.1 Return Loss (RL) of proposed patch antenna:

Return loss of rectangular patch antenna is shown in figure 2.

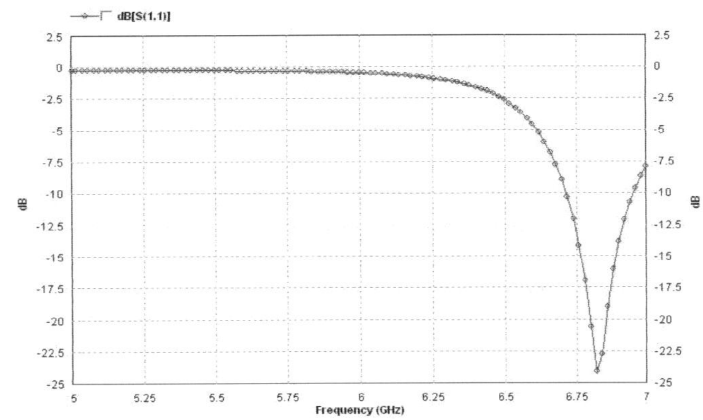

Figure 2. Return Loss of proposed patch antenna

Return Loss = -24.0412 dB
Highest Frequency (F_H) = 6.950 GHz
Lowest Frequency (F_L) = 6.716 GHz
Operating Bandwidth = F_H - F_L= 234MHz

Bandwidth of 234 MHz achieved on designing rectangular patch antenna but the operating frequency band is from 6.7 GHz to 6.9 GHz and our requirement is around 5.8 GHz.

3.2 Current distribution of proposed patch antenna:

Current Distribution of proposed patch antenna is shown in figure 3. Obtained current distribution is not satisfying our requirement for designing practical patch.

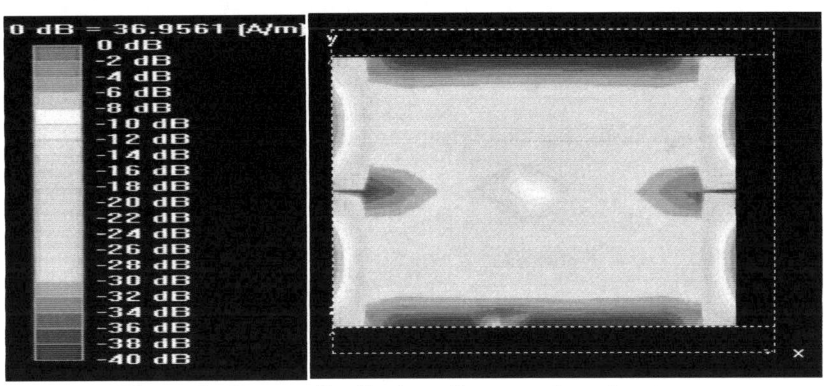

Figure 3. Current distribution of proposed patch antenna

3.3 VSWR for proposed patch antenna:
Voltage standing wave ratio for proposed patch antenna is shown in figure 4. In operating range proposed antenna VSWR is close to one. i.e. VSWR ≈ 1.

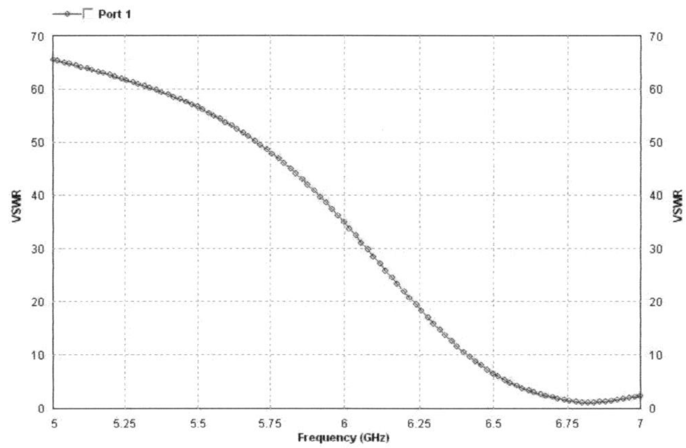

Figure 4. VSWR of proposed patch antenna

Parametric output results shows that center frequency of proposed antenna is 6.8 GHz but the required center frequency is 5.8 GHz. Further optimization in designing of patch antenna was possible. Optimization can be done by improving its design from rectangular shape to other shape like circular or triangular.

4. Optimized patch antenna design

Proposed patch antenna designed is further optimized by varying its geometry. Artificial neural network algorithm was used for calculations of dimensions variations and its effect was analyzed in wideband frequency application. Different geometry like circular, triangular etc. are simulated on electromagnetic simulator IE3D software. But our design requirement was satisfied by designing the patch that resemble to the geometry shape of English alphabet "E" [10] [11]. The pattern structure of proposed patch antenna is shown in figure 5.

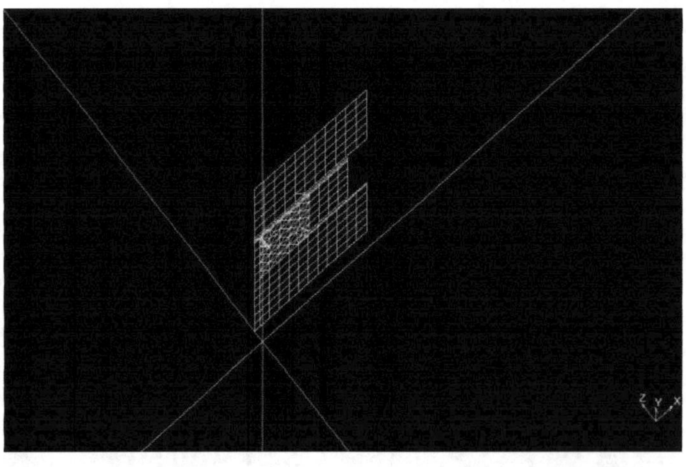

Figure 5. Pattern structure of E shaped patch antenna

Parametric outputs of designed E-shaped patch antenna after simulating it on IE3D electromagnetic simulator-

4.1 Return Loss of E-shaped Patch Antenna:
Return loss of E-shaped patch antenna is shown in figure 6. Center resonating frequency of 5.8 GHz achieved, also its operating bandwidth is 150 MHz.

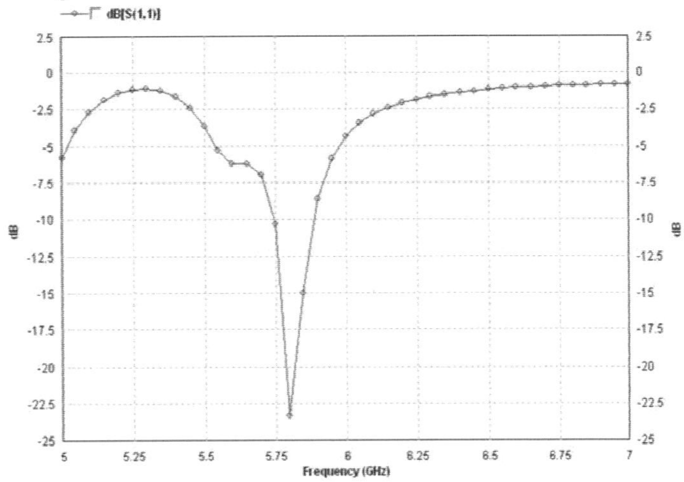

Figure 6. Return Loss of E shaped patch antenna

Return Loss = -23.3053 dB
Highest Frequency (F_H) = 5.875 GHz
Lowest Frequency (F_L) = 5.725 GHz
Operating Bandwidth = F_H - F_L= 150 MHz

4.2 Current distribution of E shaped patch antenna:
Current distribution of E shaped patch antenna is shown in figure 7. Uniform current distribution is achieved.

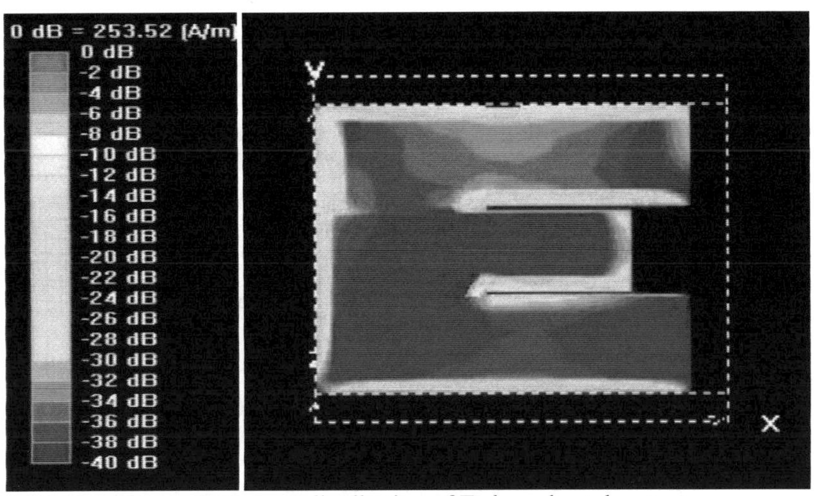

Figure 7. Current distribution of E shaped patch antenna

4.3 VSWR of E shaped patch Antenna:

Voltage standing wave ratio for designed patch antenna is shown in figure 8. In Operating range designed antenna VSWR is close to one. i.e. VSWR ≈ 1.

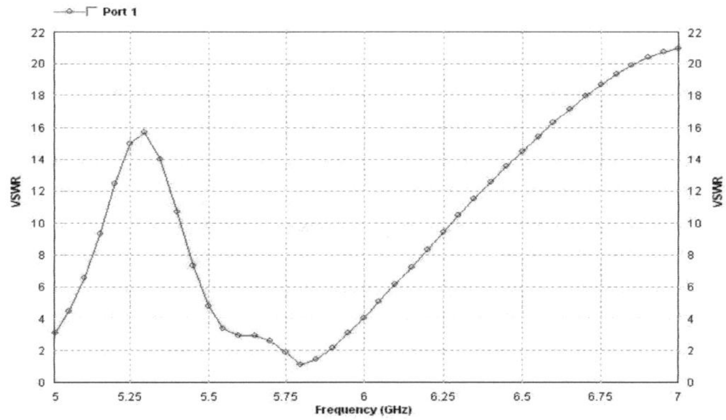

Figure 8. VSWR of E shaped patch antenna

5. Result and Conclusion

Final design of patch antenna of E shape having substrate RT Duroid of dielectric constant 2.2 on simulating on electromagnetic simulator IE3D shows the bandwidth of 150 MHz at operating frequency of 5.8 GHz which satisfy the requirement of patch antenna design as its return loss is -23 dB at operating frequency, and voltage standing wave ratio is approximately equal to one. Current distribution and near far field electric and magnetic distribution are also fulfilling the requirement. The proposed patch antenna of E-shape can be commercially used for the applications like-fixed satellite, radiolocation, amateur satellite services and mobiles.

6. References

[1] G A Deshchamps 1953 *Microstrip microwave antenna* (3rd USAF symposium on antenna)

[2] G Baissinot and H. Gutton 1955 *High frequency antenna* (French patent no 703113)

[3] J Q Howell 1975 *Microstrip Antennas* (IEEE Transaction antennas propagation) vol. **AP-23**, pages: 90-93

[4] M Kanda 1990 *A microstrip patch antenna as s standard transmitting and receiving antenna* (IEEE Transactions on Electromagnetic Compatibility) vol. **32**, Issue: 1

[5] J H Lu, C L Tang and K L Wong 2000 *Novel dual frequency and broadband designs of loaded equilateral triangular microstrip antenna* (IEEE Transaction antenna propagation) vol. **48**, pages: 1048-1055

[6] K L Wong 2002 *Compact and Broadband Microstrip Antenna* John Wiley and sons

[7] C A Balanis 2005 *Antenna theory analysis and design* John Wiley and sons

[8] A A Roy, J M Mom, D T Kureve 2013 *Effect of dielectric constant on the design of rectangular microstrip antenna* (IEEE International conference on emerging & sustainable technologies for power & ICT in a developing society NIGERCON)

[9] N P Ojaroudi 2014 *Design of microstrip patch antenna for 2.4/5.8 GHz RFID applications* (German microwave conference GeMic)

[10] S Shivani, D Tarishi, K Manoj, S Avinash 2018 *Microstrip E-shaped patch antenna for ISM Band at 5.3 GHz frequency application* (IEEE International conference on microelectronics and telecommunication engineering ICMETE)

[11] A K Rastogi, G Pravin, S Sharma 2019 *Comparative study of rectangular and E-shaped microstrip patch antenna array for X-band applications* (Chapter in advances in intelligent systems and computing)

A review on superhydrophobic materials and coating techniques

Vipul Kumar Mishra[1], R Saini[1] and N Kumar[1]

[1]Departhment of Mechanical Engineering, ABES Engineering College, Ghaziabad, India

[1]Departhment of Mechanical Engineering, ABES Engineering College, Ghaziabad, India

[1]Departhment of Mechanical Engineering, ABES Engineering College, Ghaziabad, India

Abstract- Since the recent year, we have seen the interest in scientist to mimic the natural behaviour, and they tried to mimic the nature's property of superhydrophobicity artificially and very far they also get succeeded. This superhydrophobic property is very useful in many aspects of life as well as in industries as they have an extensive property to overcame wear, corrosion, biofouling etc. There are different ways to fabricate these coatings, some are chemical and some are physical. In this review paper, we have discussed the theoretical background of superhydrophobicity and discussed the need for a surface to become superhydrophobic like surface energy and roughness and also different ways of fabrication in which some are simple and cost-effective and some are very useful for industries. Also, there are different examples of materials that can show superhydrophobic properties.

Keywords- Superhydrophobicity, Coatings, Contact Angle, Surface Roughness, Surface Energy.

1. Introduction- Scientists always inspired by the natural phenomenon and try to replicate that artificially. In this term of events, they try to replicate the different leaves, wings of some insects found in nature that have the capability to repel water from its surface example sacred lotus leaves, cicada ornis's wings etc, because they have exceptional self-cleaning property this property is associated with the superhydrophobicity [1]. Superhydrophobicity is a term we use generally for a surface that repels water and water droplet get a jump on it and roll on the surface. Also, to understand superhydrophobicity we can use the phenomenon of surface tension. As we know that molecules which are in the liquid have equal force all around it but the molecule that is on the surfaces faceless force that leads to the surface tension [2] and we know that due to surface tension the droplets always try to reduce their area and form a spherical shape. But as droplet gets in contact with the surface it changes its shape from spherical and gets distorted due to gravity [3]. Also, the cohesive force of the liquid should be more and adhesive force should be less for that surface, so for superhydrophobicity, with the help of surface roughness and surface chemistry, we try to reduce the adhesion force. In different culture, lotus leaves are known for its purity although it grows in muddy water [2]. Water on the leaves roll of from its surface including all the dirt particles along with it, this effect is called a lotus effect or self-cleaning property of lotus [2,3]. Due to this type of wetting property and self-cleaning this favoured topic for the scientists [4]. To indicate wettability, we use contact angle made by the water droplet on the surface [5]. Wetting of any surface depends on these factors, contact angle between and a droplet of water, chemical composition and surface roughness [6]. For wettability, it is said by the scientists that when the contact angle is less than 90 degree then it comes under hydrophilic surfaces and for greater than 90 degrees to 150 degrees it came under hydrophobic surfaces and contact angle greater than 150 degree comes under superhydrophobic surfaces [7,8]. Chemical composition describes the energy of the particular surface. For superhydrophobicity, the surface energy should be low. And the surface of the superhydrophobic substance should be rough [9, 10, 11]. We will discuss this property in this review further.

To get superhydrophobic surface we fabricate different layers on the surface that have to make superhydrophobic with different techniques. The main requirement for any fabrication technique is to modify the surface and head to decrease surface energy and increase surface roughness [7,12]. Superhydrophobic surfaces are very useful for daily uses and industrial purpose also, it has different applications like self-cleaning, anti-biofouling, anti-icing, an increase of corrosion resistance, anti-fogging surface etc. We see the frying pans have coatings that make them anti-sticking, we can see the hydrophobic layers in textiles industries, in different industries where high corrosion resistance is required, in ships where we need to control the biofouling. These are some places where we use these layers for different purposes. In this review, we will discuss the layers to increase corrosion resistance and to increase the anti-biofouling characteristics and self-cleaning. And also, different techniques like electrochemical deposition method, sol-gel process, chemical fabrication method, spray coating fabrication method etc of fabrication will be discussed further in this paper.

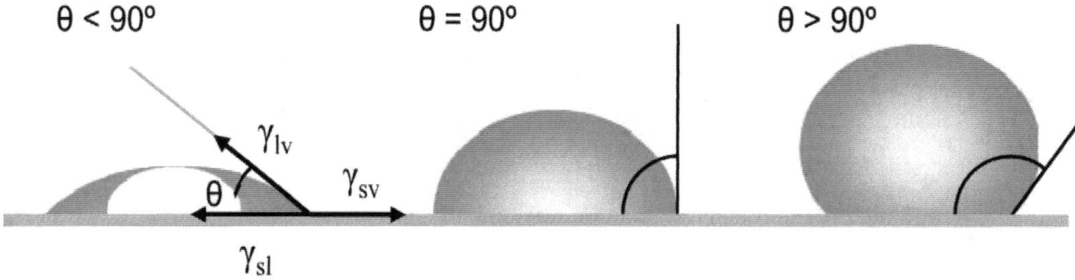

Figure 1. Water contact angles for hydrophilic hydrophobic surfaces [6].

2. Theoretical Background

2.1. Contact Angle- The definition of contact angle is that the angle which is made by the edge of the droplet and solid surface is known as the contact angle. When droplet touches the surface there is a contact point if solid, liquid and gas phases, at this point the angle between solid phase surface and tangent to liquid droplet is the measurement of contact angle [13]. So, we can say that there are three-phase and all three phases solid-liquid, liquid-air, air-solid and have their surface tension forces are γ_{SL}, γ_{LA}, γ_{SA} respectively [14] (fig 2). For contact angle measurement for wetting first model was given by Thomas Young in 1885 [15]. Derivation for contact angle by Thomas Young is in equation (1).

$$\cos \theta_Y = (\gamma_{SA} - \gamma_{SL})/\gamma_{LA} \tag{1}$$

here θY is Young's angle,

But this Young's angle is derived for the smooth surfaces and we know that in actual superhydrophobic conditions the surfaces rough and this affects the contact angle. So, to get actual angle other models were proposed we will discuss the other two models in wetting models.

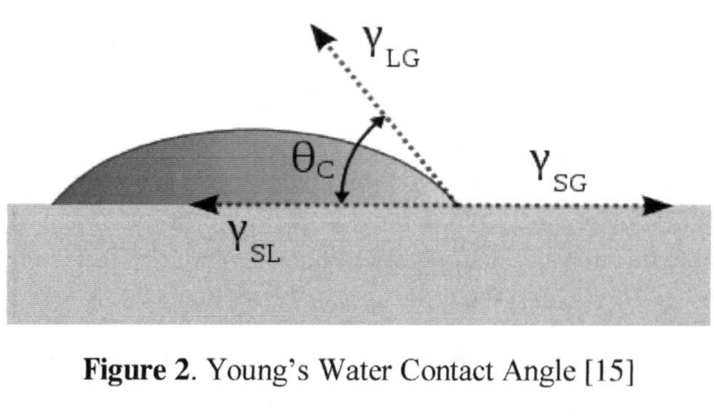

Figure 2. Young's Water Contact Angle [15]

Superhydrophobicity traditionally implies not only a high contact angle as well as a low contact angle hysteresis.
The minimal hysteresis of the superhydrophobic surface contact angle is accountable for the self-cleaning characteristics that ensure a drop of water can roll off the surface easily and eliminate dust fro m the surface Self-cleaning is one of the most essential properties of superhydrophobic surface applications and functions . [16,17].

For superhydrophobicity, the only greater contact angle is not enough but the water should be roll off from the surfaces just by tilting the surface with a small angle. This slight tilting angle of the surface is known as the sliding angle. For superhydrophobic surface, this sliding surface should be very low generally less than 5 degrees [18].

2.2. Surface Energy- We know that for superhydrophobicity the surface energy should be low. Different chemicals or surfactants are used to lower the energy of the surface. These chemicals can be coated on the surface to reduce the surface energy. The surfactants and surface roughness combine to make the surface energy so low so that it can be used for the superhydrophobic substrate. Generally, the compound which has silane group in them have a great effect to reduce the surface energy of the surface. Among silane fluoroalkysilanes are most useful for lower surface energy. But these groups due to the presence of fluorine are hazardous for the environment in nature and are also very costly. So, there is a need for the substance that can replace these and can be used as an alternative, some substance like stearic acid and polymers can be chosen as the alternative [19].

2.3. Surface Roughness- A superhydrophobicity must have a rough surface. Due to this uneven surface air get caught in the between peaks of the roughness, so when water placed on the surface due to low surface energy and this air bubble restrict water to spread at all points of the surface, and due to this surface area get reduced and this leads to lower the friction that's why the droplet gets slipped on the surface easily [20,21].

2.4. Wetting Models- Previously we have discussed young's equation for contact angle but young's equation is only valid for smooth surfaces. But we know that there are rough surfaces for the superhydrophobic surfaces and this roughness can be microstructured and nanostructured due to this roughness the contact angle gets increases. So, to calculate the contact angle for rough surfaces we use two models first is the Wenzel model and Cassie-Baxter model. Wenzel model is used for microstructure roughness. In figure 3 we can see the characteristics of water on both micro and nanostructures surface. Either Water droplet can penetrate in spikes on can be suspended over the peaks of roughness [5].

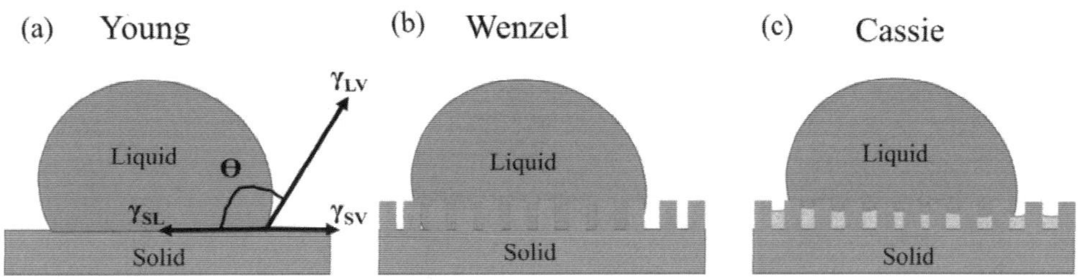

Figure 3. Wenzel And Cassie Baxter State [6]

Wenzel published his model of wettability in 1936 for a rough surface. so, Wenzel introduced the factor of roughness R that is a ratio of rough surface area to the smooth surface area, the value of R is generally greater than or equal to 1. The following equation provides the contact angle on a rough surface in the Wenzel model [22].

$$\cos \theta_W = R \cos \theta_Y \quad (2)$$

Roughness factor R = Actual surface area/Planar surface

The surface roughness can enhance both hydrophobicities as well as hydrophilicity [23].

2.5. Cassie-Baxter Model- We know that due to roughness the air can be trapped in between them, but Wenzel model does not look into it, so we can say it works only for homogeneous surfaces. For heterogeneous surfaces, Cassie-Baxter came up with another formula for contact angle. In this Cassie-Baxter says that the droplet is only contacted at the top of the roughness but cannot wet the surface in which air is trapped [6]. The formula given by Cassie-Baxter is- [24,25]

$$\cos \theta_{CB} = f(\cos \theta_W + 1) - 1 \quad (3)$$

here,

f= fraction of area which touches the water

θ_{CB}= Cassie-Baxter contact angle

from the formula, we can see that the contact angle is increased in this state.

In the Wenzel model, the water driblet gets into the roughness grooves, but in the Cassie-Baxter model, it holds on the top of the roughness grooves as the air is trapped in the grooves [23,26,27].

3. Different Methods of Fabrication to Achieve High Corrosion and Anti-Biofouling Properties- We have already discussed that two things (i)chemical composition and (ii)surface roughness is required to create the superhydrophobic layer. We can create a superhydrophobic surface by a association of the rough surface of lower energy material and by deposition of another material on low surface energy material [28,29]. So, different methods are used for the fabrication that uses base as we discussed above.

3.1. Electrodeposition Method- In the process of Electroplating, also known as electrodeposition there is a solution of ionic salts this salt is of the same material as depositing a material that has to be deposited on the conducting surface. We can use this method of fabrication to produce a thin film of metal on surfaces so that the properties of the surface can be enhanced. We can increase the corrosion resistance protection, abrasion resistance can be increased, also we can give lustrous property to surface. The main quality of this process is that we can use it in complicated parts which are not easy in some other methods. For example, in automotive part are electroplated by nickel-metal so that anti-corrosion property can be increased, also in the circuit board, we know that copper has lower resistance so it is plated on circuits so that resistance can be reduced. Another advantage of electroplating is that by this method we can electroplate alloys too on surfaces.
This method is very useful in managing the roughness morphology of coating. By this method we can produce a variety of morphologies like sheets, needles, tubes, cones, rods etc relatively fast There are several methods by which we can control both surface roughness and morphology but Electrochemical processes are way better than any other methods and are in capable of controlling these properties even on very large surfaces. [17].

Furthermore, micro-structures and nano-structures development can be easily controlled by regulating some electrodeposition parameters like current density, current type AC or DC, by using pulsed or non-pulsed current [30,31]. As an example, by using platinum anode and watts bath thin micro or

nanoparticles layer can be fabricated from a solution on the conductive surface and produce a pine cone-like hierarchical structure. [32].
Here is a fig. that shows the formation of nanostructure of ZnO by electrochemical anodization method by mixing hydrofluoric acid and methanol electrolyte on a large area surface. We can observe the difference in the formation of ZnO nanostructure that changes in nanodots, nanowires and nanoflower in less than an hour like fig "a" shows the formation after 30 seconds, "b" shows sample after 1 min, "c" shows sample after 5 min, "d" shows sample after 10 min, "e" shows sample after 15 min, "f" shows sample after 30 min [33].

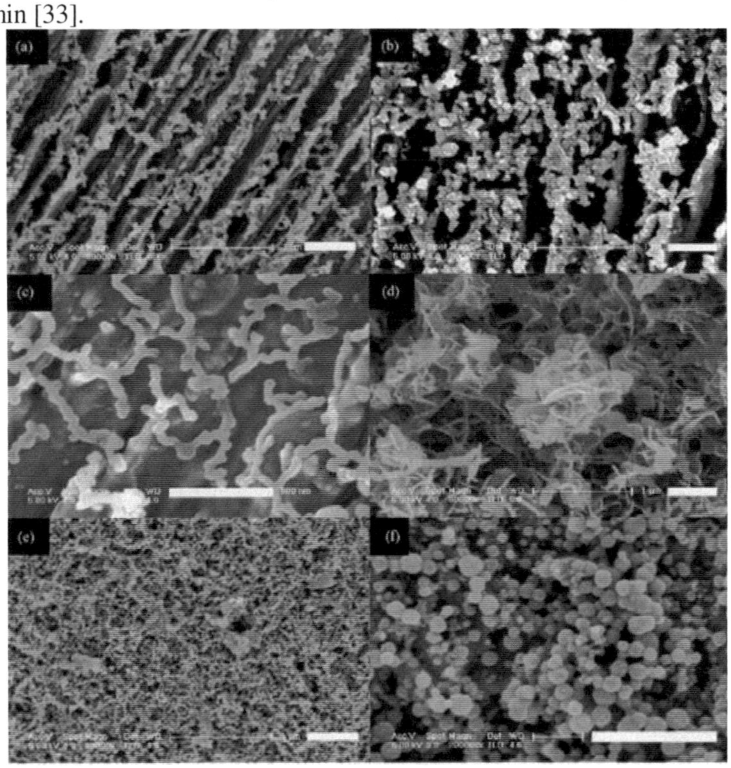

Figure 4. According to different reaction times, the growth of ZnO nanostructure on Zn foil is shown by SEM image [33].

Using direct voltage supply and one-step deposition process Huang et al. prepare the superhydrophobic copper surface. They used a dilute ethanolic stearic acid solution in which copper plates were immersed. The reaction between copper and stearic acid takes place and a surface copper electrode on anode gets transformed to superhydrophobic with flower-like nanostructure of copper stearate film. This layer exhibits the water contact angle of 153 degrees [34].

Metals such as aluminium, zinc, nickel and iron and their following alloys like brass and Zn- Fe alloy etc shows their advantage of reaction with perfluorocarboxylic acid solution, they can make a very efficient superhydrophobic surface. By controlling process time and concentration of perfluorocarboxylic acid we can create a variety of different substrate [35].

3.2. Electrospinning Method- Electrospinning is a simple process that generally used for the making of polymer nanofibers. It is a very versatile process to fabricate continuous polymer fibres in nano and micrometre sizes. This process is something like the extrusion process but this extrusion is electrically controlled [36,37]. High voltage is applied to the raw material of electrospinning, this power is required to enhance the liquid electrostatic potential. There is a relationship between electric potential and surface tension of a liquid. So, if we vary the electric potential the surface tension, the volume of fluid can get decrease or increase. But generally, oppositely surface charge to act as the electric potential get increase surface charge gets reduced and due to this change in surface charge fluid change, its shape and make the structure known as Taylor cone [38,39]. This method can be used in laboratories as well as in industries, it has a wide variety of application that's why it became the most adaptable method for fabrication of nanofibers. Some examples of electrospinning are given below to understand the variety of this process.

Carbon nanotubes have many characteristics that can be used in mechanical works like higher tensile strength, higher modulus of elasticity etc [40]. Although when CNT reinforced to make composite then it became very hard so it can't be made into string form so now this is not able to perform the mechanical properties so researchers try to synthesize CNTs into polymer nanofibers by electrospinning method [41]. A researcher found that electrospinning of CNT and polyaniline the are used as a fabrication composite and the conductivity of fibre got increases and it made a carbon sheet by electrospun method, this sheet was investigated and this layer was increasing the property of the base material [42].

Electrospinning can also use as a method to protect the wound. Fibres mat dressing can be used if the ultra-nano fibres produced using electric field and that directly spun on the wounded area of body [43,44].

Ma et al. combined PCL (polycaprolactone) by electrospinning which contains hierarchical roughness structure and PPFEMA (polymerized perfluoroalkyl ethyl methacrylate) by initial chemical vapour deposition process which has very low surface energy. So, this combination creates a very suitable state for the superhydrophobic film. This coating shows the contact angle of 175 degrees and 205 degrees of sliding angle [45].

PVDF (polyvinylidene difluoride) is well known for lower surface energy and used as a substance to lower the surface energy of the surface. It mixes with different substance like zirconia, alumina, silica etc and can make a compound that can create PVDF-ZnO, PVDF-Al_2O_3, PVDF-TiO_2 coatings that can be used by electrospinning process [46,47,48].

3.3. Sol-Gel Method- In a typical sol-gel process a series reaction like hydrolysis and poly-condensation take place on the precursor and is transfigured into a glassy material. The surface roughness can be controlled by altering the precursor mixture and conditions. The transparent superhydrophobic layer can also be fabricated by this method as this is adaptable for glass [49,50,51]. Superhydrophobic sol-gel coatings usually present good reluctance to temperature. In this process gels and nanoparticles are synthesized on the substrate. By this method different types of layers can be formed like multi-layer films, thin layer, nanocomposites, porous roughness can be easily made [52,53]. Due to its advantages like process on low temperature, lower cost of fabrication, simple technique, and better control on the process it is used widely to fabricate superhydrophobic layer [54]. In this method generally metals oxides like silica, titanium oxide, zirconia etc are used to create a superhydrophobic layer that works on the theory of Wenzel and Cassie- Baxter [55,56].

Mahadik et al. prepared silica film by sol gel method with mixture of different chemicals like methyltriethoxysilane (MTES), methoxytrimethylsilane (TMMS), trimethylchlorosilane (TMCS), methanol, hexane and Ammonia, (MTES: TMMS:CH3OH: H2O- 1:0.09:12.71:3.58) and prepared transparent film that can we useful for goggles, specs, solar panels etc. they get best result in 25 hours deposition time in which they get contact angle of 172 degrees [57].

Lu et al. prepared sol get coating with Zinc nitrate hexahydrate (Zn $(NO_3)_2 \cdot 6H_2O$) and Methenamine ($C_6H_{12}N_4$) on aluminium substrate, stearic acid used to decrease the surface energy and they get water contact of 154.8 degrees [58].

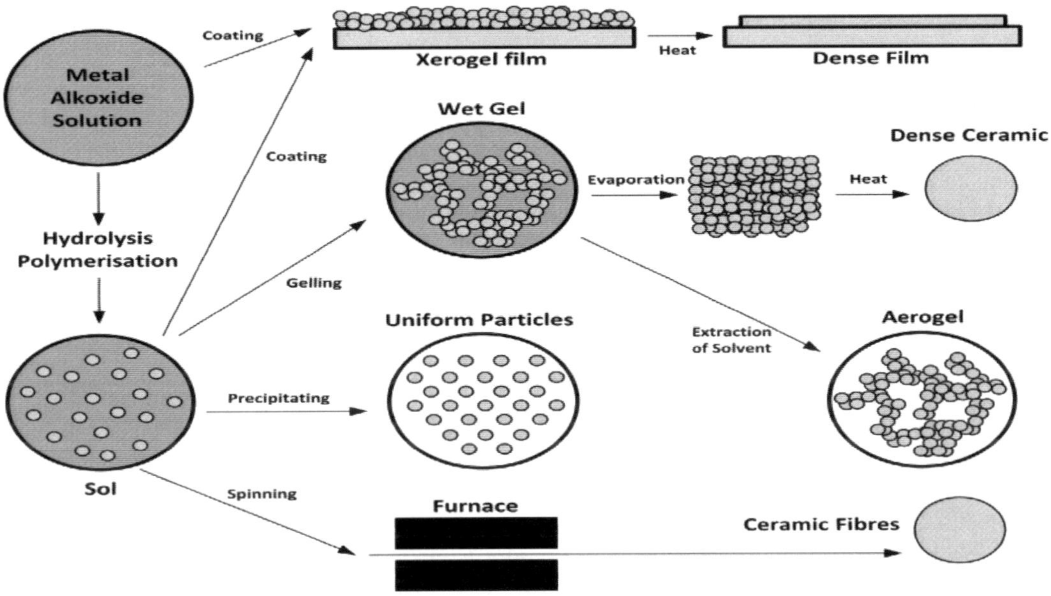

Figure 5. Sol-gel process [52].

3.4. Etching Method- Etching is a method in which superhydrophobic surface is made by increasing in roughening of the surface. The metal surface can be etcher by the different etching method that can combine with acid to increase the roughness. There is a different kind of etching processes that increase roughness can be used together with acid to increase the roughness of the surface. For inorganic materials, plasma etching techniques and optical maser etching can be used. To make multiple roughness's scales many techniques can be combined [59,60]. We can divide etching into two categories,

(a) Wet chemical Etching- in this acid or base solutions etch the surface and make the surface rougher [61,62].
(b) Dry etching- in this plasma, reaction etching like techniques are used. This generation of reactive atoms and particles takes place in gas discharge and they do an anisotropic etching of the surface [63,64]. The etching is a very simple and straight forward technique which can also be used on very large-scale surfaces but there is a demerit of this technique, due to use of corrosive acids like sulfuric acids phosphoric acids it leads to unsafe for use due to environmental issues [65].

*3.5. Plasma Treatment-*A plasma treatment of surfaces perpetually involves plasma etching. Plasma treatment of surfaces will cause a big modification within the surface micro-nanostructure attributable to the aeolotropic etching of the surface layers [66]. Plasma could be an uncomplicated and effective technique to fabricate hydrophobic surfaces. Increase in Surface roughness and reduction in surface energy both can get simultaneously. By using oxygen plasma, we can get Superhydrophobic surfaces simply. However, ageing is the main problem of superhydrophobic surfaces by plasma technique. Since the roughness of surfaces finished by plasma will be simply adjusted, this technique is a suitable way to fabricate superhydrophobic coatings with special optical properties, wherever the surface roughness of those coatings is the most vital factor [7]. Larger substrate and for continuous process plasma technique is suitable because we can treat the large surfaces easily. After all, it doesn't need vacuum instruments and can be used in inline mode rather than batch mode [67].

In one research using sound waves assisted nitric acid and cetyltrimethylammonium bromide (CTAB) Pan et al. made a rough surface on copper and made it superhydrophobic surface and get water contact angle of 155 degrees [68].
Ryu et al. created hydrophobic film by plasma method on polytetrafluoroethylene (PTFE) substrate and get spherical tips on surface that were nano structured. He get water contact angle of maximum 171 degrees [69].

3.6. Chemical Deposition- In this process generation of thin films on substrate take place. In a chemical reaction the deposition occurs and it self assembles itself and formed deposit on the substrate. It creates a crystalline structure on the substrate, we can use different materials like CdS, ZnS, CuSe etc. this deposition can be done by different processes like chemical vapour deposition, electrochemical deposition and chemical bath deposition. We can get a different type of morphologies and micro-nano structure can vary on basis of material and deposition conditions [5].

3.6.1 Chemical Vapour Deposition- In this chemical process the chemical gaseous substance(reactants) gets deposited on the substrate surface and forms a non-volatile solid film. In this process, the morphology of the layer is based on the morphology of the surface. But if change reactants and control the reaction conditions we can achieve a different kind of morphologies on the substrate [29]. Figure 7 shows by chemical vapour deposition techniques used by Mertaniemi et al. to coat FOTS (fluoro-octyl-trichloro-silane) onto the NFC (nano fibrillated cellulose) particles deposited on the surface and fabricated a superhydrophobic surface [70].

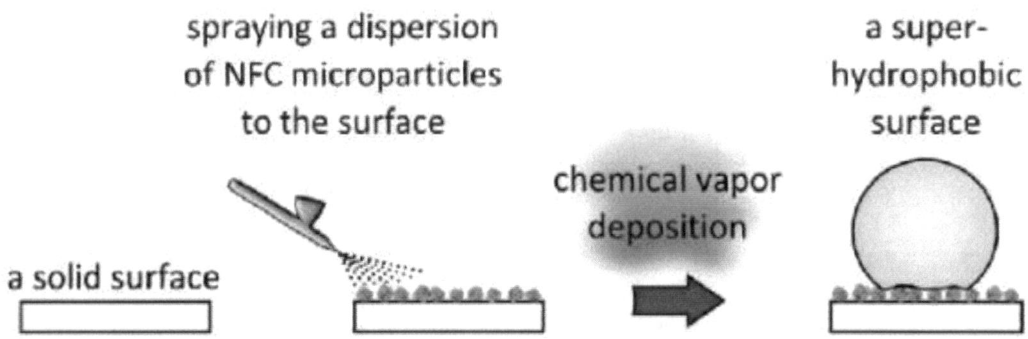

Figure 6. Generation of the superhydrophobic surface using Chemical Vapour Deposition [68]

3.6.2 Electrochemical deposition- Electrochemical deposition can create a thin layer of metal or metallic alloys on the substrate. The main benefit of this method is that we can create a large verity of morphologies and structure on the surface and also at very low cost and thickness can also vary. Coating created by this method is very stable and can be achieved at room temperature. In this method, the cations of the electrodeposited metals get reduced by an electric current from electrodes to a conductive surface. For large area production, we generally used this method in preparation of photovoltaic solar panels [13]. In this method generally ZnO, gold cluster, silver aggregate used to prepare the superhydrophobic film. In an experiment of fabrication of hydrophobic film by chemical deposition, method researchers used aqueous compound of $ZnCl_2$ and KCl and ITO glass was used as substrate. After completing of process, we found that the layer was made up of zinc, oxygen and carbon. They found the contact angle of 128 degrees and the surface was relatively rough [71]. By this method, we can generate organic coatings like PEO-polyethylene oxide, PTFF-polytetrafluoroethylene [72]. In an experiment, myristic acid was used to create a stable layer and contact angle got by layer was 154 degree and this surface was also very good for corrosion protection [6].

3.6.3 Chemical bath deposition- Generally for fabrication for semiconductors there is no method that is simpler than the chemical bath method. We can fabricate thin film on a large area very easily, but in this method, we can use only some selected materials like Ag_2S, As_2S_3, Bi_2S_3, Ag_2Se, Bi_2Se_3, Ag_2O, AgO etc. In this we use a liquid bath containing suitable reagent and the temperature can vary from 0 degrees to 100 degrees [73,74]. The object is immersed in a solution containing the precursors in chemical bath deposition method. In this method nucleation and growth of particle is the two main steps that take place so that solid structure can be fabricated on the surface from the solution. Different parameters can affect the deposition like the temperature of the solution, the pH of the solution, concentration and time of bath. For the creation of nano pin type structure from $CoCl_2$ and NH_2CO in water, Hosono et al. used borosilicate glass. In thermodynamic equilibrium condition, every metal compound deposited singly on the substrate and make a crystalline-like structure. Lauric acid was used to adjust the nano pin. Here is a figure 8 shows a structure of nano pin. By this method the maximum contact angle was formed till now that is 178 degree also, we can see that there are 166 needles structures in a $9\mu m^2$area and apex of pins are of 6.5 nanometers.[75].

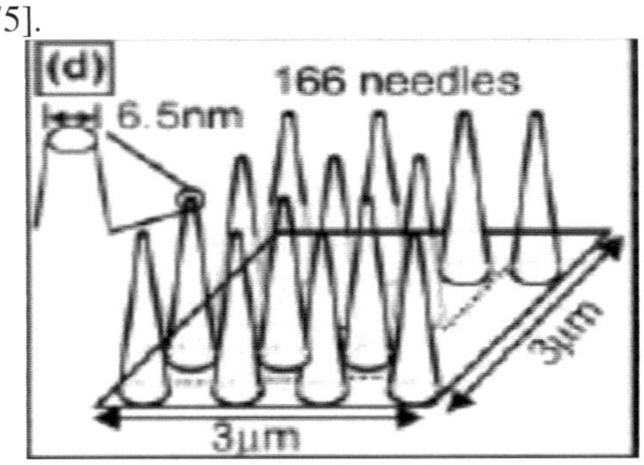

Figure 7. A nano pin structure formed by film-coated with lauric acid [73].

Hydroxide solution of a mixture of nickel sulphate, potassium persulfate and aqueous ammonia was used as a precursor for vapour bath method on Indium tin oxide coated glass. The precursor was deposited on the glass, firstly the colour was grey but after annealing, it became almost transparent and get adhere to the glass tightly. Even after annealing, the morphology was rarely changed, and it was found that this film has a porous structure [76].

Rezaei et al. used vapour deposition method to fabricate superhydrophobic layer that showed 160 degrees water contact angle by using Ammonium hydroxide (NH4OH), tetraethylorthosilicate (TEOS, P98%, Merck), vinyltrimethoxysilane (VTMS) and Ethanol on different substrates aluminium, glass and silicon slides. They also tried to fabricate the layer at minimum temperature that is 40 degrees [77].

4. Conclusion – In this review, an epitome of superhydrophobicity has been discussed with many examples. First of all, there is a discussion about natural superhydrophobicity in different natural things like flowers, insects etc. Then it covered the main aspect of wettability of surface and theoretical explanation of superhydrophobic property along with rough surface and low surface energy materials. An immense discussion of different kind of fabrication methods like electrodeposition method, electrospinning method, sol-gel method, a chemical etching method, chemical deposition methods were given. We know that there are very important and variety of applications of superhydrophobicity is there so, the properties as anti-icing, self-cleaning, anti-corrosion, anti-fogging etc. We have seen that there are many theoretical conflicts of superhydrophobicity but due to broad use of these coatings researchers and scientists showing a lot of interest in this field. With the help of micro and nanostructures, there can generate a large variety of surfaces. Now, still, we are very behind in making coatings on large areas and make it less costly so that it can be achievable to all. These coatings are also used in textile industries in fabrics to make waterproof cloths. There are some other uses of superhydrophobicity that is beyond normal superhydrophobicity like Leiden-Frost effect. In Leiden Frost effect we can see that water can go up in the opposite direction if we can create proper patterning in micro and nanostructures. There is still a lot to do in the field of superhydrophobicity so that it can be used in different fields because it can make a change in the working processes of different fields.

References-

[1] Woo Lee, Mi-Kyoung Jin, Won-Cheol Yoo, and Jin-Kyu Lee 2004 Nanostructuring of a Polymeric Substrate with Well-Defined Nanometer-Scale Topography and Tailored Surface Wettability *Langmuir* **20**, 7665-7669. doi:10.1021/la049411.

[2] AW Adamson, AP Gast. *Physical chemistry of surfaces*. A Wiley-Interscience Publication (1997).

[3] Neil J. Shirtcliffe, Glen McHale, Shaun Atherton, Michael I. 2010 Newton An introduction to superhydrophobicity. *Advances in Colloid and Interface Science* **161**(1-2),124–138. doi: 10.1016/j.cis.2009.11.001.

[4] Zen Yoshimitsu, Akira Nakajima, Toshiya Watanabe, Kazuhito Hashimoto 2002 Effects of Surface Structure on the Hydrophobicity and Sliding Behavior of Water Droplets. *Langmuir* **18**, 5818-5822 .

[5] Li, X.-M., Reinhoudt, D., & Crego-Calama, M. 2007 What do we need for a superhydrophobic surface? A review on the recent progress in the preparation of superhydrophobic surfaces. *Chemical Society Reviews* **36**(8), 1350. https://doi.org/10.1039/B602486F.

[6] Barati Darband, G., Aliofkhazraei, M., Khorsand, S., Sokhanvar, S., & Kaboli 2018 A Science and Engineering of Superhydrophobic Surfaces: Review of Corrosion Resistance, Chemical and Mechanical Stability. *Arabian Journal of Chemistry*.. doi: 10.1016/j.arabjc.2018.01.013

[7] K.Y. Law 2014 Definitions for Hydrophilicity, Hydrophobicity, and Superhydrophobicity: Getting the Basics Right. *J. Phys. Chem. Lett.* **5**, 686–688, dx.doi.org/10.1021/jz402762h

[8] Wang, Z., Shen, L., Qiu, M., Jiang, W., Chen, Y., & Zhao, J. 2019 Study on the properties of the superhydrophobic coating prepared by scanning electrodeposition on SLM substrate. *Materials Research Express* **6**(10), 106409. doi:10.1088/2053-1591/ab3682.

[9] Minglin Ma, Yu Mao, Malancha Gupta, Karen K. Gleason, and Gregory C. Rutledge 2006 Superhydrophobic surfaces. *Current Opinion in Colloid & Interface Science* **11**(4), 193–202. doi: 10.1016/j.cocis.2006.06.002

[10] Yang-Tse Cheng and Daniel E. Rodak. Is the lotus leaf superhydrophobic. *Appl. Phys. Lett.* **86**, 144101 (2005). https://doi.org/10.1063/1.1895487

[11] Manhui Sun, Chunxiong Luo, Luping Xu,Hang Ji,Qi Ouyang,Dapeng Yu,and Yong Chen 2005 Artificial Lotus Leaf by Nanocasting. *Langmuir* **21**, 8978-8981

[12] Hafiz Muhammad Ali, Muhammad Arslan Qasim, Sullahuddin Malik and Ghulam Murtaza 2018 Techniques for the Fabrication of Super-Hydrophobic Surfaces and Their Heat Transfer Applications., http://dx.doi.org/10.5772/intechopen.72820

[13] Phuong Nguyen-Tria,b., Hai Nguyen Trana, Claudiane Ouellet Plamondonc, Ludovic Tudurid, Dai-Viet N. Voe, Sonil Nandaf, Abhilasha Mishrag, Huan-Ping Chaoh,i, A.K. Bajpaij 2019 Recent progress in the preparation, properties and applications ofsuperhydrophobic nano-based coatings and surfaces: A review. *Progress in Organic Coatings* **132**, 235–256

[14] Minrui Ran, Wenyue Zheng & Huimin Wang 2019 Fabrication of superhydrophobic surfaces for corrosion protection: a review. *Materials Science and Technology*, DOI: 10.1080/02670836.2018.1560985

[15] YoungT. 1805 Anessayon the cohesionof fluids. PhilosTrans. R Soc London.; 95:65-87.

[16] Zhang, X., Shi, F., Niu, J., Jiang, Y., & Wang, Z. 2008 Superhydrophobic surfaces: from structural control to functional application. *J. Mater. Chem.* **18**(6), 621–633.

[17] Pantoja, M., Velasco, F., Abenojar, J., & Martinez, M. A. 2018 Development of superhydrophobic coatings on AISI 304 austenitic stainless steel with different surface pretreatments. *Thin Solid Films*, doi: 10.1016/j.tsf.2018.12.016

[18] Simpson, J. T., Hunter, S. R., & Aytug, T. 2015 Superhydrophobic materials and coatings: a review. *Reports on Progress in Physics* **78**(8), 086501. doi:10.1088/0034-4885/78/8/086501

[19] Hao, J.Y., Wang, Y.Y., Tong, X.L., Jin, G.Q., Guo, X.Y. 2012 Photocatalytic hydrogen production over modified SiC nanowires under visible light irradiation. *Int. J. Hydrogen Energy* **37** (20), 15038–15044. https://doi.org/10.1016/j.ijhydene.2012.08.021

[20] Oikawa, Y., Minami, T., Mayama, H., Tsujii, K., Fushimi, K., Aoki, Y., Habazaki, H. 2009 Preparation of self-organized porous anodic niobium oxide microcones and their surface wettability. *Acta Mater.* **57** (13), 3941–3946. https://doi.org/10.1016/j. actamat.2009.04.050.

[21] Cionti, C., Taroni, T., & Meroni, D. 2019 Bouncing Droplets: A Hands-On Activity To Demonstrate the Properties and Applications of Superhydrophobic Surface Coatings. *Journal of Chemical Education*, doi: 10.1021/acs.jchemed.9b00406

[22] Wenzel RN. 1936 Resistance of solid surfaces to wetting by water. *Ind Eng Chem.* **28**:988–994.

[23] Fihri A, Bovero E, Al-Shahrani A. 2017 Recent progress in superhydrophobic coatings used for steel protection: a review. *Colloids SurfA.* **520**,378–390.

[24] Cassie ABD, Baxter S. 1944 Wettability of porous surfaces. *Trans Faraday Soc.* **40**,546–551.

[25] Boinovich LB, Emelyanenko AM. 2008 Hydrophobic materials and coatings: principles of design, properties and applications. *Russ Chem Rev.* **77**:583–600.

[26] Chen, X., Yuan, J., Huang, J., Ren, K., Liu, Y., Lu, S., & Li, H. 2014 Large-scale fabrication of superhydrophobic polyurethane/nano-Al2O3 coatings by suspension flame spraying for anti-corrosion applications. *Applied Surface Science* **311**, 864–869. doi: 10.1016/j.apsusc.2014.05.186

[27] J.G. Buijnsters. 2013 Surface wettability of macroporous anodized aluminium oxide. *ACS Appl. Mater. Interfaces* **5** (8) 3224–3233.

[28] C. Liu, F. Su, J. Liang. 2014 Facile fabrication of a robust and corrosion resistant superhydrophobic aluminum alloy surface by a novel method. *RSC Adv.* **4** (98) 27,55556–55564.

[29] Darmanin, T., De Givenchy, E.T., Amigoni, S., Guittard, F. 2013 Superhydrophobic surfaces by electrochemical processes. *Adv. Mater.* **25** (10), 1378–1394.

[30] F. Su, K. Yao. 2014 Facile fabrication of superhydrophobic surface with excellent mechanical abrasion and corrosion resistance on copper substrate by a novel method, *ACS Appl. Mater. Interfaces* **6** (11) 8762–8770.

[31] F. Su. 2013 Rapid fabrication of corrosion resistant and superhydrophobic cobalt coating by a one-step electrodeposition, *J. Electrochem. Soc.* **160** (11) D593–D599.

[32] S. Khorsand, K. Raeissi, F. Ashrafizadeh 2014 Corrosion resistance and long-term durability of super-hydrophobic nickel film prepared by electrodeposition process, *Appl. Surf. Sci.* **305** 498–505.

[33] S. He, M. Zheng, L. Yao, X. Yuan, M. Li, L. Ma and W. Shen 2010 Preparation and Properties of ZnO Nanostructures by Electrochemical Anodization Method. *Applied Surface Science* Vol. 256, No. **8**, pp. 2557-2562. doi: 10.1016/j.apsusc.2009.10.104

[34] Y. Huang, D. K. Sarkar and X. G. Chen 2010 A One-Step Process to Engineer Superhydrophobic Copper Surfaces. *Materials Letters* **64** (24),2722-2724. doi: 10.1016/j.matlet.2010.09.010

[35] H. Meng, S. Wang, J. Xi, Z. Tang and L. Jiang 2008 Facile Means of Preparing Superamphiphobic Surfaces on Common Engineering Metals. *The Journal of Physical Chemistry C* **112** (30),11454-11458. doi:10.1021/jp803027w

[36] Ma, M., Mao, Y., Gupta, M., Gleason, K.K., Rutledge, G.C. 2005 Superhydrophobic fabrics produced by electrospinning and chemical vapor deposition. *Macromolecules* **38** (23), 9742–9748.

[37] Zhang, Y. Z., Venugopal, J., Huang, Z.-M., Lim, C. T., & Ramakrishna, S. 2005 Characterization of the Surface Biocompatibility of the Electrospun PCL-Collagen Nanofibers Using Fibroblasts. *Biomacromolecules* **6**(5), 2583–2589. doi:10.1021/bm050314k

[38] Taylor, G. 1964 Disintegration of water drops in an electric field. *Proc. R. Soc. Lond. Ser. A Math. Phys. Sci.* **280**(1382), 383–397.

[39] Mohammad Mirjalili, Salar Zohoori 2016 Review for application of electrospinning and electrospun nanofibers technology in the textile industry, *J Nanostruct Chem*, DOI 10.1007/s40097-016-0189-y

[40] Thostenson, E.T., Ren, Z., Chou, T.W. 2016 Advances in the science and technology of carbon nanotubes and their composites: a review. *Compos. Sci. Technol.* **1**(13), 1899–1912.

[41] Lau, A.K.T., Hui, D. 2002 The revolutionary creation of new advanced materials—carbon nanotube composites. *Compos. B Eng.* **33**(4), 263–277.

[42] Maitra, T., Sharma, S., Srivastava, A., Cho, Y.K., Madou, M., Sharma, A. 2012 Improved graphitization and electrical conductivity of suspended carbon nanofibers derived from carbon nanotube/polyacrylonitrile composites by directed electrospinning. *Carbon* **50**(5), 1753–1761.

[43] Jin, H.J., Fridrikh, S.V., Rutledge, G.C., Kaplan, D.L. 2002 ElectrospinningBombyx mori silk with poly (ethylene oxide). *Biomacromolecules* **3**(6), 1233–1239.

[44] Coffee, R.A. 1998 *A dispensing device and method for forming material.*, In. Google Patents,

[45] Ma, M., Mao, Y., Gupta, M., Gleason, K. K., & Rutledge, G. C. 2005 Superhydrophobic Fabrics Produced by Electrospinning and Chemical Vapor Deposition. *Macromolecules* **38**(23), 9742–9748. doi:10.1021/ma0511189

[46] Hashim, N.A., Liu, Y., Li, K. 2011 Preparation of PVDF hollow fiber membranes using SiO2 particles: the effect of acid and alkali treatment on the membrane performances. *Ind. Eng. Chem. Res.* **50** (5), 3035–3040.

[47] Mohamed, A.M., Jafari, R., Farzaneh, M. 2014 An optimization of superhydrophobic polyvinylidene fluoride/zinc oxide materials using Taguchi method. *Appl. Surf. Sci.* **288**, 229–237.

[48] Radwan, A.B., Mohamed, A.M., Abdullah, A.M., Al-Maadeed, M. A. 2015 Corrosion protection of electrospun PVDF-ZnO superhydrophobic coating. *Surf. Coat. Technol*, https://doi.org/10.1016/j.surfcoat.2015.12.087

[49] Shang, H., Wang, Y., Limmer, S., Chou, T., Takahashi, K., Cao, G. 2005 Optically transparent superhydrophobic silica-based films. *Thin Solid Films* **472** (1), 37–43.

[50] Tadanaga, K., Morinaga, J., Minami, T. 2000 Formation of superhydrophobic-superhydrophilic pattern on flowerlike alumina thin film by the sol-gel method. *J. Sol-Gel Sci. Technol.* **19** (1–3), 211–214.

[51] Brinker and Scherer 1990 *Sol-gel Science: The Physics and Chemistry of Sol-gel Processing*., Elsevier Inc.

[52] Xue, C.-H., Jia, S.-T., Zhang, J., & Ma, J.-Z. 2010 Large-area fabrication of superhydrophobic surfaces for practical applications: an overview. *Science and Technology of Advanced Materials* **11**(3),033002. doi:10.1088/1468-6996/11/3/033002

[53] Guo, Z., Zhou, F., Hao, J., & Liu, W. 2006 Effects of system parameters on making aluminium alloy lotus. *Journal of Colloid and Interface Science* **303**(1), 298–305.

[54] Hikita, M., Tanaka, K., Nakamura, T., Kajiyama, T., Takahara, A. 2005 Super-liquid-repellent surfaces prepared by colloidal silica nanoparticles covered with fluoroalkyl groups. *Langmuir* **21** (16), 7299–7302.

[55] Jitianu, A., Doyle, J., Amatucci, G., Klein, L.C. 2010 Methyl modified siloxane melting gels for hydrophobic films. *J. Sol-Gel Sci. Technol.* **53** (2), 272–279.

[56] Rao, A.V., Latthe, S.S., Mahadik, S.A., Kappenstein, C. 2011 Mechanically stable and corrosion-resistant superhydrophobic sol-gel coatings on the copper substrate. *Appl. Surf. Sci.* **257** (13), 5772– 5776.

[57] Mahadik, S. A., Kavale, M. S., Mukherjee, S. K., & Rao, A. V. 2010 Transparent Superhydrophobic silica coatings on glass by sol–gel method. *Applied Surface Science*, **257**(2), 333–339. doi: 10.1016/j.apsusc.2010.06.062.

[58] Lu, S., Chen, Y., Xu, W., & Liu, W. 2010 Controlled growth of superhydrophobic films by sol–gel method on aluminum substrate. *Applied Surface Science*, **256**(20), 6072–6075. doi: 10.1016/j.apsusc.2010.03.122

[59] Nimittrakoolchai, O.-U., & Supothina, S. 2008 Deposition of organic-based superhydrophobic films for anti-adhesion and self-cleaning applications. *Journal of the European Ceramic Society* 28(5), 947–952. doi: 10.1016/j.jeurceramsoc.2007.09.025

[60] R. Jafari, S. Asadollahi, M. Farzaneh 2013 Applications of plasma technology in the development of superhydrophobic surfaces. *Plasma Chem. Plasma Process* **33** (1) 177–200.

[61] N. Gupta, S. Sasikala, H.C. Barshilia 2012 Corrosion study of superhydrophobic magnesium alloy AZ31 surfaces prepared by the wet chemical etching process. *Nanosci. Nanotechnol. Lett.* **4** (8) 757–765.

[62] L. Liu, F. Xu, L. Ma. 2012 Facile fabrication of a superhydrophobic Cu surface via selective etching of high-energy facets. *J. Phys. Chem. C* **116** (35) 18722–18727.

[63] R. Jafari, L.F. Mobarakeh, M. Farzaneh. 2012 Water-repellency enhancement of nanostructuredplasma-polymerized HMDSO coatings using grey-based taguchi method, *Nanosci. Nanotechnol. Lett.* **4** (3)369–374.

[64] T. Kim, D. Tahk, H.H. Lee. 2009 Wettability-controllable super water-and moderately oil-repellent surface fabricated by wet chemical etching. *Langmuir* **25** (11) 6576–6579.

[65] L. Feng. 2014 Superhydrophobic aluminium alloy surface: fabrication, structure, and corrosion resistance. Colloids Surf. *A Physicochem. Eng. Asp.* **441**, 319–325.

[66] Subhash Latthe, S., Basavraj Gurav, A., Shridhar Maruti, C., & Shrikant Vhatkar, R. 2012 Recent Progress in Preparation of Superhydrophobic Surfaces: A Review. *Journal of Surface Engineered Materials and Advanced Technology* **02**(02), 76–94.

[67] Yung, C.S., Tomlin, N.A., Heuerman, K., Keller, M.W., White, M. G., Stephens, M., Lehman, J.H. 2018 Plasma modification of vertically aligned carbon nanotubes: Superhydrophobic surfaces with ultra-low reflectance. *Carbon* **127**, 195–201.47.

[68] Pan, L.N., Dong, H.R., Bi, P.Y. 2010 Facile preparation of superhydrophobic copper surface by HNO3 etching technique with the assistance of CTAB and ultrasonication. *Appl. Surf. Sci.* **257** (5), 1707–1711.

[69] Ryu, J., Kim, K., Park, J., Hwang, B. G., Ko, Y., Kim, H.,Lee, S. J. 2017 Nearly Perfect Durable Superhydrophobic Surfaces Fabricated by a Simple One-Step Plasma Treatment. *Scientific Reports*, **7**(1). doi:10.1038/s41598-017-02108-1

[70] Mertaniemi H, Laukkanen A, Teirfolk JE. 2012 Functionalized porous microparticles of nanofibrillated cellulose for biomimetic hierarchically structured superhydrophobic surfaces. *RSC Adv* **2**(7), 2882–2886.

[71] Mei Li, Jin Zhai, Huan Liu, Yanlin Song, Lei Jiang, and Daoben Zhu. 2003 Electrochemical Deposition of Conductive Superhydrophobic Zinc Oxide Thin Films, *J. Phys. Chem. B* **107**, 9954-9957.

[72] F. Wang, S. Arai, M. Endo. 2004 Electrochemical preparation and characterization of Nickel/Ultra-Dispersed PTFE composite films from aqueous solution, *Mater. Trans.* **45** (4) 1311–1316.

[73] S. Soundeswaran, O. Senthil Kumar, R. Dhanasekaran. 2004 Effect of ammonium sulphate on chemical bath deposition of CdS thin films. *Materials Letters* 58 2381– 2385.

[74] Hodes, G. 2007 Semiconductor and ceramic nanoparticle films deposited by chemical bath deposition. *Physical Chemistry Chemical Physics* **9**(18), 2181. doi:10.1039/b616684a

[75] Hosono, E., Fujihara, S., Honma, I., & Zhou, H. 2005 Superhydrophobic Perpendicular Nanopin Film by the Bottom-Up Process. *Journal of the American Chemical Society* **127**(39), 13458–13459. doi:10.1021/ja053745j

[76] X.H. Xiaa , J.P. Tua,, J. Zhanga , X.L. Wanga , W.K. Zhangb , H. Huangb. 2008 Electrochromic properties of porous NiO thin films prepared by a chemical bath deposition, *Solar Energy Materials & Solar Cells* **92** 628–633.

[77] Rezaei, S., Manoucheri, I., Moradian, R., & Pourabbas, B. 2014 One-step chemical vapor deposition and modification of silica nanoparticles at the lowest possible temperature and superhydrophobic surface fabrication. *Chemical Engineering Journal,* **252**, 11–16. doi: 10.1016/j.cej.2014.04.100

Review paper on removal of heavy metal ions from industrial waste water effluent

Jaya Mishra[1], R Saini[1] and D Singh[1]

[1]Departhment of Mechanical Engineering, ABES Engineering College, Ghaziabad, India

[1]Departhment of Mechanical Engineering, ABES Engineering College, Ghaziabad, India

[1]Departhment of Mechanical Engineering, ABES Engineering College, Ghaziabad, India

Abstract- Heavy metal ions is a major key of pollutions and it is one of the big environment issue today. To overcome with this problem the handling of metal ions is necessary. In recent research, there are various techniques that has been used to examin the metals ions and overview the methods also. The technologies involves are- ion-exchang, chemical precipitation, flotation, adsorption, membrane filtration, electro-chemical, coagulation-flocculation methods. It is find from experimental and literature survey article that membrane filtration, ion-exchange and adsorption are most common techniques used for the discourse of removal of meta .l ions from waste water effluent.

Keywords- Heavy metal ions, adsorption, waste water, removal, effluent.

INTRODUCTION- Environmental pollution has many forms like the water we use to drink , the air we use to breath , the ground where we cultivate our plants these are all contribute to health problems, and the major role plays heavy metals which are present in waste water . Atomic weight of heavy metal elements in between 63.8 to 200.5 and specific gravity is higher than 5.0 [1] . Here some of the common examples of metals including chromium (Cr), iron(Fe), copper(Cu), lead(Pb), zinc(Zn), mercury(Hg), nickle(Ni), cadmimum(Cd), copper(Cu), thallium(Tl), and arsenic(Ar),cobalt(Co). These heavy metals are characterize into major 3 groups: harmful metals (Pb, Hg, Zn, Cr, Ni, As, Cu, Sn, Cd, Co) useful metals (As, Pd Ag, Pt, Au) and radio metals (such as Ra, Th, U) [2]. Discharge from industry contains different organic and inorganic pollutants among these a huge amount of toxic metal ions are found and which are hafmful to human health and when it is discharge into aquatic area it also causes problem to flora and fauna [3] With the speedy development of industries like mining operations, paper industries, metal plating , fertilizer company, batteries, etc . these pollutants knowingly or unknowingly discharge into surrounding increasingly and as we know that heavy toxic metals are non- biodegradable & carcinogenic in nature. Heavy toxic metals are very harmful because this tends to bio-accumulate which sence that concentrations of chemical increase in an organism over time. Harmful metals is concern in examine of waste water effluent especially mercury, nickel, zinc, lead, copper [4]. For human health Zinc is essential but in very small quantity

Zinc is a trace element and is essential for physiological operations of alive tissue & also perform bio-chemical functions. Excess of zinc causes health issues like stomach cramps, anemia, skin diseases, nausea and vomating[5].

Cu (copper) perform important metabolic work in animals body but extreme intake of Cu brings a serious issues concern like cramps, convulsions, vomiting, sometimes it causes unproductive death also [6]. Ni (Nickel) increases its extreme level brings dangerous conditions of kidney & lung problems and also gastrointestinal distress, skin problem and pulmonary Fibrosis [7]. Hg (Mercury) is neurotoxin that distroy central nervous system. Excess amount of Hg affects the kidney function , impairment of pulmonary, chest pain and dyspnea[8] . A known environmental calamity with heavy metals is Minamata disease and which is due to Mercury(Hg) pollution in Japan.

Lead damages the Central Nervous System. Pb also affects liver, reproductive system and kidney and also brain function and cellular processes. Symptoms be like extreme headache, weakness of muscles, insomnia, anemia, dizziness, renal damage, irritation [9]. Chromium present in two states in aquatic environment that are Cr(III), Cr(VI). But we find that Cr(VI) is extreme dangerous to Cr(III). Cr(VI) damages human physiology and accumulate in the food and causes health issues and the symptoms are skin dermatitis and lung caecinoma [10].

TECHNOLOGIES ARE USED FOR TREATMENT OF HEAVY METAL IONS

1. Chemical Precipitation

This process is most efficient & common process used in companies [11] due to its simple techniques and inexpensiveness for operation. In Chemical precipitation – precipitates are formed when chemicals are combine with toxic metals ion. With the help of two process first one is filtration and second one is sedimentation the insoluble precipitates are separated. Hydroxide and sulfide chemical precipitation are the two process which comes in conventional precipitation.

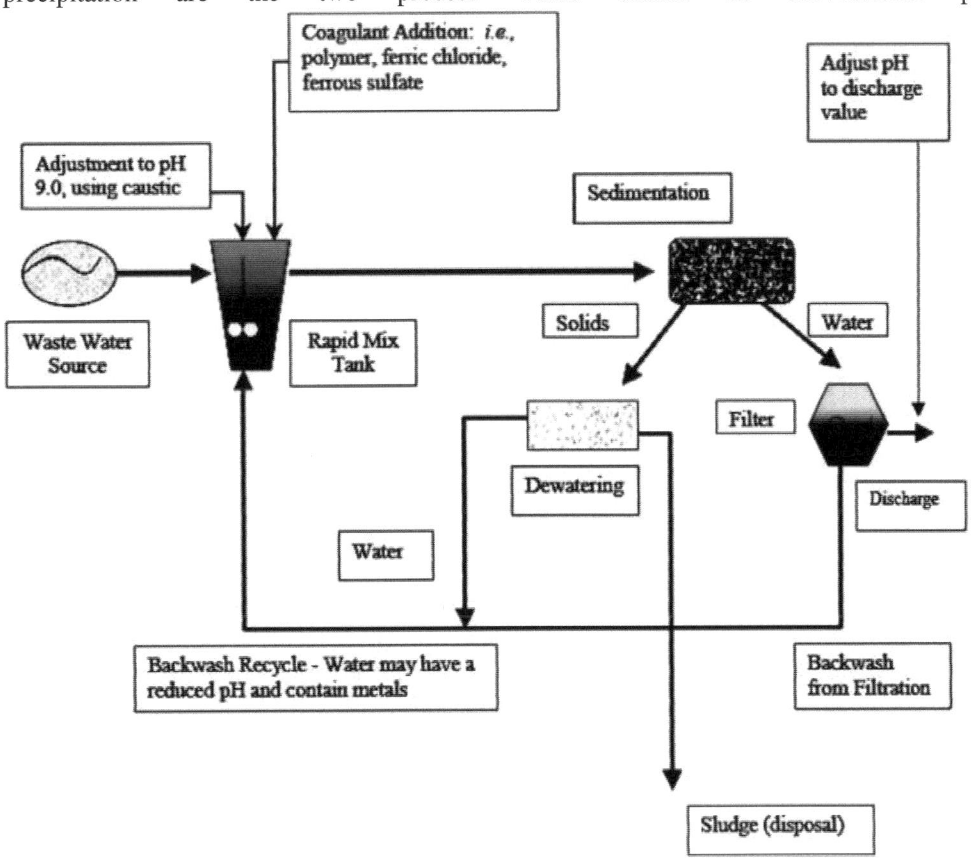

Figure 1. Removal of sludge from contaminated water by chemical precipitation.

1.1 Hydroxide chemical precipitation-The hydroxide precipitation is most common techniques used in chemical precipitation because of its control pH value and much low cost [12]. A different amount of hydroxides has used to find precipitation of toxic ions from contaminated water because of its low cost and simple handling, lime is best suitated choice for hydroxide chemical precipitation in companies [13].

Hydroxide chemical precipitation method takes Ca(OH)2 & NaOH for eliminating Cr(VI) & Cu(II) ions from contaminated water [14]. To increase the property of lime precipitation , the seed material we used is fly ash [15]. This carbonation treatment enhance the size of particle and gives better efficiency of toxic metals ion elimination. Although this process is large in use but it also has some major drawback first one is it creates disposal & dewatering problem because HCP gives high volume & small density muck [16] and the second thing is some hydroxide metals are amphoteric and compound metal cause some problems.

1.2 Sulfide precipitation-The most widely used method is Sulphide chemical precipitation , it is very effective for heavy metal ions treatment. The biggest advantage of this process is that its solubilities is lower than HCP process and it is not amphoteric in nature. So that's why it removal efficiency of heavy metal ions is much more than hydroxide chemical precipitation. Ozverdi and Erdem (2006) examine for removal of Cd2+, Pb2+ and Cu2+ pyrite sulfide iron is used. Chemical precipitation exist when its pH is smaller than 3.0 because of H2S formation (Equ (a) and (b)) and adsorption process occurs when pH varies from range(3.1- 6.0)

>[17] Dr. Mahmood M. Brbootl et al,.used hydroxide precipitation for the removal of Iron (III), chromium (III), copper (II), Lead(II), Cadmium(II) and Nickel (II) from the aqu solution with the help of jar tester magnesia used as a precipitator.

>[18] The pH value range is 9.6 – 10 with MgO precipitant

The optimum value of MgO precipitant is 1.6 – 3.0 g/l .

2.0 Ion- exchange

Considering too much advantage second method is ion-exchange . This is widely used method for the treatment of industrial wastewater effluent. These advantages are its removal efficiency, treatment capacity is high, and high kinetics[19]. Resin of ion-exchange method , either natural solid or synthetic resin. It possesses some special capacity to replace its cations to the metal ions in the contaminated water. Synthetic resin is best used materials among all the material take part in ion-exchange method because it is very effective to eliminate the toxic metal ions from effluent [20].

The cation exchanger are acidic group of sulfonic(-S03H) with stronge acidic group and acidic group of carboxyl (-COOH) with week acidic resin. Heavy metals present in the solution are go through the cations col where heavy metal-ions are replaces for the H+ on the resins with below reaction process:

$$nR - SO3H + M^{a} \longrightarrow (R- SO3^{-})n\ M^{a} + nH+ \qquad (a)$$

$$nR- COOH + M^{a} \longrightarrow (R- COOH^{-})n\ M^{a} + nH+ \qquad (b)$$

Lots of researcher find that Zeolite possesses stronge cation replace capability for toxic ions under variant circumstances [21]. In recent work studies some researcher also find that for better exchange capability of clinoptilolite the surface of it surrounded with amorphous iron-oxide [22]. Research studies shows that for eliminating metal ions an ion exchange resin is used as zeolities and montmorillonites. But as compared to synthetic resins they are small in stock.

>[23] Alyuz and Veli initially well defined a most efficient techniques for the removal of unwanted metal ions from the contaminated water and that is Ion-exchange method with a better efficiency.

>[24] Zagorodni describes the ion-exchange method for the elimination of heavy metal ions with the use of inorganic ion-exchanger material that is zeolite

>[25] Rathor et al,. describes metal sulphide used as ion-exchanger for the better efficiency in removing heavy metal ions from the waste water effluent.

>[26] Fathima and Pandith et al describes hybrid techniques in which organic and inorganic are hybrid for high potential of removing heavy metal ions from the waste water effluent.

3.0 Adsorption

Considering economic technique for removal of heavy metal ions we go through the adsorption process , it is very effective method for treatment of wastewater effluent. The use of this technique is very flexible in nature. Sometimes adsorption is a reversible process, and for the desorptions process the adsorbent can be re-generated and here there are several adsorbents present for removal of heavy metal ions

3.1 Activated carbon adsorbents- The most common adsorbent used for removal of heavy metal ions is activated carbon adsorbents (ACA). It contains large volume of mesopore and micropore which gives large surface area this is the main advantage of this adsorbent. Many researchers find that activated carbon is the best adsorbent for eliminating toxic metal ions [27]. Due to exhaustion of coal the price of AC is increased so without affecting this factor we will use activated carbon composite and additives in place of ACA, Tannic acid[28] , surfactants [29] , additives of alginate [30] , magnesium [31] and most important activated carbon composite can be effectively absorbents for heavy metal contaminants. Due to expensive source of AC adsorbent there is conversion of carbonaceous material into activated carbon for heavy metal removal.

3.2 Carbon nanotube adsorbents- In 1991 Iijima discovered CNT (carbon nanotube). He describes the properties of CNTs and also give basic idea of its application. Carbon nanotube gives the best results for eliminating toxic metal ions like cadmium [32] , copper [33] , lead [34] , chromium [35] , nickel [36] from the contaminated water. And carbon nanotube adsorption gives the excellent result of removal of these heavy metals. Carbon nanotubes are characterizied into two groups (a) SWCNT single walled carbon nanotubes adsorbent) & (b) MWCNT (multi walled carbon nanotubes adsorbent [37]. The procedure of carbon nanotube adsorbent are very complex in nature so the mechanism go through the chemical-interaction , electrostatic-attraction and sorption precipitation in-between the suface of functional-group and the heavy metal ions of the carbon nanotubes adsorbents [38] Due to low capacity of sorption of metals it oxidized through KMnO4, HNO3 and NaClO solution to increase its capacities of removing toxic metal ions [39].

3.3 Low cost adsorbents- From the research it is found that insteed of being expensiveness of activated carbon it is widely used adsorbent for removing metal ions. So, the researcher continue studied to find the low cost adsorbent . More than hundreads of articles has been published for the low cost adsorbent and they are by-product of industrial waste , natural material and agricultural-waste for the treatment of toxic metal ions. In 2008 Gupta & Bhattachar survey the two kaolinite and montmorillonite adsorbent for eliminating the toxic metal ions . Also in 2008 Sud et al. survey that agricultural by product waste is used as adsorbent for treatment of heavy metal ions through aqueous solutions. In 2008 Wan Ngah & Hanafiah survey that chemically plant waste used as adsorbent for eliminating of toxic metal ions. Many researchers examine that by product of industrial waste like diatomite (40) lignite (41) , natural-zeolite (42) , lignin (43) , clino pyrrhotite (44) , peat (45) , aragonite- shells (46) , kaolinite (47) . In 2010 a researcher Jiang examine that from Longyan we find the kaolinite-clay and it is used for elimination of toxic metals from the contaminated water and the time is approx. 29-30 minutes was observed. In 2009 the two researcher Navia & Agoubordea investigated that sediments & saw-dust and sometimes the mix of both material used as adsorbent for the eliminations of Zn & Cu from the aqu soln. The adsorption capacities are 4.84 , 2.57 & 5.58 mg per grm for Zn and for Cu it is 4.70 , 2.30 , 4.32 mg per gram .

3.4 Bio-adsorbents- The very new adsorbent is discovered for removal of toxic metals from aqua solu is bioadsorbent and the process is known as bioadsorption. This adsorbent is inexpensive in comparion to other adsorbents and it is highly effective for removal of metal ions these are the advantages of this adsorbent. It is best suitable adsorbent for the treatment of contaminated water. Bio-adsorbent are basically divided into three groups (48) they are ; (a) microbial biomass { such as yeast , bacteria , fungi } , (b) non-living biomass { such as squid , shrimp , bark , krill , lignin , crab shell etc } and (c) alga biomass. There are different in-expensive bioadsorbents are sawdust [49] , egg-shell [50], coffee-husk[51] , citrus peel [52] , potato peel [53] , black gram husk [54] , seed shells [55] and sugar beet pectin gels [56] etc for removal of heavy toxic metal ions. Algae is a renewable biomass which present in large amount in the earth zone which is used naturally for the treatment of waste water effluent. This low cost adsorbent possesses a lot of advantages like it is widely available in nature secondly it is low cost in nature and have high metal ions sorption capability [57]. Removal of heavy metal ions through bacteria from aqua solutions are Pseudomonas aeruginosa [58] and Escherichia coil [59] etc. Fungi & Yeast produces large biomass and they are easy to grow, Fungi bio-adsorbents are saccharomyces cerevisiae [60] , Rhizopus arrhizus [61] , Lentinus edodes [62], Lentinus edodes [63] . Renu , Madhu Agarwal et al., describes that for the removal of toxic metal ions the best suitable method is adsorption and the charactertised it into two different parts that is (1) commercial adsorbent (2) bioadsorbent for removing copper , chromium , and cadmium .

>[64] In 2011 Fu and Wang discovered the most efficient techniques used for removing heavy metal ions that is adsorbtion.

>[65] Gopalakrishnan et al., in 2015 describes the quality of nanomaterial which possesses the functional group , surface area is high and enhance active site due to these qualities the removing of heavy and unwanted metals ions can remove easily.

>[66] Renu et al ., in 2017 describes adsorbtion techniques for the removal of toxic metal ions with the use of commercial adsorbent and bio-adsorbent wich increase the removing capacity .

>[67] Sabry M. Shaheen et al . discover that with the help of zeolite in adsorption techniques the elimination of heavy metal ions will take place with better efficiency .

>[68] K.Singh et al., found that adsorbent like composite of carbon nanotubes , alumina , grapheme sand composite , eggshell and many more are used to eliminate the toxic metal ions like copper , chromium , cadmium from the waste water effluent.

More Case Study Related To Adsorptions

[69] Laxmipriya Panda et al., do excellent work towards removal of heavy metal ions with the use of waste ash from the industry and utilizes in the elimination of cd2+ , Ni2+ , Co2+ , Pb2+. And that ash is dolochar which generates the geopolymer due to its volume and size enhancement it gives total 99% removal by archiving temperature = 343.3 K , ph = 7.9 and ion concentration = 9.9- 10 ppm

[70] Bao-Lin Zhang gives knowledge about two adsorbent that is MoFs MIL-100(Fe) and second is Fe-BTc which works on removal of cd and pb . The performance rate of MoFs is higher than Fe-BTc with ph change from 2-7 . These two adsorbent removes pb more than 99.5% and cd approx 99.2%. The noticeable thing is that here endothermic phenomena take place that is spontaneously change in entropy.

[71] Researcher review here for the removal of Cr(VI) metal ions because of its high toxicity with the use of activated carbon such as Bentonite which plays excellent work towards removal of Cr(VI) and gives result like equilibrium time rate was 119-120 min with temperature rate is 35-40 degree celcius and ph is 2 . Here concentration of chromium (VI) was found 200 mg/L.

4.0 Membrane filtration

Membrance filtration method have different types of membrance for treatment of waste water effluent . It is easy to operate with space saving and also possess high efficiency of metal removal. The membrane which used to eliminate ions from the contaminated water are reverse osmosis , electrodialysis , ultra-filtration , nano-filtration.

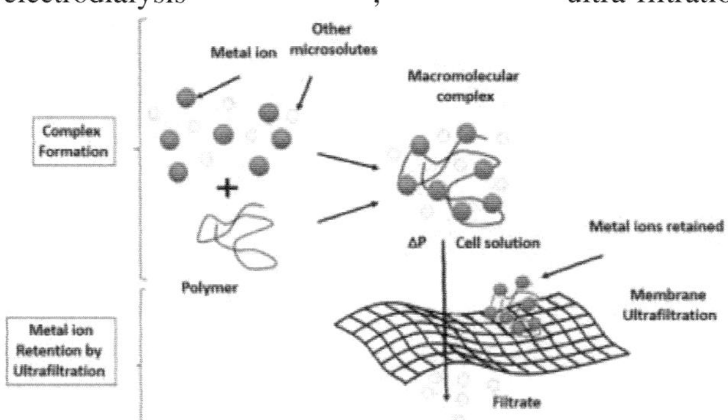

Figure 2. Filtration of contaminated water through membrane filtration

4.1 Electrodialysis- ED is the type of membrane filtration for the ions separation from solution by electrice discharge through solu . Ion exchange membrane is used in electrodialysis process . The two exchanger works here one is anion exchanger and second one is cation exchanger . For the purpose of drinking water and process-water from the sea-water and briny water electro-dialysis is widely used. Electrodialysis gives good result for sepeartion of useful metal ions from the waste effluent and sea water [72] . Electrodialysis also gives the effectively resul for the treatment of toxic metal ions removal. In 2007 the researcher Nataraj gives a working model for elimination of hexa-valent Cr ions by electrodialysis method using ion exchanger membrane , and the outcome gives the satisfactory result. In 2009 the researcher Cifuentes discover that electrodialysis is very effective for the elimination of copper and iron from the aqua solution . From a lot of experience were performed and then after researcher found that for improving the performance of cell the temperature and voltage will be increased and it is also find that with increase of flow rate the separation percentage will be decreased.

4.2 Reverse Osmosis -In this process the solution which is being purified will pass through the semi permeable membrane via eliminating the unwanted materials. Reverse osmosis is useful techniques for eliminates a large quantity of dissolved contaminats from water. 20 percent of universe desalination capacity is performed by reverse osmosis[73].

In 2009 the two researcher Dialynas & Diamadopoulos studied that efficiencies of elimination metal ions will increase when reverse osmosis is combine with bio-reactor membrance. But it has some disadvanges also like it consume high power because of pressure of pumping and second thing is its membrane restoration.

4.3 Nanofiltration -Between Reverse Osmosis and Ultra-Filteration the in-between process happens and that we known as NF (Nano-filteration) . Nano-filteration is a worthfull techniques for the elimination of toxic metal ions like Cr[74] , As [75] , Ni [76] , and Co [77] from contaminated water . Nanofilteration possess a lot of advantages like it consumes less power supply , its efficiency of eliminating toxic metal ions is comparatively high with other filteration process, it is easy yo operate and more reliable in nature [78].

In 2010 a researcher Figoli reviewed that with the help of 2-commercial nano-filtration membranes { NF90 & N30F }the elimination of pentavalent As from the synthetic water occurs. And at the same time they found that pH is increases while temperature of operation is decrease gradually. Murthy & Chaudhari takes a year to examine the elimination process of toxic ions with the help of nano-filtration method and they reported the use of polyamide nano-filteration membrane for the removal of Ni ions from aqua solution of contaminated water [79] . and they also observe that the max removal of Ni ions is 97.9% and 92.0 % for feed-concentration of 6 and 249 mg per L respectively and they also research on Cd and Ni rejection capability using nano-filtration membrane for removal of toxic heavy metal ions [80]. The outcome for the Cd and Ni rejection are 98.93 % and 82. 68% respectively for feed-concentration of 5.0 mg per L.

4.4 Ultrafiltration- For the rejection of colloidal and dissolved material from aqua solution a membrane techniques work on low trans-membrane pressure and that is known as UF (ultra-filtration). For obtaining high efficiency of metal removal the MEUF (micellar en-hanced ultra-filtration) and second one is PEUF (polymer en-hanced ultra-filtration) was come.

In 1980s a researcher Scamehorn et al., was initially discovered Micellar enhanced ultrafiltration for the treatment of dissolved and multi-valent toxic metal ions from aqua solution [81] . Efficiency of metal rejection through micellar enhanced ultrafiltration totally dependent on the following parametears like concentration of metals to surfactants , pH of solu , ionic-strength and terms synchronize to membrane operation .

>[82] Hani Abu-Qudais use the part of membrance filtration that is reverse osmosis and nano-filtration for the removal of cadmium and copper from the waste water effluent.

>[83] Blocher , Dorda J et al., describes the hybrid techniques of separation of heavy metal ions from the contaminated water.

5.0 Flotation

Extensively used method for the removal of waste water treatment is flotation.

For the separation of toxic ions from the aqua solution with the help of attachment of bubble , mineral originated process we use flotation techniques . The main process of flotation are dissolved air-flotation , ion-flotation and precipitation are used for the treatment for waste water effluent.

Dissolved air flotation permit to bubbles of air to stuck on suspended-particle in the aqua solution , making a thin film of low density than solution where the toxic metal ions eliminate as a sludge formation [84] . dissolved air flotation proves that for the removal of ions it is the widest used techniques in 1990s [85].

Ion- flotation method is also very effective for the removal of metal ions from the contaminated water. Ion flotation process totally depend on im-parting metal ions in contaminated hydrophobic with the help of surfactants and through air bubbles these ionic-metals eliminates from the contaminated solutions [86].

In 2008 the researcher Yuan et al., examined the capacity of ion-flotation to eliminate the cd , cu and pb from the dilute aqua soltn from the bio-derived tea (saponin) . The max elimination of cu2+ , pb2+ and cd2+ are 81.12 % , 89. 94% and 71.16% respectively.

Another process that is used for the removal of metal ions from the waste water effluent is Precipitate flotation it is type of flotation process with the help of air bubbles in the formation of precipitate. Under the observation of metal soltn concentration the formation of metal hydro-oxide occurs due to precipitation formation [87]. Elimination of Cr (III) from the dil aqua solution is carried because of precipitation flotation process and here ethanol is used as an-ionic collector [88]. And the excellent result come with accuracy of ph 7.99 and max removal of metal ions is 96.1%.

>[89] Mavrov V et al., discover the hybrid of flotation process for the removing of cadmium and copper toxic metal ions from the waste water effluent.

>[90] Blocher et al. in 2003 has also given the hybrid theory for removal process in which he combine the flotation and membrane filtration for the removal of zinc , copper , nickel from the waste water effluent .

>[91] In 2002 Rubio et al., used collector such as SOS and MIBC for the removal of cadmium and it gives result from 89.3% to 97.6% .

6.0 Electrochemical Treatment

In this method on a cathode surface the metallic ionc are spread and by this procedure useful metals are recover in elemental state. But it has major disadvantages like it is very expensive techniques due to large electric consumptions. So, this process is not more in use but from last 2 decades it is widely used because of its modify basic properties [92]. And here we studied that different form of this technologies such as electroflotation (EF) , electro-coagulation(EC) and electro-deposition (ED).

EC takes electrodes of AL or Fe to generate coagulate by mixing electrically either Al or Fe [93]. At the anode side there is generation of ions of metal take place and at the cathode side there is formation of hydrogen ion take place and these hydrogen gas (ions) help to flash out the flocculated particles from the aqua [94] . In 2008 the two researcher Heidmann and Calmano studied and examine the removing capability of Zn2+ , Ni2+ , Cu2+ , Ag+ with the help of electro-coagulation . For the examination of Hg2+ concentration of synthetic solution the electro-flotation method is alsoused [95]. The removing efficiency is greater than 99.7% when the gap between the electrode is nearly 3 cm and the current density varies from 2.6 to 3.123 Adm-2 and pH value of mercury ions varies from 3.1 to varies accordingly

>[96] David Hasson et al., describes the removal techniques of phosphate with the use of electrochemical treatment is also described calcium and magnesium hardness. Electrochemical techniques is known for its high potential for the elimination of hard metal from the industrial wastewater.

>[97] Chaloempan Petsriprasit et al., describes electrochemical process with lots of advantage and disadvantage , it is the best suitable process for the removal of heavy metal ions from the industrial waste water effluent .

7.0 Electroflotation

EF (electro-flotation) is a process of liquid/solid separation in which pollutant comes at the surface of water solution in the form of gas bubbles of oxygen and hydrogen. Electro-flotation gives the good result in the removal process of toxic metal ions .The researcher survey the electro-flotation techniques with the help of Al electrode for the waste water treatment [98]. Using this method the metal ions like Ni , Zn , Fe , Co , pb can easily removes from waste water. And it gives 99.9% removal rate which is actually a excellent result of removal. In electro-flotation process with the use of hybrid electrode of Fe and with the use of micro- filtration , filter-paper and ultra-filtration test operate for increasing efficiency of Ni with or without use of any external source of oxygen [99].

Electro-deosition is a method which is used for the recovery of useful metal ions from contaminated water and it is totally different process. In this method there is no extra residue added for the separation of metal ions so, researcher say that it is very clean process for metal recovery [100]. The researcher found that ED is very useful technique for the recovery of useful metal ions from the waste water effluent [101].

8.0 Coagulation & Flocculation

Involving two process sedimentation & filtration to eliminate metal ions from contaminated water and the whole process is known as Coagulation and Flocculation method. The main work of coagulation is disturbing the colloid particles by making the forces neutral which keep them apart. There are example of some coagulants like ferrous sulphate , aluminium and ferric chloride which gives the excellent work of removing impurities and metal ions from waste water effluent and also form the precipitate of amorphous metal of hydroxide. In a research it is found that when coagulants like poly-aluminium chloride and ferric chloride mix with impure water of industries then this coagulants removes the impurities and toxic metal ions from the effluent [102].

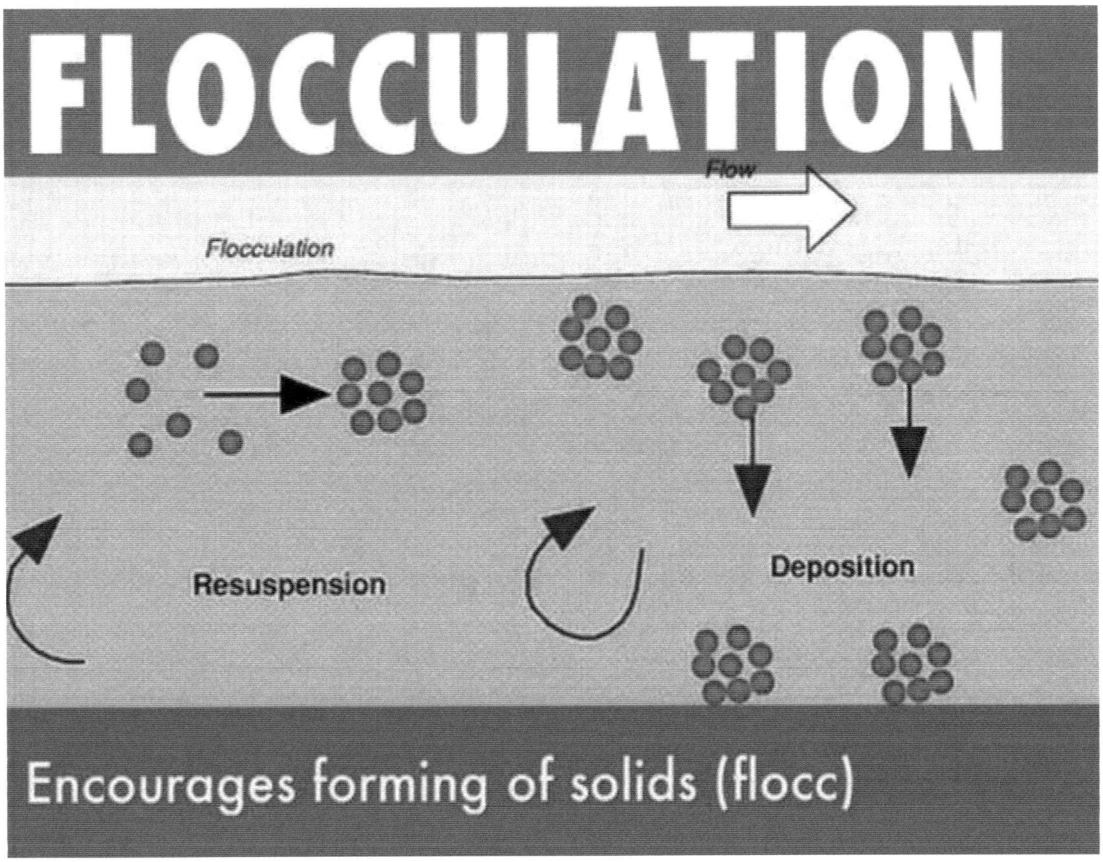

Figure 3. Removal of heavy metal ions from waste water effluent through flocculation

Coagulation is very effective process for the removal of metal ions from the contaminated sources , suspended particle and hydrophobic collides are the major two objective of the coagulations. Sodium xanthogenate group works effectively in the eliminations of metal ions from waste water effluent [103] . The turbidity decreases and Ni2+ removal increases when the pH of water sample is increases.

Flocculation is a process in which we add attorney and due to this impurities which are present in aqua solution comes out in the form of flocs. When the suspended-particles are flocculated into big particle then the elimination of these large particle can be easily performed by a various techniques like flotation , filtration , straining. There are a lot of flocculants are available and they are macromolecule , polyaluminium chloride , poly-acrylamide , poly-ferric sulphate are most common used for the elimination of toxic metal ions from the waste water effluent. A research is performed in which a new macro-molecule flocculent (mercaptoacetyl-chitosan) is invented through which not only turbidity remove but also metal ions also remove when this chitosan is react with mercaptoacetic acid [104]. The flocculants are formed to remove metal ions from contaminated water and they are ; N-carboxy-ethylated Chitosns two new { derivate of poly- ampholyte-chitosan} [105] and the second one is co-sodium xanthate { derivate of poly-acrylamide} [106]. A new method flocculation techniques comes in which without any additional process like filtration and centrifugation to seprate the metal ions from effluent with the use of thermosenstive polymer [107]. And a new tannic type flocculant has been discovered for the elimination of metal ions [108].

From a research it is find that heavy metal ions are not completely eliminate from the wastewater through coagulation and flocculation method [109]. So, there is need of some other techniques to removes metal ions completely and that can be precipitation , flocculation and coagulation process to eliminate tungsten from the industrial waste water using ferric chloride as a agent [110]. With the use of aluminium composite (micro-alloyed) for the spontaneous reduction process take place and it also give the good result for removal of metal ions [111]

Summary of treatment of heavy metal ions removal

As we can see at the above techniques there is many different process which is used for removal of useless metal ions and impurities with many advantages and contain limitations also.

When the removal of metal ions from aqua solution then we go through the chemical precipitation process because of its easy operation and also low cost but this process is useful only for high concentration waste water and it is not effective for low concentration of waste water. Due to non-economical property of chemical precipitation it generates large amount of sludge which causes difficulties in metal removal process.

Ion-exchange is another techniques which is also used for the removal of impurities which is present in waste water but it can not properly work when the water concentration is low . It is expensive in nature and also cause pollution because of additional of resins so it is not used in large scale purpose.

Adsorption is widely used method for the removal of toxic metal ions from industrial waste water effluent. There are many varities of low cost adsorbent present in market through which removal techniques may be performed in better way. It is very important point that what type of adsorbent is used for the removal process because the efficiency of removal process is totally depend upon the type of adsorbent is used. In adsorption method there is new type of adsorbent is introduce and that is bio-adsorbent.

Membrane-filtration is useful for high efficiency removal of metal ions but it consist of high cost, complexity in operation and sometimes generate a large amount of flux so their use is in limitated for removal process of metal ions.

Electro-chemical process take forward with the use of chemicals due to this agent it working properties improves and it does not produce sludges . It is very useful for the removal of heavy metal ions but it is not well developed technologies because it capital cost is high and consumes a large power capacity.

Flotation is also used for the elimination of toxic metal ions from the waste water effluent but it consisa lot of advantages like low detention time , efficiency of removal is high comparative to other techniques , metal sensitivity is high , overflow rate is high and operating cost is low but the more concentrated sludge is formed [112]. And also consist some disadvantage like high initial cost, maintance cost is also high and operational cost is also high. Go through the coagulation-flocculation process for the removal of heavy metal ions the sludge produced in this process is good sludge and is useful for the elimination of metal ions using chemical consputions. Although all the process is used for the removal of metal ions from the contaminated water. The selection of process depends upon the following parameters like first check what is the concentration of solutions, what is the initial cost of method and what is the capital-investment of the process some more parameters like flexibility and reliability of the plant and the last one which is more important to know what is the environmental impact and considering all these things then after a suitable technique is used for the removal of toxic metal ions [113].

Conclusions

As we can see that day by day there is increase in environmental pollution and one of the reason is heavy metal ions polloution of waste water effluent. To overcome with these environment problem we adopt removal techniques like membrane filtration , ion-exchange , chemical precipitation , adsorption , flotation , coagulation-flocculation for the elimination of toxic metal ions from the contaminated water sources. But it is observe that adsorption process is the best suitable method among all the above techniques and it is examine under 180-200 articles survey. Adsorbent of low cost and bio-adsorbent is also very useful for the removal of heavy metal ions , adsorption is very effective and less expensive as compare to other process, it is more economical method for low concentration of treatment of waste water. Adsorption process gives 98-99.8% removal rate of heavy metal ions with maintain ph value and appropriate temperature rate.

References

[1] AnasRaklami Abdel-ilahTahiri NouraBechtaoui El GharmaliAbdelhay 2021 Restoring the plant productivity of heavy metal-contaminated soil using phosphate sludge, marble waste, and beneficial microorganisms .Journal of Environmental SciencesVolume 99

[2] V.Balaram 2020 Microwave plasma atomic emission spectrometry (MP-AES) and its applications – A critical review .Microchemical Journal Volume 159, 105483

[3] Mehvish Hameed,Zulaykha Khurshid Dijoo,Rouf Ahmad Bhat,Irteza Qayoom .Concern and Threats of Heavy Metals' Contamination on Aquatic Ecosystem .Bioremediation and Biotechnology, Vol 4 pp 1-19

[4] Summaira Sharaf, Muti-ur-Rehman Khan, Asim Aslam, Masood Rabbani, Aisha Sharf, Misbah Ijaz, Ahsan Anjum & Naqi Hussain .2020 Environmental Science and Pollution Research (2020)

[5] Amjad Ali Maitlo, Wahid Bux Jatoi, Almas Fatima Memon, Amjad Hussain Soomro & Muhammad, Siddique Bhayo 2020 .Biological Trace Element Research (2020) Published: 15 August 2020.Assessment of Zinc, Lead, Chromium, and Cobalt in Commonly Consumed Herbal Medicines in Sindh,

[6] Abdelghany Hefnawy Abdelghany, Heba mohamed Elkhaiat 2015 The importance of copper and the effects of its deficiency and toxicity in animal health , January 2015 ,International Journal of Livestock Research DOI: 10.5455/ijlr.20151213101704

[7] Giuseppe Genchi,[1] Alessia Carocci,[2,*] Graziantonio Lauria,[1] Maria Stefania Sinicropi,[1] 2020 Human Health and Environmental Toxicology , Int J Environ Res Public Health. 2020 Feb; 17(3): 679. Published online 2020 Jan 21. doi: 10.3390/ijerph17030679

[8] A.N.TriantafillouMD ,G.A.PattersonMDM.S.PohlRN ,P.A.DeloneyRN ,R.S.SundaresanMD C.L.RoperMD. 1995 Bilateral pneumectomy (volume reduction) for chronic obstructive pulmonary disease .The Journal of Thoracic and Cardiovascular Surgery Volume 109, Issue 1, January 1995, Pages 106-119

[9] Jayanta Kumar Biswas , Mahendra Rai , Monojit Mondal , Avinash P. Ingle 2018 The Flop Side of Using Heavy Metal(oids)s in the Traditional Medicine: Toxic Insults and Injury to Human Health.Biomedical Applications of Metals pp 257-276

[10] Qiao YiChen[1]AnthonyMurphy[1]HongSunMaxCosta . Molecular and epigenetic mechanisms of Cr(VI)-induced carcinogenesis 2019 Toxicology and Applied PharmacologyVolume 377, 15 August 2019, 114636

[11] HaimingHuangJiahuiLiuPengZhangDingdingZhangFamingGao. 2017 Investigation on the simultaneous removal of fluoride, ammonia nitrogen and phosphate from semiconductor wastewater using chemical precipitation .Chemical Engineering JournalVolume 307, 1 January 2017, Pages 696-706

[12] QuanyuanChen·YuanYao·XinyingLi·JunLu·JuanZhou ,ZhaoluHuang 2018 Comparison of heavy metal removals from aqueous solutions by chemical precipitation and characteristics of precipitates. Journal of Water Process EngineeringVolume 26, December 2018, Pages 289-300

[13] W. Ashane M.Fernando· I.M.S.K.Ilankoon·Tauqir H.Syed·MohanYellishetty 2018 Challenges and opportunities in the removal of sulphate ions in contaminated mine water: A review. Minerals EngineeringVolume 117, March 2018, Pages 74-90

[14] QuanyuanChen·YuanYao·XinyingLi·JunLuJuanZhou·ZhaoluHuang 2018 Comparison of heavy metal removals from aqueous solutions by chemical precipitation and characteristics of precipitates Volume 26, December 2018, Pages 289-300

[15] WeiZhou ,XiangLu ,ChongchongQi ,MengYang 2020 Utilisation of ultrasonic treatment to improve the soil amelioration property of coal fly ash .Journal of Environmental ManagementVolume 276, 111311

[16] Zhiquan Huang ,Lila R. Dahal,Sylvain Marsillac,Nikolas J. Podraza 2020 Real Time and Mapping Spectroscopic Ellipsometry of Hydrogenated Amorphous and Nanocrystalline Si Solar Cells . Spectroscopic Ellipsometry for Photovoltaics pp 255-315

[17] EbrahimTilahun· ErkanSahinkaya , BarişÇalli 2018 A hybrid membrane gas absorption and bio-oxidation process for the removal of hydrogen sulfide from biogas. International Biodeterioration & BiodegradationVolume 127, Pages 69-76

[18] XuChen Yong SikOk· DineshMohan· Charles U.PittmanJr.· XiaominDou 2017 The stability and removal of water-dispersed CdSe/CdS core-shell quantum dots from water. Chemosphere Volume 185, Pages 926-933

[19] XuhuiLi· WeigangZhu· GengjianMeng· ChaoshengZhang , RuichaoGuo 2020 Efficiency and kinetics of conventional pollutants and tetracyclines removal in integrated vertical-flow constructed wetlands enhanced by aeration. Journal of Environmental ManagementVolume 273, 111120

[20] Hao Peng , Jing Guo 2020 Removal of chromium from wastewater by membrane filtration, chemical precipitation, ion exchange, adsorption electrocoagulation, electrochemical reduction, electrodialysis, electrodeionization, photocatalysis and nanotechnology: a review Published: 23 July 2020

[21] MeiHong· LingyunYu· YandingWang· JianZhang· ZhuwenChen· LeiDong· QijieZan· RuiliLi 2019 . Heavy metal adsorption with zeolites: The role of hierarchical pore architecture. Chemical Engineering JournalVolume 359, 1 March 2019, Pages 363-372

[22] SurasesWiriyathamcharoen· SudiptaSarkar· PijitJiemvarangkul· Trung ThanhNguyen· WantanaKlysubun· SurapolPadungthon 2020 Synthesis optimization of hybrid anion exchanger containing triethylamine functional groups and hydrated Fe(III) oxide nanoparticles for simultaneous nitrate and phosphate removal. Chemical Engineering JournalVolume 381, 1 February 2020, 122671

[23] Jiyeon Choi 2020 Won Sik Shin. Removal of Salicylic and Ibuprofen by Hexadecyltrimethylammonium-Modified Montmorillonite and Zeolite

[24] B. Senthil Rathi & P. Senthil Kumar 2020 Electrodeionization theory, mechanism and environmental applications. A review . Published: 20 April 2020

[25] XianghongFang· ZhonghuiXu· YaodongLuo· LiRen· WeiHua 2016 Removal of Radionuclides from Laundry Wastewater Containing Organics and Suspended Solids Using Inorganic Ion Exchanger

[26] WengFu· GuozhaoJi· HuihuangChen· SiyuanYang ·BaoGuo· HongYang ·ZhiqiangHuang. 2020 Molybdenum sulphide modified chelating resin for toxic metal adsorption from acid mine wastewater

[27] Mahmoud O. Abd El-Magied 2018 Removal of nickel (II) ions from aqueous solutions using modified activated carbon: A kinetic and equilibrium study. Journal of Dispersion Science and Technology Volume 39, 2018 - Issue 6

[28] Somayeh Karandish 2019 An efficient solid phase extraction of Pb^{2+} using tannic acid-coated cerium oxide nanoparticles followed by electrothermal atomic absorption spectrometry . Separation Science and Technology Volume 54, 2019

[29] JunZhao , XiangLiu , YapanWu· Dong-ShengLi· QichunZhang 2019 Surfactants as promising media in the field of metal-organic frameworks .Coordination Chemistry Reviews Volume 391, 15 July 2019, Pages 30-43

[30] KishoreGinjupalli , RamakrishnaAlla ,TusharShaw 2018 Comparative evaluation of efficacy of Zinc oxide and Copper oxide nanoparticles as antimicrobial additives in alginate impression materials .materialstoday: PROCEEDINGSVolume 5, Issue 8, Part 3, 2018, Pages 16258-16266

[31] Fang-Zhen Teng 2017 Mineralogy and Geochemistry (2017) 82 (1): 219–287.

[32] XingXuanJae Y.Park 2018 A miniaturized and flexible cadmium and lead ion detection sensor based on micro-patterned reduced graphene oxide/carbon nanotube/bismuth composite electrodes. Sensors and Actuators B: ChemicalVolume 255, Part 2, February 2018, Pages 1220-1227

[33] Anthony B. Dichiara Michael R. Webber William R. Gorman Reginald E. Rogers* 2015 Removal of Copper Ions from Aqueous Solutions via Adsorption on Carbon Nanocomposites , Publication Date:June 30, 2015 //doi.org/10.1021/acsami.5b04974

[34] Yan-Hui Li , Shuguang Wang , Jinquan Wei 2002 Lead Adsorption on Carbon Nanotubes , DOI: 10.1016/S0009-2614(02)00502-X

[35] Muataz Ali Atieh,1,2 Omer Yahya Bakather,1 Bassam S. Tawabini,3 Alaadin A. Bukhari,4 Mazen Khaled,5 Mamdouh Alharthi,1 Mohammed Fettouhi,5 and Faraj Ahmad Abuilaiwi2,6 2010 Removal of Chromium (III) from Water by Using Modified and Nonmodified Carbon Nanotubes , Volume 2010 |Article ID 232378 | doi 10.1155/2010/232378

[36] Chungsying Lu , Chunti Liu 2006 Removal of nickel(II) from aqueous solution by carbon nanotubes , Journal of Chemical Technology & Biotechnology 81(12):1932 - 1940

[37] YiminHuang , XinqingLee ,Florika C.Macazo ,MatteoGrattieri· RongCai ,Shelley D.Mintee 2018 Fast and efficient removal of chromium (VI) anionic species by a reusable chitosan-modified multi-walled carbon nanotube composite . Chemical Engineering JournalVolume 339, 1 May 2018, Pages 259-267

[38] Abu Nasar, Fouzia Mashkoor 2019 Application of polyaniline-based adsorbents for dye removal from water and wastewater—a review Published: 05 January 2019

[39] Rajeev Kumar , Nazrul Haq 2014 Application of Carbon Nanotubes in Heavy Metals Remediation. Environmental Science and Technology Volume 44, 2014 - Issue 9

[40] AliErgün 2011 Effects of the usage of diatomite and waste marble powder as partial replacement of cement on the mechanical properties of concrete . Construction and Building MaterialsVolume 25, Issue 2, February 2011, Pages 806-812

[41] Ioannis K.Oikonomopoulos ,MariaPerraki ,NikolaosTougiannidis , TheodoraPerraki ,Haino UweKasper , MarcusGurk 2015 Clays from Neogene Achlada lignite deposits in Florina basin (Western Macedonia, N. Greece): A prospective resource for the ceramics industry . Applied Clay ScienceVolume 103, January 2015, Pages 1-9

[42] R.Miandad , M.A.Barakat , M.Rehan 2017 Plastic waste to liquid oil through catalytic pyrolysis using natural and synthetic zeolite catalysts .Waste ManagementVolume 69, November 2017, Pages 66-78

[43] Atul V.Maldhure[1]J.D.Ekhe 2013 Pyrolysis of purified kraft lignin in the presence of $AlCl_3$ and $ZnCl_2$. Journal of Environmental Chemical EngineeringVolume 1, Issue 4, December 2013, Pages 844-849

[44] Md. Juned K.AhmedM.Ahmaruzzaman 2016 A review on potential usage of industrial waste materials for binding heavy metal ions from aqueous solutions . Journal of Water Process Engineering Volume 10, April 2016, Pages 39-47

[45] T.ViraraghavanFlorde Maria Alfaro 1998 Adsorption of phenol from wastewater by peat, fly ash and bentonite . Journal of Hazardous Materials Volume 57, Issues 1–3, January 1998, Pages 59-70

[46] JutikaBoro ,DhanapatiDeka , Ashim J.Thakur 2012 A review on solid oxide derived from waste shells as catalyst for biodiesel production . Renewable and Sustainable Energy Reviews Volume 16, Issue 1, January 2012, Pages 904-910

[47] JutikaBoro· DhanapatiDeka · Ashim J.Thakur 2012 A review on solid oxide derived from waste shells as catalyst for biodiesel production . Renewable and Sustainable Energy ReviewsVolume 16, Issue 1, January 2012, Pages 904-91

[48] YingjieDai , QiyaSunWensiWang , LuLuMei, LiuJingjin, LiShengshuYang 2018 Utilizations of agricultural waste as adsorbent for the removal of contaminants: A review. ChemosphereVolume 211, November 2018, Pages 235-253

[49] Tushar KantiSen , SharmeenAfroze ,H.M.Ang 2014 Dye and its removal from aqueous solution by adsorption: A review . Advances in Colloid and Interface ScienceVolume 209, July 2014, Pages 172-184

[50] Sharmeen Afroze , Tushar Sen, 2018 A Review on Heavy Metal Ions and Dye Adsorption from Water by Agricultural Solid Waste Adsorbents, June 2018 Water Air and Soil Pollution 229(7) DOI: 10.1007/s11270-018-3869-z

[51] Hernández Rodiguez, M., Yperman, J., Carleer, R., Maggen, J., Dadi, D., Gryglewicz, G., … Otero Calvis, A 2018 Adsorption of Ni(II) on spent coffee and coffee husk based activated carbon. Journal of Environmental Chemical Engineering, 6(1), 1161–1170. doi:10.1016/j.jece.2017.12.045

[52] Zhou, Y., Cao, S., Xi, C., Li, X., Zhang, L., Wang, G., & Chen, Z 2019 A novel Fe3O4/graphene oxide/citrus peel-derived bio-char based nanocomposite with enhanced adsorption affinity and sensitivity of ciprofloxacin and sparfloxacin. Bioresource Technology, 121951.

[53] S. Lairini1,2, K. El Mahtal1,2, Y. Miyah1 , K. Tanji1 , S. Guissi3 , S. Boumchita1 , F. Zerrouq1 2017 The adsorption of Crystal violet from aqueous solution by using potato peels (Solanum tuberosum): equilibrium and kinetic studies, JMES, 2017 Volume 8, Issue 9, Page 3252-3261

[54] Jirekar, D. B. ; Ghumare Pramila ; Mazahar Farooqui 2015 Kinetics and isotherm studies on crystal violet dye adsorption onto Black gram seed husk, International Journal of ChemTech Research 2015 Vol.7 No.1 pp.427-434 ref.22

[55] Yusuff, A. S. 2018 Optimization of adsorption of Cr(VI) from aqueous solution by Leucaena leucocephala seed shell activated carbon using design of experiment. Applied Water Science, 8(8). doi:10.1007/s13201-018-0850-3

[56] Mata, Y. N., Blázquez, M. L., Ballester, A., González, F., & Muñoz, J. A. 2009 Optimization of the continuous biosorption of copper with sugar-beet pectin gels. Journal of Environmental Management, 90(5), 1737–1743. doi:10.1016/j.jenvman.2008.11.028

[57] Abdel-Raouf, N., Al-Homaidan, A. A., & Ibraheem, I. B. M. 2012 Microalgae and wastewater treatment. Saudi Journal of Biological Sciences, 19(3), 257–275. doi:10.1016/j.sjbs.2012.04.005

[58] Chellaiah, E. R. 2018 Cadmium (heavy metals) bioremediation by Pseudomonas aeruginosa: a minireview. Applied Water Science, 8(6). doi:10.1007/s13201-018-0796-5

[59] Ahmad Khosravi1 · Marzieh Javdan2 · Ghazal Yazdanpanah3 · Mohammad Malakootian, 2020 Removal of heavy metals by Escherichia coli (E. coli) bioflm placed on zeolite from aqueous solutions (case study: the wastewater of Kerman Bahonar Copper Complex), Applied Water Science (2020) 10:167 https://doi.org/10.1007/s13201-020-01257-5

[60] Shafiquzzaman Siddiquee1 *, Kobun Rovina1 , Sujjat Al Azad2 , Laila Naher3 , Saallah Suryani1 and Pasicha Chaikaew4 2015 Heavy Metal Contaminants Removal from Wastewater Using the Potential Filamentous Fungi Biomass, J Microb Biochem Technol 2015, 7:6 DOI: 10.4172/1948-5948.1000243

[61] Manisha Shakya1; Pratibha Sharma2; Syeda Shaima Meryem3; Qaisar Mahmood4; and Arun Kumar5 2015 Heavy Metal Removal from Industrial Wastewater Using Fungi: Uptake Mechanism and Biochemical Aspects May 2015Journal of Environmental Engineering 2015(9) DOI: 10.1061/(ASCE)EE.1943-7870.0000983

[62] Bayramoglu and Arica in the year 2008 , low cost techniques used here which is bio-adsorption for the removal of heavy metal ions from the waste water , Bio-resource technologies 99.2766-2777 .

[63] Adriana Jazmín Legorreta-Castañeda, Carlos Alexander Lucho-Constantino , Rosa Icela Beltrán-Hernández, Claudia Coronel-Olivares and Gabriela A. Vázquez-Rodríguez, 2020 Biosorption of Water Pollutants by Fungal Pellets , Water 2020, 12, 1155; doi:10.3390/w12041155

[64] Fenglian Fu a,*, Qi Wang 2011 Removal of heavy metal ions from wastewaters , Journal of Environmental Management 92 (2011) 407e418Issue 12 doi-10.1080/02786826.2015.1109053

[65] Ranganathan Gopalakrishnan, Peter H. McMurry & Christopher J. Hogan, Jr, The Bipolar Diffusion Charging of Nanoparticles 2015 A Review and Development of Approaches for NonSpherical Particles, Aerosol Science and Technology_Volume 49, 2015 -

[66] Renu; Madhu Agarwal; K. Singh 2017 Heavy metal removal from wastewater using various adsorbents 2017 Journal of Water Reuse and Desalination (2017) 7 (4): 387–419. doi.org/10.2166/wrd.2016.104

[67] Sabry M. Shaheen , Aly Soliman Derbalah , Farahat S. Moghanm 2012 Removal of Heavy Metals from Aqueous Solution by Zeolite in Competitive Sorption System , January 2012 International Journal of Environmental Science and Development DOI: 10.7763/IJESD.2012.V3.248

[68] Renu; Madhu Agarwal; K. Singh , Heavy metal removal from wastewater using various adsorbents 2017 Journal of Water Reuse and Desalination (2017) 7 (4): 387–419. doi.org/10.2166/wrd.2016.104

[69] Laxmipriya Panda, Sandeep K. Jena, Swagat S. Rath & Pramila K. Misra 2020 Heavy metal removal from water by adsorption using a low-cost geopolymer, *Environmental Science and Pollution Research* volume 27, pages24284–24298(2020)

[70]Bao-LinZhang[a] WeiQiu[a]Pan-PanWang [a]Yu-LeiLiu[b] JingZou[c] LuWang[a] JunMa[a], 2020 Mechanism study about the adsorption of Pb(II) and Cd(II) with iron-trimesic metal-organic frameworks , Chemical Engineering Journal Volume 385, 1 April 2020, 123507

[71]Sujatha[a]R.Sivarethinamohan[b] 2020 A critical review of Cr(VI) ion effect on mankind and its amputation through adsorption by activated carbon, Materialstoday Available online 23 July 2020

[72] Luigi Gurreri , Alessandro Tamburini * , Andrea Cipollina and Giorgio Micale , 2020 Electrodialysis Applications in Wastewater Treatment for Environmental Protection and Resources Recovery: A Systematic Review on Progress and Perspectives, Membranes 2020, 10, 146 ; doi:10.3390/membranes10070146

[73] Mathhar Bdour 1,* , Zakariya Dalala 1 , Mohammad Al-Addous 1 , Atef Kharabsheh 2 and Hadi Khzouz 1, Mapping RO-Water Desalination System Powered by Standalone PV System for the Optimum Pressure and Energy Saving, Appl. Sci. 2020, 10, 2161; doi:10.3390/app10062161 w

[74] Jinyue Yang, Baohong Hou,* Jingkang Wang, Beiqian Tian, Jingtao Bi, Na Wang, Xin Li, and Xin Huang*, 2020 Nanomaterials for the Removal of Heavy Metals from Wastewate, Nanomaterials (Basel). 2019 Mar; 9(3): 424, doi: 10.3390/nano9030424

[75] TA Siddique , Naba K. Dutta * and Namita Roy Choudhury 2020 Nanofiltration for Arsenic Removal: Challenges, Recent Developments, and Perspectives, Nanomaterials 2020, 10, 1323; doi:10.3390/nano10071323

[76] Tiago da Silva Arouche 2020 Rosely Maria dos Santos Cavaleiro,[1,2] Phelipe Seiichi Martins Tanoue,[1] Tais Sousa de Sa Pereira,[1] Tarciso Andrade Filho,[3] and Antonio Maia de Jesus Chaves Neto , Heavy Metals Nanofiltration Using Nanotube and Electric Field by Molecular Dynamics, Volume 2020 |Article ID 4063201 | 10.1155/2020/4063201

[77] Kwang-Ho Choo ,Dae-Joong Kwon ,Kwang-Won Lee and Sang-June Choi 2002 Selective Removal of Cobalt Species Using Nanofiltration Membranes , *Environ. Sci. Technol.* 2002, 36, 6, 1330–1336 Publication Date:February 7, 2002 doi.org/10.1021/es010724q

[78] Cristina-VeronicaGherasim , PetrMikulášek 2014 Influence of operating variables on the removal of heavy metal ions from aqueous solutions by nanofiltration , Desalination ,Volume 343, 16 June 2014, Pages 67-74

[79] B.A.M. Al-Rashdi, Daniel James Johnson, Nidal Hilal 2013 Removal of Heavy Metal Ions by Nanofiltration, April 2013 Desalination 315:2–17 DOI: 10.1016/j.desal.2012.05.022

[80] Murthy, Z.V.P., Chaudhari, L.B., 2009. Separation of binary heavy metals from aqueous solutions by nanofiltration and characterization of the membrane Using Spieglere Kedem model. Chem. Eng. J. 150, 181e187

[81] Mahnaz Afifi , Hosein Alizadeh Golestani , Shadi Sharifi , Shirin Kiani 2014 Wastewater treatment of raisins processing factory using micellar-enhanced ultrafiltration , January 2014 Desalination and water treatment 52(1-3) , DOI: 10.1080/19443994.2013.782254

[82] Hani Abu-Qudais , Hassan Moussa 2004 Removal of Heavy Metals from Wastewater by Membrane Processes: A Comparative Study, April 2004 Desalination 164(2):105-110

[83] C Blöcher , J Dorda , V Mavrov , H Chmiel 2003 Hybrid flotation - Membrane filtration process for the removal of heavy metal ions from wastewater , September 2003 Water Research 37(16):4018-26 DOI: 10.1016/S0043-1354(03)00314-2

[84] Shahin Ahmadi1* and Ferdos Kord Mostafapour2 2017 Survey of Efficiency of Dissolved Air Flotation in Removal Penicillin G Potassium from Aqueous Solutions , 15(3): 1-11, 2017

[85] George Z. Kyzas , Kostas A. Matis 2018 Flotation in Water and Wastewater Treatment, p*rocesses* 2018, *6*(8), 116; doi.org/10.3390/pr6080116

[86] Hurriyet Polat , D Erdogan 2007 Heavy Metal Removal from Waste Waters by Ion Flotation , October 2007 Journal of Hazardous Materials 148(1-2):267-73 , DOI: 10.1016/j.jhazmat.2007.02.013

[87] Jacek Namieśnik & Anna Rabajczyk , The speciation and physico-chemical forms of metals in surface waters and sediments , ISSN: 0954-2299 (Print) 2047-6523

[88] B.Y. Medina , M.L. Torem , L.M.S. de Mesquita 2005 On the kinetics of precipitate flotation of Cr III using sodium dodecylsulfate and ethanol , February 2005 Minerals Engineering 18(2):225-231DOI: 10.1016/j.mineng.2004.08.018

[89] V. Mavrov , St. Stamenov , Ekaterina Ivanova Todorova , H. Chmiel 2006 New hybrid electrocoagulation membrane process for removing selenium from industrial wastewater , November 2006 Desalination 201(1-3):290-296DOI: 10.1016/j.desal.2006.06.005

[90] C Blöcher, J Dorda , V Mavrov , H Chmiel 2003 Hybrid flotation - Membrane filtration process for the removal of heavy metal ions from wastewater , September 2003 Water Research 37(16):4018-26DOI: 10.1016/S0043-1354(03)00314-2

[91] Rubio, Renu Bisht , Madhu Agarwal , Kailash Bahadur Singh 2017 Methodologies for removal of heavy metal ions from wastewater , October 2017 Interdisciplinary Environmental Review 18(2):124-142 DOI: 10.1504/IER.2017.10008828

[92] Thiện Khánh Trần , Hoang-Jyh Leu , Kuo-Feng Chiu , Chiu-Yue Lin 2017 Electrochemical treatment of heavy metal containing wastewater, March 2017 Journal- Chinese Chemical Society Taipei 64(5) DOI: 10.1002/jccs.201600266

[93] Mohammed Chafi , Gourich Bouchaib , Abdel hafid Essadki , Ch. Vial 2011 Comparision of Electrocoagulation Using Iron and Aluminium Electrodes With Chemical Coagulation for the Removal of a High Soluble Acid Dye , October 2011 Desalination 281(1):285-292 DOI: 10.1016/j.desal.2011.08.004

[94] Chen, G.H., 2004. Electrochemical technologies in wastewater treatment. Sep. Purif. Technol. 38, 11e41.

[95] Nanseu-Njiki, C.P., Tchamango, S.R., Ngom, P.C., Darchen, A., Ngameni, E., 2009. Mercury(II) removal from water by electrocoagulation using aluminium and iron electrodes. J. Hazard. Mater. 168, 1430e1436.

[96] David Hasson , Georgiy Sidorenko , Raphael Semiat 2011 Low electrode area electrochemical scale removal system , July 2011 Desalination and water treatment 31(1):35-41 DOI: 10.5004/dwt.2011.2389

[97] Chaloempan Petsriprasit , Jirachaya Namboonmee, Mali Hunsom 2010 Application of the electrocoagulation technique for treating heavy metals containing wastewater from the pickling process of a billet plant , May 2010 Korean Journal of Chemical Engineering 27(3):854-861 DOI: 10.1007/s11814-010-0145-3

[98] Belkacem Merzouk , Khodir Madani , Sekki Abdelkrim 2008 Treatment characteristics of textile wastewater and removal of heavy metals using the electroflotation technique , August 2008 Desalination 228(1-3):245-254 DOI: 10.1016/j.desal.2007.10.013

[99] José Adilsonde Castro[a]Ruide Góes Casqueira[b]Angelo Gomesde Oliveira Junior[b] 2015 Study of electroflotation method for treatment of wastewater from washing soil contaminated by heavy metals , Volume 4, Issue 2, April–June 2015, Pages 109-113

[100] D.-T. Chin 2002 Metal recovery from wastewater with an electrochemical method , March 2002 Chemical Engineering Education 36(2):144-149+155

[101] Anna Antonyová *1 and Peter Antony 2017 Electro-Flotation as the Method of the Water Treatment in Recycling Process of Contaminated Water as an Integral Part of the Reverse Logistics System , 21 December 2017 (03:55:44 CET), 10.20944/preprints201712.0147.v1

[102] Salman Hussein Abbas1*; Wail Hassan Ali2 2018 Electrocoagulation Technique Used To Treat Wastewater: A Review , American Journal of Engineering Research (AJER). Volume-7, Issue-10, pp-74-88

[103] Siti Nor Aishah Mohd-Salleh , Nur Shaylinda Mohd Zin , Norzila Othman 2019 A Review of Wastewater Treatment using Natural Material and Its Potential as Aid and Composite Coagulant , February 2019 Sains Malaysiana 48(1):155-164 DOI: 10.17576/jsm-2019-4801-18

[104] Cuiling Zhang , Cuiling Zhang, Qing Chang 2013 Preparation of mercaptoacetyl chitosan and its removal performance of copper ion and turbidity , December 2013 Desalination and water treatment 53(7):1-8 DOI: 10.1080/19443994.2013.870743

[105] Bratskaya, S.Y., Pestov, A.V., Yatluk, Y.G., Avramenko, V.A., 2009. Heavy metals removal by flocculation/precipitation using N-(2-carboxyethyl)chitosans. Colloid Surf. 339, 140e144.

[106] Duan, J.C., Lu, Q., Chen, R.W., Duan, Y.Q.,Wang, L.F., Gao, L., Pan, S.Y., 2010. Synthesis of a novel flocculant on the basis of crosslinked Konjac glucomannan-graftpolyacrylamide- co-sodium xanthate and its application in removal of Cu2þ ion. Carbohydr. Polym. 80, 436e441

[107] Chee Yang Teh† ,Pretty Mori Budiman† ,Katrina Pui Yee Shak† ,Ta Yeong Wu*† 2016 Recent Advancement of Coagulation–Flocculation and Its Application in Wastewater Treatment, *nd. Eng. Chem. Res.* 2016, 55, 16, 4363–43 , March 4, 2016

[108] J.Beltrán Heredia[1]J.Sánchez Martín 2009 Removing heavy metals from polluted surface water with a tannin-based flocculant agent , Volume 165, Issues 1–3, 15 June 2009, Pages 1215-1218

[109] Kingsley Ogemdi Iwuozor 2019 Prospects and Challenges of Using Coagulation-Flocculation method in the treatment of Effluents , January 2019 Advanced Journal of Chemistry-Section A 2(2):105-127 10.29088/SAMI/AJCA.2019.2.105127

[110] Plattes, M., Bertrand, A., Schmitt, B., Sinner, J., Verstraeten, F., Welfring, J., 2007. Removal of tungsten oxyanions from industrial wastewater by precipitation, coagulation and flocculation processes. J. Hazard. Mater. 148, 613e615

[111] Bojic, A.L., Bojic, D., Andjelkovic, T., 2009. Removal of Cu2þ and Zn2þ from model wastewaters by spontaneous reductionecoagulation process in flow conditions. J. Hazard. Mater. 168, 813e819

[112]ShangyongLin[ab]RunqingLiu[ab]MeirongWu[ab]YuehuaHu[ab]WeiSun[ab]ZhizhongShi[c]HaishengHan[ab] 2020 Minimizing beneficiation wastewater through internal reuse of process water in flotation circuit , Volume 245, 1 February 2020, 118898

[113] Kurniawan, T.A., Chan, G.Y.S., Lo, W.H., Babel, S., 2006. Physicoechemical treatment techniques for wastewater laden with heavy metals. Chem. Eng. J. 118, 83e98

Comparison of wear behaviour of polymer spur gears using FDM process

K.Mohan Harsha[1],Y. Seetharama Rao[2] and DVNJ Jagannadha rao[3]

[1,2,3]DEPARTMENT OF MECHANICAL ENGINEERING, GAYATRI VIDYA PARISHAD COLLEGE OF ENGINEERING (Autonomous)

Abstract. The goal of this study is to explore the potential of three different thermoplastic materials used in plastic gearing applications, such as ABS(Acrylonitrile butadiene styrene), PLA (Polylactic acid) and NYLON. The gears are created by the process of 3D printing.

In order to find the torque at various speeds, experimental investigations are conducted using a gear test rig.

Key words: Polymer-based gear, Injection moulding process, Acrylonitrile Butadiene Styrene (ABS), High Density Polyethylene (HDPE) and Polyoxymethylene (POM).

INTRODUCTION

Due to its lower weight and low manufacturing cost, part weight and quiet performance compared to metal gears, polymer composite gears are finding growing applications. However, compared with metals, polymeric materials suffer from low mechanical strength and thermal resistance. Reinforced polymers are ideal for structural/load bearing and have high mechanical strength and thermal resistance and are suitable for structural/load bearing applications.

Short fiber-reinforced polymers allow complex formed products to be manufactured economically using the injection-molding process. Many studies have been done on the performance problems of polymer and polymer composite gears. In these papers, the effect of reinforcement on polymer gears' fatigue efficiency and wear resistance is well discussed.

Because many polymer-based gear designs are derived from the method of designing metal gears, polymer-based gears require complex heat build-up considerations and their related effects that occur in operation.

Prashant et al. Investigated on the potential of three different thermoplastic materials used for plastic gearing applications are Acrylonitrile Butadiene Styrene (ABS), High Density Polyethylene (HDPE) and Polyoxymethylene (POM).The gears are manufactured by injection moulding process.

Akant et al.studied the performance and durability of the polymer gears is an important aspect about their application. This paper investigates the various techniques employed for the enhancement of the performance of polymer gears.

asuggested the significance of polymer gears to transmit power and motion is increasing continuously due to their inherent characteristics.Polymer gears have established themselves as attractive alternatives to traditional metal gears in plethora applications. They are light in weight, have lower inertia, and run noiseless than their metal counterparts.

W. Li et al.Performed the experiment on polymer gear (acetal and nylon) friction and wear behaviour. First, a unique test method for polymer gear wear will be described in brief and later used in the extensive investigation of acetal and nylon gear wear. Initial tests were performed using acetal pinions with acetal gears, and nylon pinions with nylon gears, with further investigation carried out using dissimilar polymer gears.

P. Langlois et al. Examined on machine cut acetal gear wear and thermal mechanical contact behaviour. The results for machine cut acetal gears will be compared to previously published results obtained for polymer gears manufactured through injection moulding.
K. Mao. Employed a new design method for polymer composite gear has been proposed in the current paper. This design method is based on the link between polymer gear wear rate and its surface temperature. It has been found from the tests that the polymer (acetal) gear wear rate will be increased dramatically when the load reaches a critical value for a specific geometry.
S. Senthilvelan et al.Experimented onpolymer based gears replace metal gears in many light duty power and/or motion transmission applications due to their noiseless operation even under unlubricated conditions. Visco elastic behaviour of the polymer, which is mainly responsible for the sound absorption, is altered by the addition of short fibers.
S.M. Evans et al.Investigated a new methodology to predict the transient operational temperature of a polymer–steel gear pair under loaded running is presented.
R. Gnanamoorthy et al.studied unreinforced nylon 6/6 and 20 per cent carbon-fibre-reinforced nylon 6/6 gears were injection moulded in the laboratory and tested for transmission efficiency using an in-house developed power absorption test rig.
K D Dearn et al. Observed the role of materials in the generation of noise made by plastic gears.
T.J. Hoskins et al.suggested the sound frequency spectrum is influenced by the various polymeric gear materials and operating conditions.

POLYMER SPUR GEAR

A great deal of modern technology depends on more conventional equipment to get the job done, considering the proliferation of electronic devices. For instance, in many important machines, from wind farms to heart pumps, gears still play a large role. Several gears are made of plastic to save cost and weight.
Gears made of plastic tend to be lighter in weight than similar gears made out of metal.Plastic gears are, in general, less expensive to produce than those made of metal.Plastic gears can have their teeth worn down from repeated use. While this is true of any gear, plastic as a material isn't as durable as metals, though manufacturers can add coatings for the teeth.

POLYMER SPUR GEAR

A great deal of modern technology depends on more conventional equipment to get the job done, considering the proliferation of electronic devices. For instance, in many important machines, from wind farms to heart pumps, gears still play a large role. Several gears are made of plastic to save cost and weight.

Gears made of plastic tend to be lighter in weight than similar gears made out of metal.Plastic gears are, in general, less expensive to produce than those made of metal.Plastic gears can have their teeth worn down from repeated use. While this is true of any gear, plastic as a material isn't as durable as metals, though manufacturers can add coatings for the teeth.

MATERIALS AND METHODS

The materials used to manufacture the polymer spur gear are ABS (Acrylonitrile butadiene styrene),NYLON and PLA (Polylactic Acid).

Tensile Strength(MPa)	40.96
Flexural Strength(MPa)	45.44
Impact Strength(MPa)	22.11
Melting Temperature °c	220 – 260
Density g/cm^3	1.07

Table 1 :ABS material properties

Tensile Strength(MPa)	62.63
Flexural Strength(MPa)	65.02
Impact Strength(MPa)	4.28
Melting Temperature °c	190 – 220
Density g/cm^3	1.25

Table 2 : PLA material properties

Tensile Strength(MPa)	50 – 55
Flexural Strength(MPa)	85 – 90
Impact Strength(MPa)	354
Melting Temperature (°c)	220 – 260
Density g/cm^3	1.15

Table 3 : NYLON material properties

FUSED DEPOSITION MODELING PROCESS :

Fused deposition modeling (FDM) is a method of additive manufacturing that belongs to the extrusion class of materials. An object is constructed in FDM by selectively depositing melted material into a layer-by-layer pre-determined path.

The production was performed in think-3D,Visakhapatnam,

Fig1(a and b) FDM

These materials takethe form of plastic threads, or filaments, during printing, which are unwound from a coil and fed through an extrusion nozzle. The nozzle melts and extrudes the filaments onto a foundation, often referred to as a building platform or table. A computer controls both the nozzle and the foundation, which transforms an object's dimensions into X, Y and Z nozzle coordinates and base to follow during printing.

In a standard FDM device, the extrusion nozzle travels horizontally and vertically across the building platform, drawing a cross section of an object onto the platform. This thin plastic layer cools and hardens, binding to the layer underneath it instantly. The base is lowered after a layer is finished, normally by approximately one-sixteenth of an inch, to make room for the next layer of plastic.

Printing time is dependent on the size of the produced item. Small objects, only a few cubic inches, and tall, thin objects print easily, while printing takes longer for larger, more geometrically complex objects.

Once an object is removed from the FDM printer, the support materials are either removed by soaking the object in a solution of water and detergent or snapping the support material off by hand in the case of thermoplastic supports. In order to enhance their function and appearance, objects can often be sanded, milled, painted or plated.

Once an object is removed from the FDM printer, in the case of thermoplastic supports, the support materials are either removed by soaking the object in a water and detergent solution or snapping the support material off by hand. Objects may frequently be sanded, milled, painted or plated to improve their function and appearance.

FABRICATION OF POLYMER SPUR GEARS

3D PRINTING PROCESS

In the production of a three-dimensional object through additive manufacturing, a variety of processes, equipment and materials are used.

Some techniques melt or soften the layer content, such as fused deposition modeling (FDM), selective laser melting (SLM), selective laser sintering (SLS), fused filament manufacturing (FFF), etc. Thin layers are cut to form and joined together (e.g., paper, polymer, metal).

Each process has its own advantages and disadvantages, which is why some businesses have the material used to create the object with a choice between powder and polymer. In order to create a robust prototype, others often use regular, off-the-shelf business paper as the construction material.

Speed, costs of the 3D printer, the printed prototype, choice and cost of the materials, and color capabilities are usually the key considerations in selecting a computer.

Fig 2. Nylon gear

Fig 3. PLA gear

IOP Conf. Series: Materials Science and Engineering 1168 (2021) 012028 doi:10.1088/1757-899X/1168/1/012028

Fig 4. ABS gear

MODELING OF THE SPUR GEAR IN POLYMER:

Polymer gears give engineers more versatility and different cad software such as CATIA, PRO-E and solid-works etc. can design the gear and the part can be drawn in a specific dimension where the part can be drawn.

A gear component built by CATIA V5 R21.

You must save this part in the form of the STL format. This file is used for the process of fabrication.

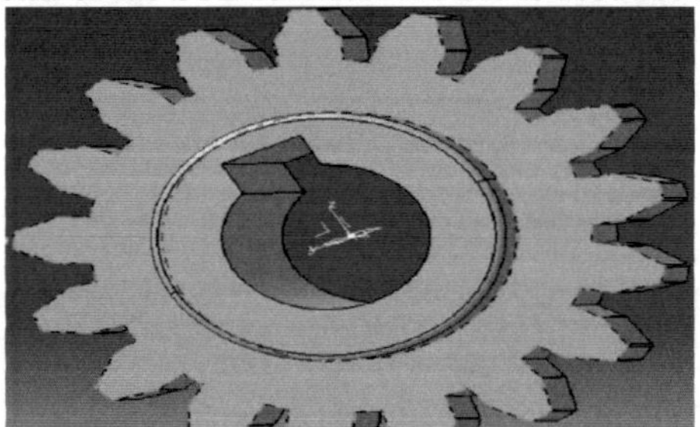
Figure 5: Polymer Spur gear using catia V5

POLYMER GEAR DIMENSIONS:

	Test Gear	Mating Gear
Material	ABS PLA NYLON	Mild steel
Manufacturing Process	3D Printing Process	Gear hobbing
Tooth profile	Involute	Involute
Type	Spur Gear	Spur Gear
Pressure angle	20^0	20^0
Module	2	2
Diameter of Pitch circle (mm)	34	34
Diameter of Tip circle (mm)	38	38
Diameter of Root circle (mm)	29	29
No. of teeth	17	17
Face width (mm)	6	6
Diameter of Hub (mm)	18	18
Diameter of Bore (mm)	12	12

Table 4: Dimensions of polymer spur gear and mating metal gear

RESULTS AND DISCUSSION:

Polymer gear test rig: The test rig is a system used on the components or assemblies to carry out such tests. It's a way of exploring and obtaining results from an experimental test.

Polymer gears are built of polymers such as nylon etc. These gears are to be checked on a test rig to provide further design approval based on the results of the tests.

These rigs typically include components such as structural frame, engines, gear mount fixtures, data output sensors such as rpm, acceleration, temperature, Show of user interfaces etc.

A motor, generator and a rheostat are part of the RAMSON polymer gear test rig. The speed of the motor and generator can be changed.

The DC motor starts spinning once the speed is determined, which gives rotational motion to the mild steel gear that meshes with the polymer spur gear attached to the DC generator. A rheostat is mounted between the engine and the generator, which provides the DC generator with a load of 1 kW at intervals of 250W, 250W and 500W.

The corresponding values are given by the temperature sensor and the speed sensors present in the apparatus. From the user interface display, all these will be noted.

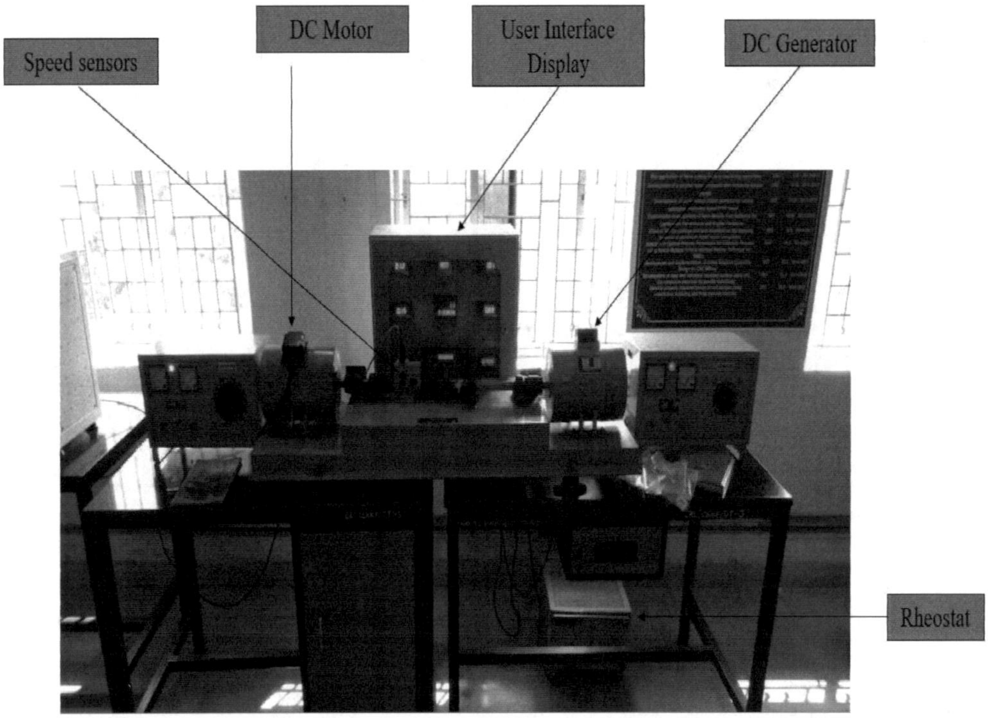

Fig 6. Polymer gear test-rig (Gitam College Visakhapatnam)

Experimentation is carried out for three different material at different speeds (600rpm, 800rpm, 1100rpm) and specific wear rate is calculated using standard formula.

$$Ws = \frac{Wv}{2zmbNT}$$ [1]

Ws= specific wear rate
Wv = wear volume ($mm3$)
z= number of gear teeth
m= module (mm)
b= tooth face width (mm)
NT= total number of revolution (rev)

$$W_S = \frac{W_v}{2zmbN_t}$$ [1].

$W_v = 34.78261mm^3$.

$z = 17$.

$m = 2mm$.

$b = 6mm$.

$N_t = 371579$.

IOP Conf. Series: Materials Science and Engineering 1168 (2021) 012028 doi:10.1088/1757-899X/1168/1/012028

S.no	Days	Speed (rpm) N	Motor		Generator		Torque (N-m)	No. of rotations per day
			Voltage (V)	Current (amp)	Voltage (V)	Current (amp)		
1	1st							123912
2	2nd	600	101	0.99	33	0.40	1.364	123882
3	3rd							123785
4	1st							123581
5	2nd	800	147	0.97	15	0.36	1.371	123423
6	3rd							123350
7	1st							123682
8	2nd	1100	175	0.98	60	1.03	1.388	123211
9	3rd							123146

Table 5: Experimental values of Nylon Gear at different speeds.

Fig 7. Nylon gear at 600 rpm

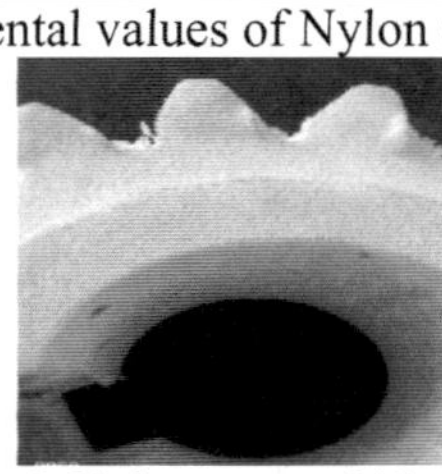
Fig 8. Nylon gear at 800 rpm

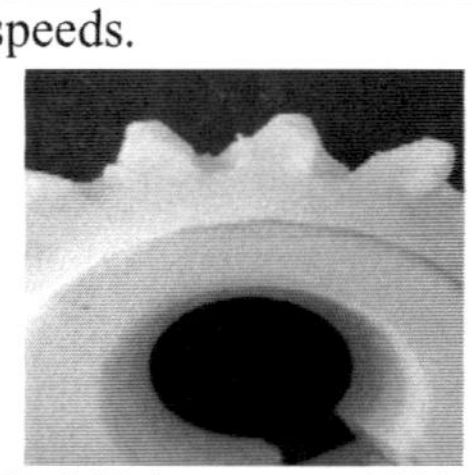
Fig 9.Nylon gear at 1100rpm

IOP Conf. Series: Materials Science and Engineering 1168 (2021) 012028 doi:10.1088/1757-899X/1168/1/012028

Table 6: Experimental values of ABS Gear at different speeds

S.no	Days	Speed (rpm) N	Motor		Generator		Torque (N-m)	No. of rotations per day
			Voltage (V)	Current (amp)	Voltage (V)	Current (amp)		
1	1st							123756
2	2nd	600	101	0.99	33	0.40	1.364	123621
3	3rd							123512
4	1st							123461
5	2nd	800	147	0.97	15	0.36	1.371	123245
6	3rd							123152
7	1st							123402
8	2nd	1100	175	0.98	60	1.03	1.388	123164
9	3rd							123069

Table 6: Experimental values of ABS Gear at different speeds

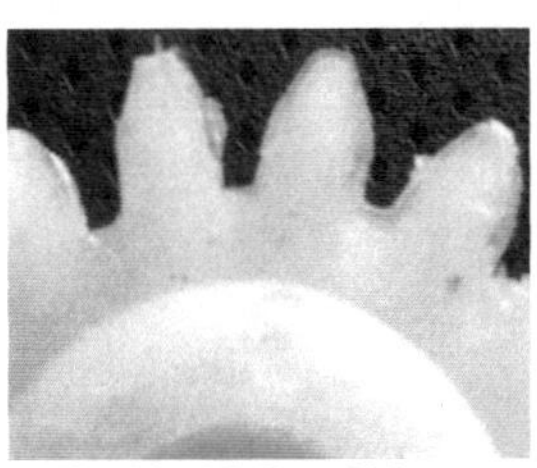
Fig 10. ABS at 600 rpm

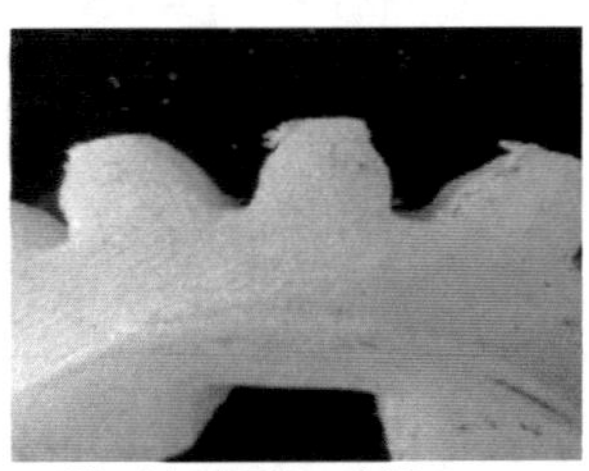
Fig 11. ABS at 800 rpm

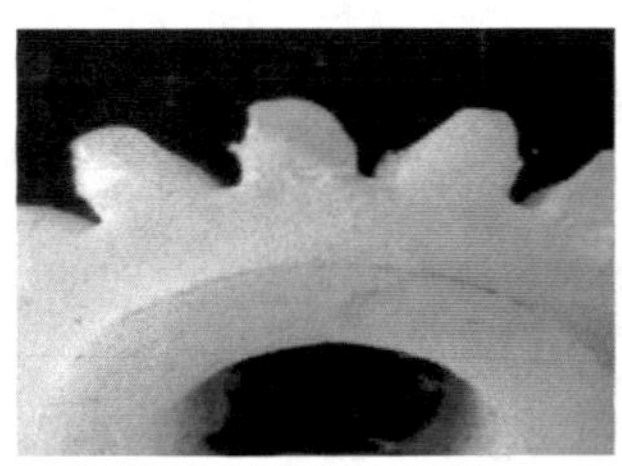
Fig 12. ABS at 1100 rpm

IOP Conf. Series: Materials Science and Engineering 1168 (2021) 012028 doi:10.1088/1757-899X/1168/1/012028

S. no	Days	Speed (rpm) N	Motor		Generator		Torque (N-m)	No. of rotations per day
			Voltage (V)	Current (amp)	Voltage (V)	Current (amp)		
1	1st	600	101	0.99	33	0.40	1.364	123692
2	2nd							123522
3	3rd							123348
4	1st	800	147	0.97	15	0.36	1.371	123458
5	2nd							123231
6	3rd							123123
7	1st	1100	175	0.98	60	1.03	1.388	123021
8	2nd							122983
9	3rd							120341-Break

Table 7: Experimental values of PLA Gear at different speeds

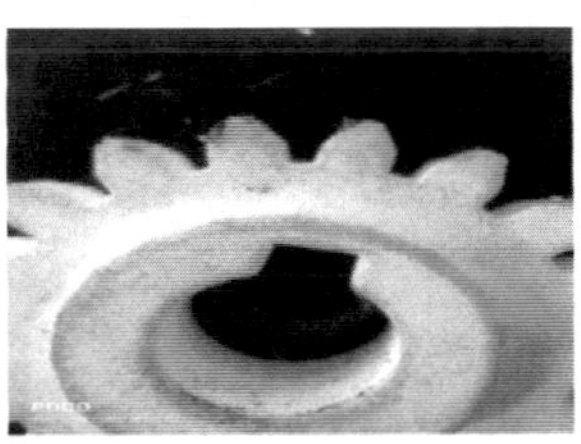
Fig 13. PLA at 600 rpm

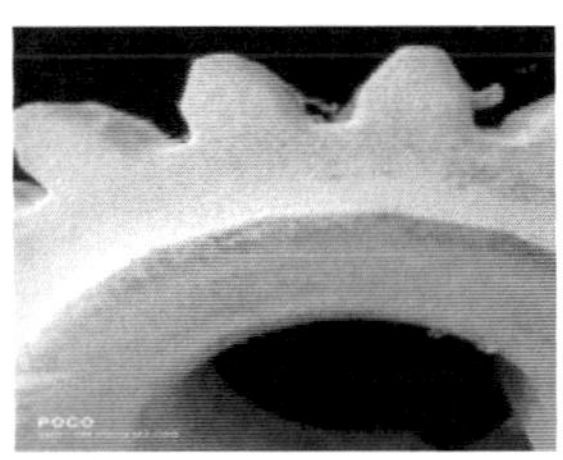
Fig 14. PLA at 800 rpm

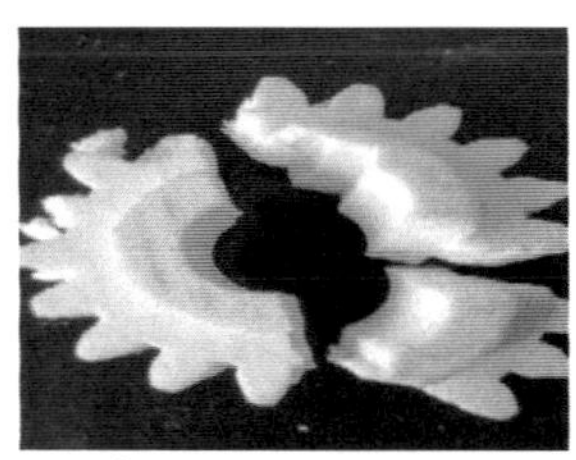
Fig 15. PLA at 1100 rpm

Speed	Weight loss. (g)	Densi y (g/cm^3)	Weight volume (mm$^{3)}$	No of rotations,(N)	Specific Wear $\times 10^6$ mm^3/mm-mm-rev
600 rpm	0.040	1.15	34.783	371579	0.229
800 rpm	0.092		80.000	370354	0.529
1100 rpm	0.140		121.739	370039	0.806

Table 8: Experimental values of specific wear rate for Nylon gear.

Speed	Weight loss. (g)	Densit y (g/cm^3)	Weight volume (mm^3)	No of rotations,(N)	Specific Wear $\times 10^6$ mm^3/mm-mm-rev
600 rpm	0.08	1.07	74.766	370889	0.494
800 rpm	0.14		130.841	369858	0.867
1100 rpm	0.22		205.607	369635	1.363

Table 9 : Experimental values of Specific wear rate for ABS gear.

Speed	Weight loss. (g)	Densit y (g/cm^3)	Weight volume (mm^3)	No of rotations,(N)	Specific Wear $\times 10^6$ mm^3/mm-mm-rev
600 rpm	0.05	1.25	40	370562	0.264
800 rpm	0.14		112	367011	0.748

Table 10: Experimental values of Specific wear rate for PLA gear.PLA gear breaks at 1100 rpm.

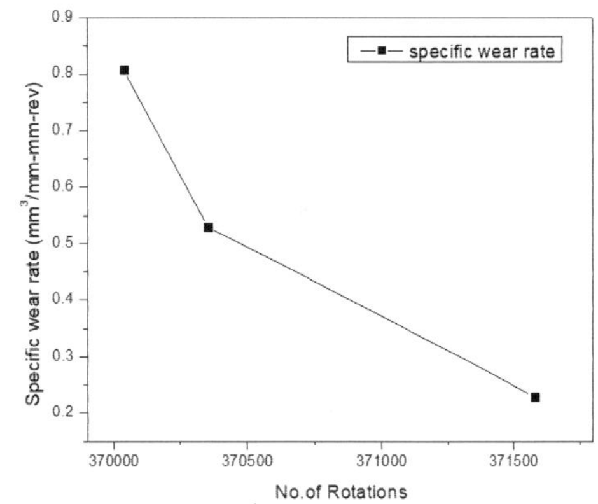

Fig 16. Plot between Specific wear rate and No. of Rotations for Nylon gear.

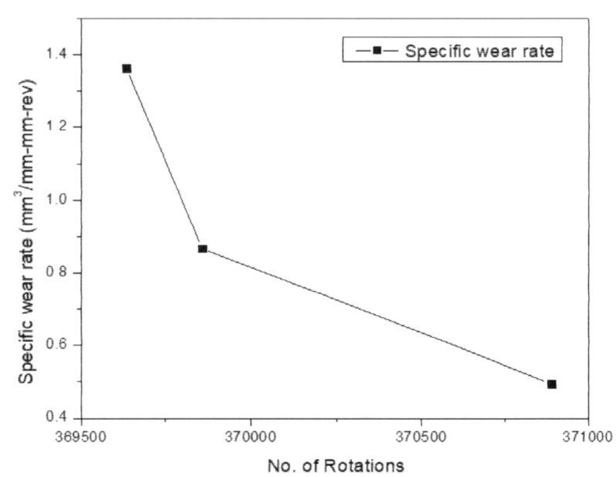

Fig 17. Plot between Specific wear rate and No. of Rotations for ABS gear.

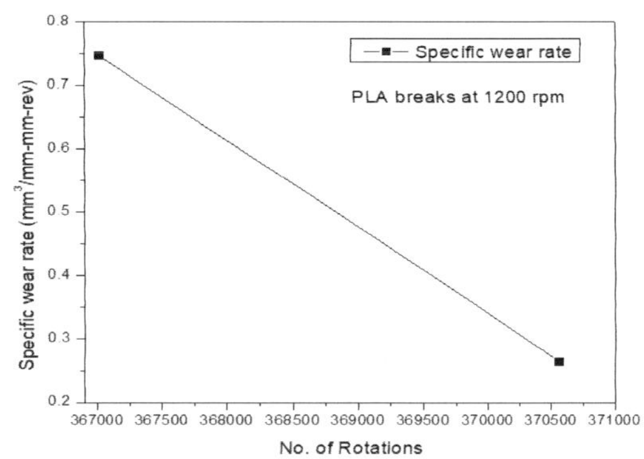

Fig 18. Plot between Specific wear rate and No. of Rotations for PLA gear

CONCLUSIONS

- For plastic gearing applications, the potential of three different polymeric gears (Nylon, ABS and PLA) created by the FDM process (3D printing) is examined.
- With increasing rotation speed, the real wear rate for polymer equipment increases.
- The results show that ABS gear has the maximum specific wear rate (at 600 rpm 0.494, at 800 rpm 0.867 and at 1100 rpm 1.363), while nylon has the minimum specific wear rate (at 600rpm 0.229 ,at 800 rpm 0.529 and at 1100 rpm at 0.806).
- It is clear that the nylon gear shows the best results while compared to other material gears.
- Polymer gear failure is investigated at a torque of 1.388Nm and a speed of 1200 rpm. PLA gear has an unacceptable wear rate and breaks after ABS and Nylon gears have completed 3L rotations at various speeds.

References

[1]P. K. Singh, Siddhartha, and A. K. Singh, "An investigation on the thermal and wear behavior of polymer based spur gears," *Tribology International*, vol. 118, pp. 264–272, Feb. 2018.

[2] P. K. Singh, Siddhartha, and A. K. Singh, "An investigation on the effects of the various techniques over the performance and durability of polymer gears," *Materials Today: Proceedings*, vol. 4, no. 2, pp. 1606–1614, 2017.

[3] A. K. Singh, Siddhartha, and P. K. Singh, "Polymer spur gears behaviors under different loading conditions: A review," *Proceedings of the Institution of Mechanical Engineers, Part J: Journal of Engineering Tribology*, vol. 232, no. 2, pp. 210–228, Feb. 2018.

[4] K. Mao, W. Li, C. J. Hooke, and D. Walton, "Friction and wear behaviour of acetal and nylon gears," *Wear*, vol. 267, no. 1–4, pp. 639–645, Jun. 2009.

[5] K. Mao *et al.*, "The wear and thermal mechanical contact behaviour of machine cut polymer gears," *Wear*, vol. 332–333, pp. 822–826, May 2015.

[6] K. Mao, "A new approach for polymer composite gear design," *Wear*, vol. 262, no. 3–4, pp. 432–441, Feb. 2007.

[7] S. Senthilvelan and R. Gnanamoorthy, "Damping characteristics of unreinforced, glass and carbon fiber reinforced nylon 6/6 spur gears," *Polymer Testing*, vol. 25, no. 1, pp. 56–62, Feb. 2006.

[8] S. M. Evans and P. S. Keogh, "Efficiency and running temperature of a polymer–steel spur gear pair from slip/roll ratio fundamentals," *Tribology International*, vol. 97, pp. 379–389, May 2016.

[9] S. Senthilvelan and R. Gnanamoorthy, "Efficiency of injection-moulded polymer composite spur gears," *Proceedings of the Institution of Mechanical Engineers, Part J: Journal of Engineering Tribology*, vol. 223, no. 6, pp. 925–928, Jun. 2009.

[10] W. Li, A. Wood, R. Weidig, and K. Mao, "An investigation on the wear behaviour of dissimilar polymer gear engagements," *Wear*, vol. 271, no. 9–10, pp. 2176–2183, Jul. 2011.

[11] T. J. Hoskins, K. D. Dearn, S. N. Kukureka, and D. Walton, "Acoustic noise from polymer gears – A tribological investigation," *Materials & Design*, vol. 32, no. 6, pp. 3509–3515, Jun. 2011.

[12] C. Gurunathan, S. Kirupasankar, and R. Gnanamoorthy, "Wear characteristics of polyamide nanocomposite spur gears," *Proceedings of the Institution of Mechanical Engineers, Part J: Journal of Engineering Tribology*, vol. 225, no. 5, pp. 299–306, May2011.

Effect of swirl flow on characteristics of the annular diffuser

Hardial Singh[a*], B.B. Arora[b]
[a]Department of Mechanical Engineering, Delhi Technological University,Delhi, and Postcode-110042, India
[b] Department of Mechanical Engineering, Delhi Technological University,Delhi, and Postcode-110042, India
Corresponding author Email id: hardialsingh3@gmail.com

Abstract. The current work deals with the axial flow annular diffuser's performance characteristics with divergent casing angle 9° with an area ratio of 3. We performed the numerical simulation to examine the inlet swirl's effect on the annular diffusers. The velocity profiles were measured along the span of diffusers at a number of locations. Analysis of the results shows that the inlet swirl improves the stability at outlet flow, gains the recovery of pressure, and delays the flow separation on the casing. It is found that static pressure recovery increases by 5% with swirl flow. The results prove that the best performance is observed in the unequal hub and diverging casing diffuser with an equivalent cone angle of 15° and the inlet swirl angle of 12°.

Keywords: annular diffuser; swirl angle; longitudinal velocity profile; static pressure recovery coefficient

1. Introduction

The diffuser is the expanding cross-section, which recovers the fluid's highest possible static pressure by reducing its flow velocity. The flow gets decelerated when it passes across the diffuser, as part of the fluid's kinetic energy is transformed into potential energy of pressure. The effective diffuser is that which changes maximum pressure energy from kinetic energy. In the gas turbine, the central core shaft is required, and the necessity is fulfilled with the annular diffuser. Many researchers have been investigated swirl flow through the diffuser and examine the flow behaviour inside the diffusers [1-2]. Stevens and Williams [3] carried out the work on annular diffusers having uniform diameter centre bodies and diverging casing walls. It was found that increasing the turbulence level at the inlet improvement in the pressure recovery and stable flow at the outlet. Adkins [4] studied the optimum design of annular diffuser geometries, but those methods do not play a vital role in the designing of diffusers because of the complex geometry and flow field. Annular diffusers with swirl flow at inlet have very less literature available, but an extensive range of literature is available for conical diffusers of different geometries. Arora et al. [5] have performed the numerical study of the swirling flow on the annular diffuser by applying the RNG k- ε turbulence model. They have found that flow separation is completely removed from the casing wall by using swirling flow. The paper's primary focus is to analyze the diffuser's performance with fully developed turbulent swirl and non-swirl flow.

1.1 Geometric model

The annular diffuser with casing wall angle 9° and hub wall 5° of area ratio (AR) 3 have been selected for present investigations. The fully developed turbulent flow with swirl angle 0°- 25° is to be chosen to generate swirl velocity type distribution at the diffuser's inlet. Based on the existing literature optimized geometric values of the diffusers are used to carry out the numerical analysis [6, 7, 8]. The annular diffuser's geometric value has been optimized for maximum static pressure recovery and minimum total pressure loss coefficient. Figure 1 and Table 1 show the geometrical configuration details of the unequal hub and diverging casing axial annular diffuser parameters.

Table 1. Geometrical dimension of Unequal Hub and Diverging Casing (UHDC) annular diffuser.

Geometrical parameters	Dimensions
Area ratio (AR)	3
Diffusing axial length (L), cm	37.55

Equivalent cone angle (2Θ)	15°
Hub inlet diameter, cm	7.6
Casing inlet diameter, cm	15.5
Hub wall angle (Θ_h)	5°
Casing wall angle (Θ_c)	9°

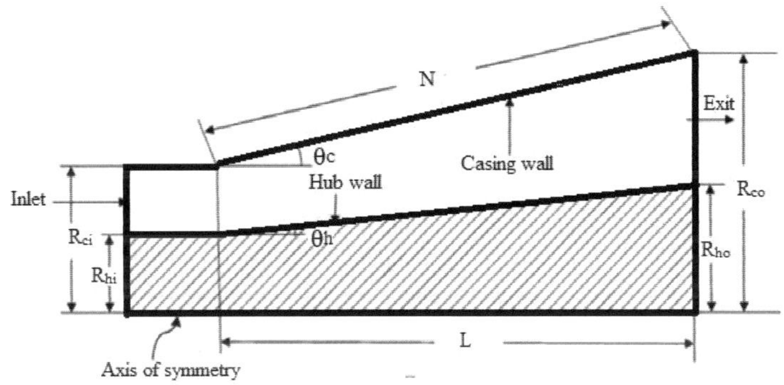

Figure 1. Diffuser geometrical parameter

2. Turbulence model

Equations (1) and (2) are the governing equations for k and ε, and equations (3) to (4) give the auxiliary relations for the RNG k-ε turbulence model. The transport equations are given below:

$$\rho u_i \frac{\partial k}{\partial x_i} = \frac{\partial}{\partial x_j}\left[\alpha_k \mu_{eff} \frac{\partial k}{\partial x_j}\right] + G_k - \rho\varepsilon + S_k \tag{1}$$

$$(\rho u_i)\frac{\partial \varepsilon}{\partial x_i} = \frac{\partial}{\partial x_j}\left[\alpha_\varepsilon \mu_{eff} \frac{\partial \varepsilon}{\partial x_j}\right] + C_{1\varepsilon}\frac{\varepsilon}{k}G_k - C_{2\varepsilon}\frac{\varepsilon^2}{k} - R_\varepsilon + S_\varepsilon \tag{2}$$

$$\mu_{eff} = \rho C_\mu \frac{k^2}{\varepsilon} \tag{3}$$

$$\mathrm{R}_\varepsilon = \frac{C_\mu \rho \eta^3 \left(1 - \eta/\eta_0\right)}{1+\beta\eta^3}\frac{\varepsilon^2}{k} \tag{4}$$

Here α_k represents the counter-effective Prandtl number for k and α_ε represents the counter effective Prandtl numbers for ε. S_k and S_ε are the source terms defined by the user. The term $\eta = sk/\varepsilon$ and value of the constants are considered in the equation as follows. $C_{1\varepsilon}$=1.42; $C_{2\varepsilon}$ =1.68; C_μ= 0.0845; η_0 =4.38; and β =0.012.

Previous studies conducted on the annular diffusers by (Singh et al., [9], Mohan et al., [7], Arora, [10]) have considered different two-equation turbulence models, such as standard k-ε (Sadasivan et al.,[11]), and RNG k-ε (Arora, [10]) to analyze the flow field variables. Here the RNG k-ε turbulence model has been chosen due to its better performance in predicting the velocity profile against the experimental results of Arora and Pathak [12].

3. Grid independence test

The grid study has been conducted for the unequal hub and diverging casing annular diffuser. The grid's size depends upon the diffuser's geometrical parameters, but different diffusers will have varied grid sizes. Four grids cell counts between 0.9 to 1.9 million are created (coarse grid=0.9×10^5, medium grid=1.2×10^5, fine grid=1.6×10^5, and finer grid =1.9×10^5). The numerical analysis has been performed with swirl flow using the RNG k-ε turbulence model at the Reynolds number $2.5\text{x}10^5$. It is observed that static pressure coefficient profiles for the fine grid and finer grid match with each other, as shown in Figure 2. Thus fine grid size (1.6×10^5 cells) has been selected to conduct the numerical study.

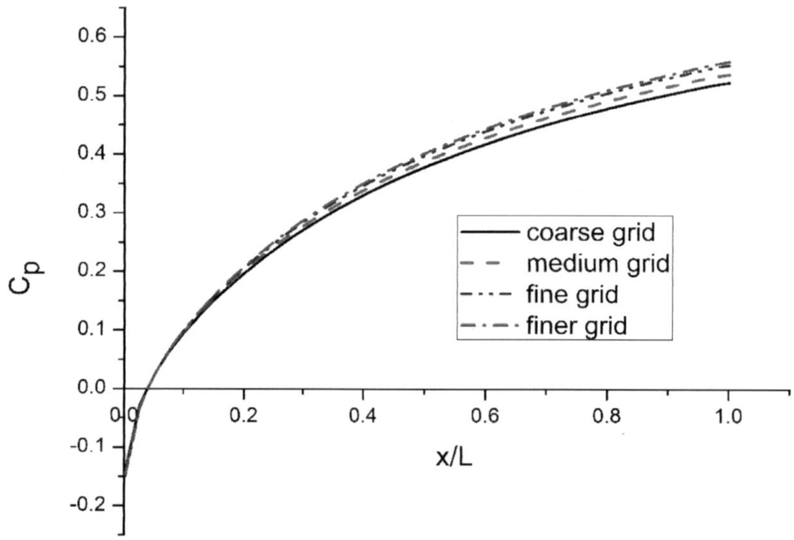

Figure 2.Grid independence test

4. Parameters of performance

The most important parameters are evaluated and are given below:-

Static pressure recovery (C_p):

The amount of kinetic energy transformed into pressure energy in terms of static pressure rise from exit to the inlet section of the diffuser due to diffusing action [13].

$$C_p = \frac{P_{ex} - P_{in}}{0.5\rho u_{avg}^2}$$

Here P_{ex} represents the mass average static pressure at the exit, and P_{in} represents the mass average static pressure at the inlet section of the diffuser while u represents the average velocity of flowing fluid.

Diffuser effectiveness (η):

It is the ratio of actual recover of the pressure of the diffuser to the ideal recover of the pressure of the diffuser for the similar area ratio.

$$\eta = \frac{C_p}{C_{pi}}$$

$$C_{pi} = 1 - \frac{1}{AR^2}$$

Where AR represents the area ratio.

Total pressure loss coefficient (C_{TL}):
The amount of total pressure loss due to turbulent intermixing of fluid due to viscous forces to the inlet dynamic head.

$$C_{TL} = \frac{P_{tin} - P_{tex}}{0.5\rho u_{avg}^2}$$

Where P_{tin} represents the mass average total pressure at the inlet, and P_{tex} represents the mass average total pressure at the exit section of the diffuser.

5. Analysis of velocity distribution

In the current work, the effect of swirling and non-swirling flow is evaluated. Moreover, the impact of the recirculation zone (RZ) on the performance is analyzed. The detailed analysis of streamline contours of UHDC diffuser at swirl angle 0°-25° is presented in Figure 3. At swirl angles 0°-12° of UHDC diffuser, velocity contours show symmetry of flow between the walls. As observed from the figure, there is no flow separation and reversal of flow from the walls. The primary intention behind the correct selection is the geometrical parameters and the inlet swirl angles.

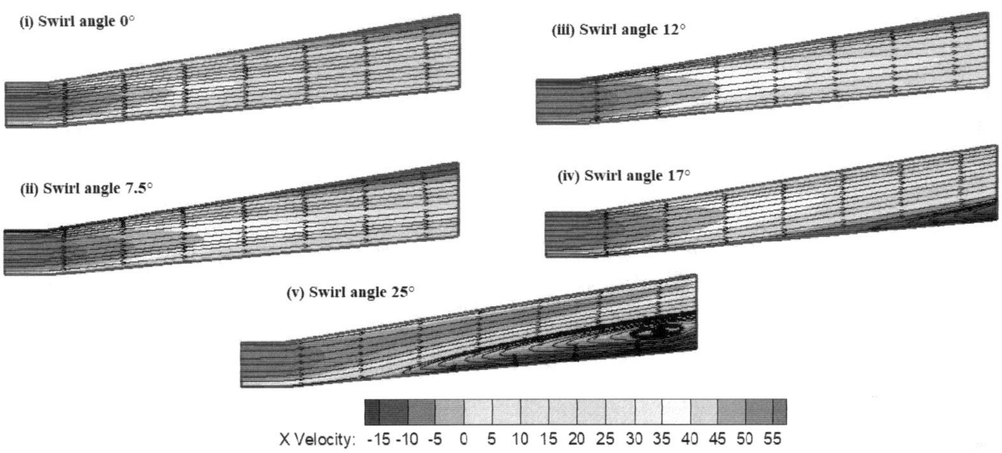

Figure 3. Streamline contours of UHDC diffuser at inlet swirl angles (i) 0°, (ii) 7.5°, (iii) 12°, (iv) 17°, and (v) 25°.

The longitudinal velocity distribution of UHDC annular diffuser at swirl angles 0°, 12°, and 25° with area ratio 3 is presented in Figure 4. The diffusion takes place continuously along the diffuser length. For the 0° inlet swirl angle, most of the flow close to the hub wall and separation occurs at a swirl angle 25°, as seen in Figure 4. The reverse flow is occurred on the inner wall at the inlet swirl angle 25° for x/L=0.3-0.9, respectively. The swirl velocity distribution is shown in Figure 5 for the inlet swirl angles 12°- 25°. These plots indicate that swirl velocity increases toward the casing wall of the diffuser. Due to the forced vortex nature, swirl velocity near the outer wall is higher in the case of a 25° inlet swirl. The high swirl velocity near the casing indicates that the flow separations from the hub. The above phenomena indicate that swirl decay as the flow moves in the downward direction of the diffuser.

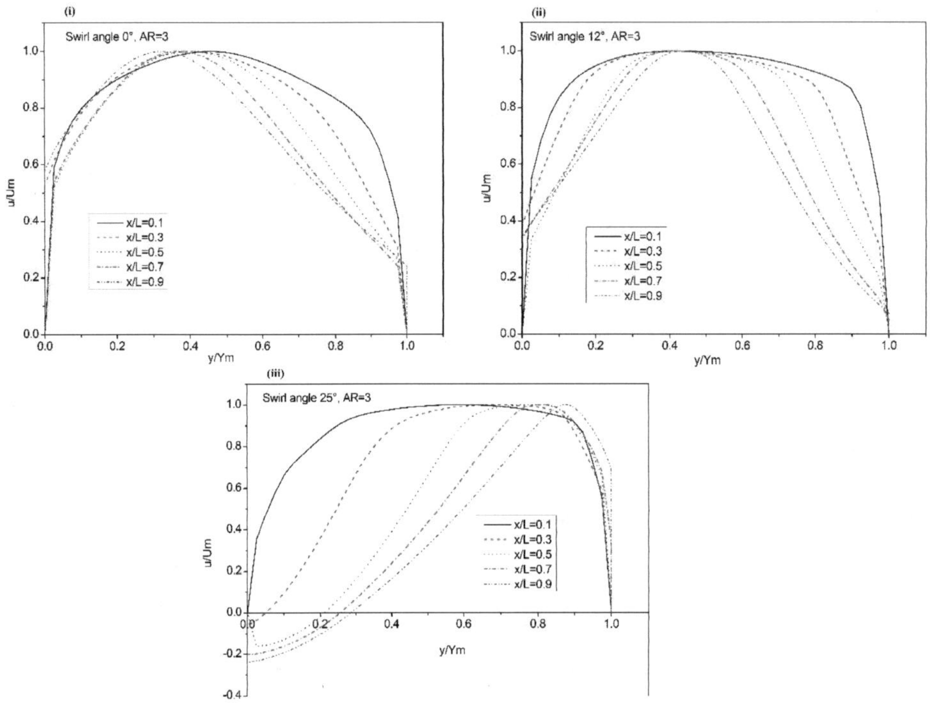

Figure 4. Longitudinal velocity profiles at the different axial location of UHDC diffuser (i) Swirl angle 0°; (ii) Swirl angle 12°; (iii) Swirl angle 25°.

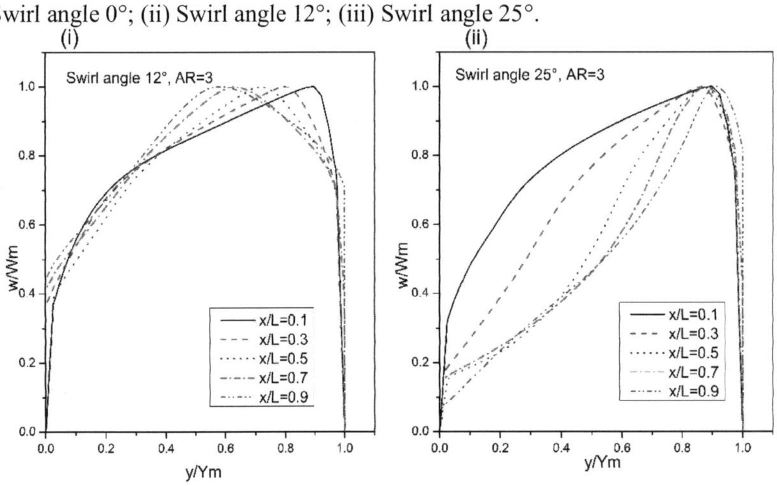

Figure 5. swirl velocity profiles at the different axial location of UHDC diffuser (i) Swirl angle 12°; (ii) Swirl angle 25°.

The flow distribution of velocity profiles very well matched with the Caladipietro et al. [14] and Hoadley [15] with swirl flow, who attributed this behavior to centrifugal effect force. The centrifugal force pushes the flow toward the diffusers' casing wall, delayed the flow separation, stabilizes the flow, and escalates the flow separation from the hub wall. The wall angle increases, then velocity profiles are distorted, as seen in the velocity profiles. The performance of the diffuser is susceptible to the wall angle as well as the swirl angle.

6. Pressure coefficient at casing

The mass average coefficient of pressure recovery has been determined in the annular diffuser with inlet swirl angles 0°-25° at the casing wall of the diffuser. It is found that there is an increase in pressure recovery using swirl flow in comparison to non-swirl flow in the diffuser. The pressure recovery coefficient increases from 0° to 12°swirl angle in the UHDC diffuser, whereas the further increase of swirl angle results in a decrement of the pressure recovery, as shown in Figure 6. The improvement of pressure recovery up to a 12° swirl angle is very distinct. For 17° to 25° swirl angle, drop in Cp due to RZ exist at the hub wall. A sudden drop in pressure recovery is observed for the 25°swirl angle after x/L=0.18 due to the sharp change in the velocity profile.

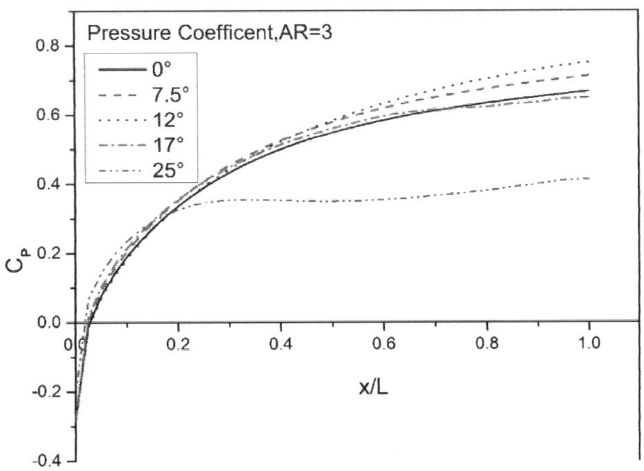

Figure 6. Pressure recovery coefficient at the casing wall of UHDC diffuser

7. Conclusion

- By introducing swirl, the flow shifts from the hub wall to the casing wall due to which no flow separation occurs on the casing wall.
- Due to the boundary layer development velocity profiles have different shapes along with the downstream of the diffuser passage.
- The UHDC diffuser's performance is degraded due to the presence of a recirculation zone (RZ) on the hub wall at the inlet swirl angles 17° and 25°.
- The best performance is observed on the optimized swirl angle 12° for the UHDC diffuser.

References

[1] Kline SJ. On the nature of stall. Journal of basic Engineering. 1959 Sep 1;81(3):305-19.

[2] Singh H, Arora BB. Effect of area ratio on flow separation in annular diffuser. In Advances in fluid and thermal engineering 2019 (pp. 297-305). Springer, Singapore.

[3] Stevens, S. J., and Williams, G. J. The influence of inlet conditions on the performance of annular diffusers. Trans. ASME J. Fluids Eng., 1980, 102, 357–363.

[4] Adkins RC. A simple method for designing optimum annular diffusers. American Society of Mechanical Engineers; 1983 Mar 27.

[5] Arora BB, Kumar M, Maji S. Analysis of flow separation in wide angle annular diffusers. International Journal of Applied Engineering Research. 2010; 5(20):3419-28.
[6] Singh, S. N., Agarwal, D. P., Sapre, R. N., and Malhotra, R. C. Effect of inlet swirl on the performance of wide-angled annular diffusers. Indian J. Eng. Sci., 1994, 22(6), 305–313.
[7] Mohan R, Singh SN, Agrawal DP. Optimum inlet swirl for annular diffuser performance using CFD. Indian Journal of Engineering and Materials Sciences, Vol. 5, pp. 15-21.
[8] Singh H, Arora BB. Optimization of Inlet Swirl for Flow Separation in Annular Diffuser. In Advances in Electromechanical Technologies 2020 (pp. 409-419). Springer, Singapore.
[9] Singh SN, Seshadri V, Saha K, Vempati KK, Bharani S. Effect of inlet swirl on the performance of annular diffusers having the same equivalent cone angle. Proceedings of the Institution of Mechanical Engineers, Part G: Journal of Aerospace Engineering. 2006 Feb 1; 220(2):129-43.
[10] Arora BB. Performance analysis of parallel hub diverging casing axial annular diffuser with 20 equivalent cone angle. Australian Journal of Mechanical Engineering. 2014 Jan 1;12 (2):179-94.
[11] Sadasivan S, Arumugam SK, Aggarwal MC. Numerical simulation of diffuser of a gas turbine using the actuator disc model. J Appl Fluid Mech. 2019 Jan 1; 12:77-84.

[12] Arora, B. B., Pathak, B. D. Effect of Geometry on the Performance of Annular Diffuser. International Journal of Applied Engineering Research. 5(20). 2009 pp. 2639-2652, ISSN 0973-4562.
[13] Singh H, Arora BB. Effects of casing angle on the performance of parallel hub axial annular diffuser. International Journal of Turbo & Jet-Engines. 2020 Aug 19;1(ahead-of-print).
[14] Coladipietro R, Schneider JH, Sridhar K. Effects of inlet flow conditions on the performance of equiangular annular diffusers. Transactions of the Canadian Society for Mechanical Engineering. 1975; 3(2):75-82.
[15] Hoadley, D. 1970, "Three Dimensional Turbulent Boundary Layers in an Annular Diffuser", Ph.D. Thesis, Department of Engineering, University of Cambridge, London

High-efficiency thermodynamic cycles for Kalina power generation systems: A comprehensive review

Alok Manas Dubey and Manoj Sharma
Department of Mechanical Engineering, Greater Noida Institute of Technology, Greater Noida, Uttar Pradesh, 201310, India

alokmanas28@rediffmail.com, manojunited1987@gmail.com

Abstract: In this paper, different Kalina power cycles have been identified and presented. The main endeavor of this review paper is to propel a complete understanding of simple and complex Kalina power generation systems. Additionally present status emphasizing some enhanced performance in Kalina power systems has been detailed. A hypothetical study of diverse Kalina power systems, incorporating an ejector, distillation column, sliding condensation, split cycle, double pressure and variable composition has been made. These modifications to the fundamental cycle decrease expansion losses using diverse principles and they require dissimilar mechanical hardware of special intricacies and expenditures. Kalina cycle with an ejector has maximum potential for efficiency improvement followed by the sliding condensation pressure method. Generally, the alternative options present several benefits to Kalina power systems such as decreased losses, improved performance and reduced energy expenditure.

1. Introduction

It is vital to reduce fossil fuel consumption and greenhouse gas emissions for a sustainable future. Organic Rankine and Kalina power systems are two promising options which use low grade energy and have been extensively discussed in the available literature. The striking characteristic of the Kalina cycle is the increase in efficiency due to heat exchange processes occurring at varying temperatures. The increase results due to higher net work output. Efficiency is augmented by means of recuperator exchangers. The increase occurs as a consequence of the distinctive variable boiling and condensing feature of the work medium. The heat addition and rejection to the ammonia– water mixture takes place at varying temperatures. The varying temperature through the heat-exchange processes decreases the thermal unavailability. Moreover thermal pinch in a boiler is also reduced. Scientists and engineers have been investigating for the performance optimization of Kalina cycle. Researchers are making ever increasing efforts to compute thermodynamic properties of ammonia-water mixture precisely and analyze diverse Kalina cycles to improve thermal and exergy efficiencies.

2. Kalina cycle with a distillation column

Performance of the Kalina power cycle can be improved by using a distillation column. The DCSS configuration comprises of a flash separator, a flash preheater, two reheaters, an absorber, a feedwater heater and a condenser. The working fluid exiting from the turbine (15) is refrigerated in the flash preheater (16) and in the reheaters (18). It is mixed with a poor ammonia mixture (6) approaching from the separator so as to obtain a lower concentration of ammonia and hence a higher boiling temperature. This concentration (19) is called the basic composition.

After being condensed in the condenser (20) the mixture is pumped (21) to the second reheater (22) and heated. A portion of the liquid is extracted (23) to weaken the ammonia-rich stream (8) incoming via the separator so as to refurbish the composition of mixture (9). The remaining mixture (24) is exchanges heat in both first reheater (1) and flash preheater (2). The stream is divided into a liquid stream (3) with a low ammonia fraction and a vapor stream (7) with a high ammonia fraction in the flash separator. The temperature of both the streams reduces while heat is exchanged in the first reheater and the second reheater (5). After throttling the stream (6) is then mixed with the incoming stream from the turbine. The liquid (23) is mixed with the vapor stream (8)in order to refurbish the composition of the working fluid. Next the mixture (9) is cooled (10) in the feedwater heater, condensed (11) in the condenser and pumped (12) again to the feedwater heater (13) before entering the vapor generator.

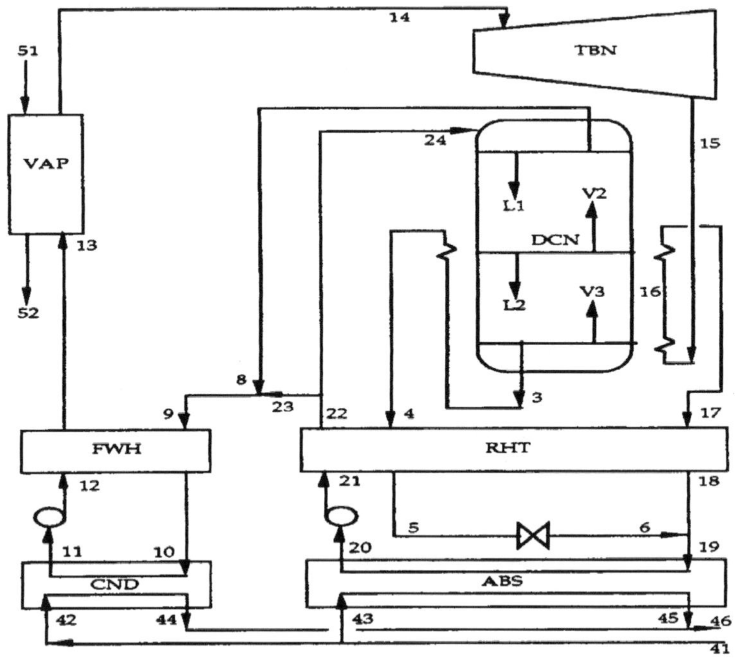

Fig.1. Layout of Kalina power system with a distillation column [1]

Glasare et al. [1] investigated and compared three dissimilar arrangements of the Kalina bottoming cycle by changing the diverse compositions in the cycles. The cycle energy efficiency and the second law efficiency have also been calculated. In addition to the rate of exergy loss in each unit was also computed. However the three configurations differ very little in efficiency as depicted in the results. This configuration is a novel alternative to the arrangement investigated by Chuang and Ishida [2] and temperature pinches in the heat exchanger are less intense in comparison to the original cycle described by El-Sayed and Tribus[3]. Chuang and Ishida observed that a distillation column reduces the intensity for pinch points because the maximum pinch was established in reheater 1.

The energy and exergy efficiencies differ with the working and the basic compositions. The maximum value of energy and exergy efficiencies are found at a concentration of 0.645 kgNH_3/kg,tot The exergy efficiency is 34.10% and the exergy efficiency is 69.19% at this optimum composition.

3. Kalina cycle with sliding condensation pressure

The efficiency of the KSG-1 cycle can be improved by utilising the sliding condensation pressure technique for low grade waste energy obtained from industry. The layout is shown in Fig. 2. The operating medium is a mixture of ammonia-water. The functioning of the system is detailed as follows. The saturated liquid at state 1 is pumped from the tank into the recuperator. The subcooled liquid at state 2 is heated to state 3 by the recuperator. Next, the basic solution extracts heat from the brine solution in the evaporator and passes it to the separator at state 4.The basic solution cannot be completely vaporized since the heat source is brine at 120°C and hence a separator is utilized to split the two-phase fluid into a rich ammonia mixture of state 5and a lean ammonia mixture of state 6. The expansion pressure ratio in the turbine can be adjusted using sliding pressure control method whilst the vapor mixture is changed to state 7. Alternatively, the liquid mixture is throttled through the expansion valve to state 8. The exhaust gas from the turbine is mixed with the low-pressure liquid in the mixer. Subsequently the mixture is passed to the recuperator where it rejects heat to the saturated liquid coming from the tank. Next this mixture is condensed in the air-cooled condenser and converted again to saturated liquid.

Wang et al. [4] established a hypothetical model based on sliding condensation pressure method and subsequently framed a mathematical program to investigate the cycle performance. The condensation pressure alteration according to the varying ambient temperature has been mathematically confirmed with different ammonia-water mixture concentrations.Hu et al. [5] investigated the off-design performances of an organic Rankine cycle system with three diverse control techniques together with sliding pressure control. Generally, sliding pressure control is used to cater to the changeable demand in load by varying evaporation pressure. Modi et al. [6] and Li et al. [7] performed an off-design performance analysis of the Kalina cycle with sliding pressure control. Usman et al. [8] carried out experimental analysis of an organic Rankine cycle using sliding pressure control method. In comparison to the ORC with a clean working fluid, a zeotropic working fluid like an ammonia-water mixture is used in the Kalina cycle. This provides flexibility as far as the ammonia mass fraction is concerned.

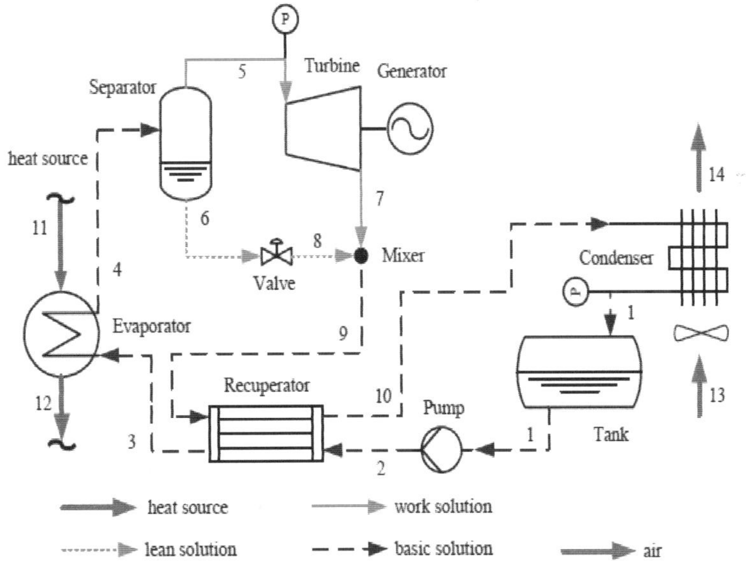

Fig.2. Layout of the Kalina power system with sliding condensation pressure [4]

4. Kalina split-cycle power system

The effectiveness of Kalina power system could be also enhanced by using split system. Fig. 3 is the schematic representation of the Split-cycle method. To concentrate on the unique split stream boiler, the Split-cycle arrangement was configured with fewer components required for evaluating the concept. Two streams a rich stream (25) and a lean stream (31) enter the boiler. The rich stream is completely vaporized before it mixes with the lean stream in the mixer (4) heated to its boiling point. This configuration helps in lowering the temperatures as liquid (25, 31) is converted to vapor (2). An auxiliary mixing subsystem is necessary to generate the two streams with required concentrations and mass flow rates. This comprises of 3 splitters and 2 mixers and the 3 splitters split the incoming streams (11, 12 and 18) as desired. In most cases, desired flow rates and concentrations of the streams can be achieved from a series of splitting fractions, however there may be situations when streams cannot be formed as needed. The slope of the evaporation temperature curve can be attuned to the temperature gradient of the heating source by choosing an optimum composition for both streams, as shown in Fig. 3.

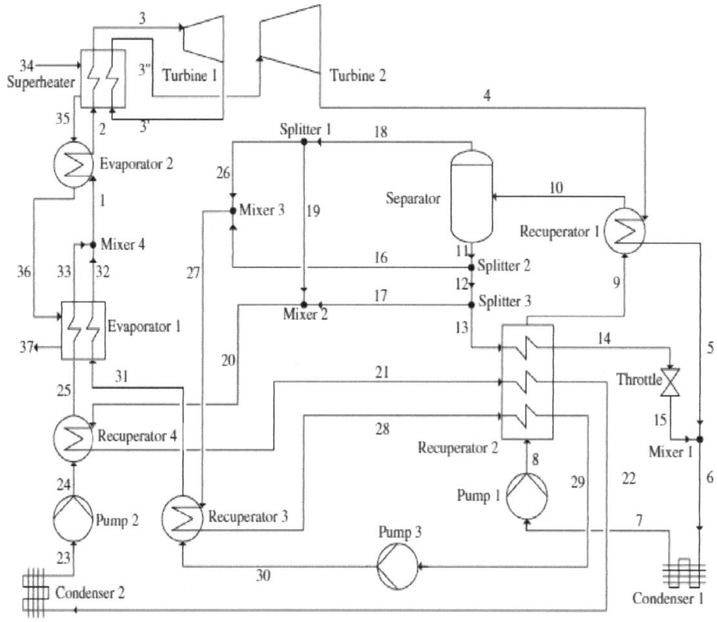

Fig.3. Schematic representation of Kalina split system [9]

Larsen et al. [9], concluded that the components that produce maximum impact on process efficiency are separator, recuperator, boiler and turbine. Additionally, it was suggested that the most significant variables that affect the performance of the system were the concentration of ammonia and temperature of the cooling water.

Moreover an economic study of the Kalina split-cycle was also conducted and it was concluded that payback time of this configuration is comparable to that of a conventional Kalina cycle. The maximum thermal efficiency is 23.2% with reheating option which was about 3.4-5.9% more in comparison to the reference cycle. The maximum thermal efficiency was obtained as a result of optimization performed using the genetic algorithm technique. Nguyen et al. [10] devised a technique to optimize this unique Kalina system and confirmed that in comparison to a typical Kalina power system the split cycle process enhances the thermal efficiency from 20.1% to 21.5%. Nevertheless this optimization technique is confined to the boiler and turbine components only. Bombarda et al. [11] thermodynamically compared the performance of Kalina cycle with that of an ORC cycle, with hexamethyldisiloxane working fluid. The waste heat was recovered from two Diesel engines both with a power of 8900 kW. Assuming a LMTD of 50°C for heat exchanger, an electric power of 1615 kilowatts in the Kalina cycle and 1603 kilowatts for the ORC cycle was estimated. Marston [12] theoretically investigated the Kalina cycle by varying different parameters. The turbine inlet composition and separator temperature were recognized as the significant variables for optimization. These results were acknowledged by Nag and Gupta [13], who conducted an exergetic study of the Kalina cycle, and confirmed that the turbine inlet temperature and composition, in addition to the separator temperature, as producing the maximum effect on the thermal efficiency of the system. Dubey et al. [14] critically investigated a CO_2/propylene based transcritical cascaded system for simultaneous cooling and heating. Energy utilization factor (EUF) enhancement of 8 % was confirmed by the addition of a split unit. A multilinear regression analysis was carried out for the transcritical cascaded refrigeration system to optimize EUF, intermediate temperature and mass flow ratio.

5. Double pressure Kalina power system

Yet another method to improve the performance of Kalina cycle is the use of a double pressure system as shown in Fig.4. A cycle's stream (17) absorbs some heat from the heat source and leaves the LP evaporator as it is a two-phase mixture with both liquid and vapor (1). Hence the phases can be divided in separator 1, from where the ammonia rich stream passes to the turbine (4) after further heating in the HT regenerator. The residual liquid stream (3) with lower concentration of ammonia is pumped to the HP evaporator where it absorbs heat. The two phases are again divided in separator 2, from where the ammonia rich stream passes to the turbine (12). The lean stream then enters the pressure valve (8) via the HT and LT regenerator. The stream's pressure drops and it mixes with the exhaust of the turbine (14) in the mixer (14). The product of the mixer (10) condenses in the condenser. The condensate (15) is then pumped (6) to be preheated by the separator's mentioned liquid extract and enters the LP evaporator (17) via the economizer.

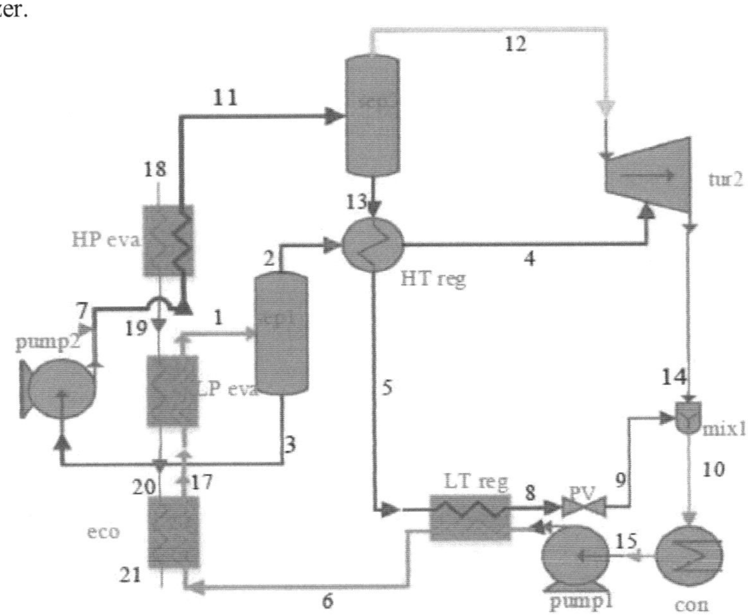

Fig.4. Schematic representation of Kalina cycle with double pressure [15]

Rasool et al. [15] conducted an optimization study for four different arrangements of double pressure Kalina cycle system all of which are innovations of Kalina cycle system 11. Exergy efficiency is considered as the objective function and the modified double pressure cycles were compared to the base Kalina cycle. Findings illustrate that the Kalina cycle system 112b has the maximum efficiency at the base condition. It was further demonstrated that exergy efficiency and the cost of equipment rose by escalating the heat source temperature at the optimum condition however the payback cost of electricity decreases. Marson and Hyre performed a comparative analysis of single stage and triple stage Kalina cycle with a triple pressure steam cycle [16]. Sadeghi et al. [17] projected a modified Kalina cycle with two turbines which may be synchronized appropriately with a heat source operating in a range of 80–200°C. Guo et al. compared the base Kalina cycle [18] with a double pressure evaporation arrangement. The results confirmed that the novel arrangement increased the efficiency of the base cycle by 17 percent.

6. Kalina cycle with ejector

In an EKalina cycle expansion energy is recovered and fluid pressure is increased by replacing the throttle valve and the absorber with an ejector. The main components of an EKalina cycle include a regenerator, vapor-liquid separator, expander, evaporator, condenser, fluid pump and an ejector. The Ammonia solution leaving the evaporator is divided into saturated rich vapor and saturated weak liquid in the separator. The saturated vapor flows to the expander while the saturated liquid flows to the ejector via the regenerator. The exhaust from the expander mixes with the cooled liquid coming out from the regenerator. Further the mixture coming out from the ejector is condensed in the condenser and pressurized back to the evaporator after absorbing heat in the regenerator.

Li et al. [19] projected a Kalina cycle with ejector as a replacement for the throttle valve and the absorber. The back pressure of the expander is reduced by the ejector thus escalating the power produced as well as thermodynamic efficiency of the cycle. For values of X_B in the range of 0.5-0.9, X_V of 0.92-0.98 and water temperature of 110°C, the maximum and minimum improvement in power of the EKalina cycle in comparison to KCS 11 are 39.55%, and 0.75% respectively. The increases in efficiency are 39.13%, and 0.72% respectively. Rashidi et al. [20] introduced an ejector before the evaporator in a novel configuration of Kalina combined power and cooling cycle. It was concluded during investigation of energy efficiency that under similar operating conditions the novel cycle with an ejector provides a performance enhancement without significantly increasing system intricacy.

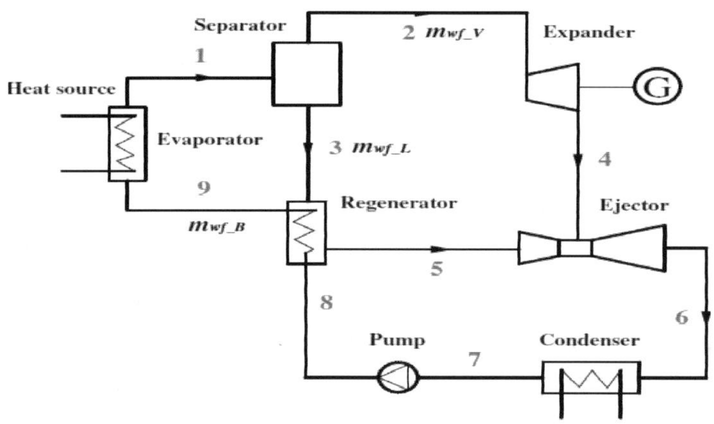

Fig.5. Layout of the Kalina cycle system with ejector [19]

Seckin et al. [21] suggested an arrangement of a cogeneration cycle wherein an ejector cooling system is combined with the Kalina power system to recuperate heat from weak ammonia mixture which exits from the separator at an elevated temperature and pressure and is discharged as a waste. The thermal efficiency of the cogeneration cycle augmented with rise in strength of weak NH_3-H_2O solution. However the efficiency reduced with both increasing condenser exit temperature and pressure of heat recuperator. Thermodynamic modeling and simulation of the novel cogeneration cycle was performed with the viewpoint of the energy and exergy. The exhaust of the turbine was separated in two streams one to feed the condenser and the other to feed flow heater. Dubey et al. [22] suggested a numerical model for investigating the performance of a CO_2/propylene based transcritical cascade refrigeration-heat pump system. The expansion valve was replaced by a vortex tube for recuperating expansion work. An improvement of 6 % in the COP of the suggested system was calculated for the specified values of temperatures and pressures.

7. Composition-adjustable Kalina cycle

Another option to reduce the expansion losses is the use of separators which have potential for improving efficiency. As shown in Fig. 6, the basic solution a saturated liquid flows to Tank 1. A density sensor is positioned at its exit to determine the density of the working fluid to calculate its composition and sent as a feedback signal for monitoring the composition. The solution at state 3 is pressurized and converted to subcooled liquid. This subcooled liquid absorbs heat in the recuperator and is then heated further in the evaporator to convert into a two-phase mixture by geothermal brine. The temperature of the brine water is set as 120°C in this study so that operating variables are similar to that in literature [25] for ease of comparison.

The basic solution is not fully converted into vapor due to lower temperature of brine. Therefore a separator is used to split the two phase mixture into a rich stream of saturated vapor and a lean stream of saturated liquid. The high temperature high pressure vapor mixture expands in the turbine to produce useful work. The low pressure liquid at state 10 is mixed with the low temperature low pressure exhaust from the turbine in a mixer. This two-phase mixture then passes to the heat exchanger, where the temperature of this mixture reduces again after losing its heat to the basic solution from pump1. The ammonia-water mixture from the recuperator is split into two streams in a separator to facilitate the condensation process. The incoming liquid stream from Tank 2 is subsequently pumped and subsequently sprayed in Mixer 2, condensing the concentrated ammonia vapor stream further. The mixture exchanges heat from the atmospheric air and is converted into saturated liquid at state 1.

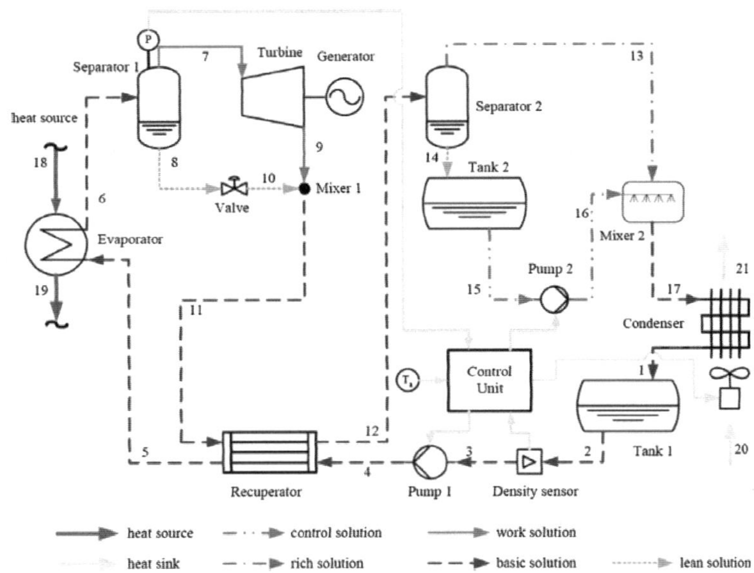

Fig.6. Configuration of Variable Composition Kalina cycle [23]

Wang et al. [23] proposed a numerical model to simulate the operating code of a variable composition Kalina power system. A mathematical program is subsequently designed to investigate the performance of the cycle with changing climatic conditions. It was established that better thermal efficiency could be attained by adjusting the composition of Kalina power system in comparison to a conventional system. Although, such an increase in thermal efficiency essentially depends on the system variables and climatic conditions. The thermal efficiency of Cycle B is steady at 6.12%. Nevertheless, the thermal efficiency of the other cycle A augmented from 6.12% to 9.24% with decrease in ambient temperature. The ambient air temperature can be synchronized with the condensation temperature of working fluid to increase the power output of the turbine considerably.

Hua et al. [24] examined the transient performance of a Kalina cycle for high-temperature waste heat recuperation, which can control the concentration of the working fluid. This method controls the on/off state of two valves to maximize power generation with varying temperature of the waste heat source. The inlet pressure of the turbine is altered by regulating the concentration of the working fluid. With a variation of flow rate ranging between With a variation of flow rate in the range of 100% to 80%, the Kalina power system efficiency by means of concentration control is 12.8% more in comparison to valve throttling method. A thermal efficiency of 21.25% was established for this cycle.

Nasruddin et al. [25] numerically investigated a KCS-34 Kalina cycle with Cycle Tempo 5.0 software and validated the results with the experimental data of the Husavik power plant which presented a good agreement. Hettiarachchi et al. [26] investigated the efficiency of KCS-11 power system for harnessing renewable geothermal energy. An optimum value for ammonia concentration was established for a given turbine inlet pressure. Sun et al. [27] investigated the performance of a KCS-11 Kalina cycle for research in solar energy. It was confirmed that the ammonia mass fraction was a significant system variable and its optimization would decrease irreversibility of the system.

8. Conclusions

Extensive reviews of Kalina Power cycle modifications such as use of a distillation column, sliding condensation pressure, split unit, ejector, double pressure and composition variation have been carried out in the paper. The below mentioned conclusions can be drawn from the above review.

- Configuration with a distillation column in the DCSS reduces efficiency due to higher exergy losses in the DCSS owing to reduced heat transfer. However there is an advantage that temperature profiles are smooth which are achievable with counter current heat transfer. Moreover an improved design of the distillation column would attain additional exergy efficient heat transfer.
- The estimated thermal efficiency of the Kalina cycle by sliding condensation pressure is 8.21%, enhanced by 33.5% in comparison to the conventional Kalina cycle.
- An average thermal efficiency of 23.2% can be attained by split cycle while utilizing reheat in contrast to 20.8% without reheat. Reheat augments the thermal efficiency of conventional cycle by 3.4-5.9%.
- Maximum improvements in the exergy efficiencies of double pressure cycle as compared to the conventional cycle are 16.7, 23.6 and 10.2 % for the 3 temperature levels, 383.15 K, 413.15 K and 443.15 K, respectively.
- For values of X_B in the range of 0.5-0.9, X_V of 0.92-0.98 and water temperature of 110°C, the maximum and minimum improvement in net power output of the EKalina cycle in contrast to KCS 11 are 39.55%, and 0.75% respectively. The increases in efficiency are 39.13%, and 0.72% respectively.
- With a variation of flow rate in the range of 100% to 80%, the Kalina power system efficiency by means of concentration control is 12.8% more in comparison to valve throttling method. A thermal efficiency of 21.25% was established for this cycle.

9. References

[1] Glasare K B, Olsson E K, Spakovsky M R, Gunnar Svedberg. Analysis of Kalina cycle designs with modifications including a distillation column. Presented at the International Gas Turbine and Aeroengine Congress and Exposition Cincinnati, Ohio, 1993,May 24-27.
[2] Chuang C, Ishida M, Graphic. Exergy Analysis of the Kalina Power cycles. Analysis of Thermal and Energy Systems, Greg. Foundas, Athens, Greece, 1991, 93-104.
[3] El-Sayed Y M, Tribus M A. Theoretical Comparison of the Rankine and Kalina Cycles. ASME publication, AES 1, 1985, 97-102.
[4] Wang E, Yub Z, Zhang F. Investigation on efficiency improvement of a Kalina cycle by sliding condensation pressure method. Energy Conversion and Management, 2017, 151:123-135.
[5] Hu D, Zheng Y, Wu Y, Li S, Dai Y. Off-design performance comparison of an organic Rankine cycle under different control strategies. Applied Energy, 2015, 156:268–79.
[6] Modi A, Andreasen J G, Karn M R, Haglind F. Part-load performance of a high temperature Kalina cycle. Energy Conversion and Management, 2015, 105:453–61.
[7] Li H, Hu D, Wang M, Dai Y. Off-design performance analysis of Kalina cycle for low temperature geothermal source. Applied Thermal Engineering, 2016, 107:728–37.
[8] Usman M, Imran M, Lee D H, Park B S. Experimental investigation of off-grid organic Rankine cycle control system adapting sliding pressure strategy under proportional integral with feed-forward and compensator. Applied Thermal Engineering, 2017, 110:1153–63.
[9] Larsen U, Nguyen T V, Knudsen T, Haglind F. System analysis and optimisation of a Kalina-split cycle for waste heat recovery on large marine diesel engines. Energy, 2013, 64:484-94.
[10] Nguyen T V, Knudsen T, Larsen U, Haglind F. Thermodynamic evaluation of the Kalina split-cycle concepts for waste heat recovery applications. Energy, 2014, 7:277-288.
[11] Bombarda P, Invernizzi C M, Pietra C. Heat recovery from diesel engines: a thermodynamic comparison between Kalina and ORC cycles. Applied Thermal Engineering, 2010, 30(2-3):212-219.
[12] Marston C H. Parametric analysis of the Kalina cycle. J Eng Gas Turbines Power, 1990, 112(1):107-16.
[13] Nag P, Gupta A. Exergy analysis of the Kalina cycle. Applied Thermal Engineering, 1998, 18(6):427-39.
[14] Dubey A M, Agrawal G D, Kumar S. Thermodynamic analysis of a transcritical CO2/propylene cascade system with split unit in HT cycle. J Braz. Soc. Mech. Sci. Engineering, 2015, 37:1365-1378.
[15] Rasool B, Ali B. Thermodynamic optimization and thermoeconomic analysis of four double pressure Kalina cycles driven from Kalina cycle system 1. Energy Conversion and Management, 2017, 152: 110-123.
[16] Marston C H, Hyre M . Gas turbine bottoming cycles: triple-pressure steam versus Kalina. J Eng Gas Turbines Power, 1995, 117(1):10–5.
[17] Sadeghi S, Saffari H, Bahadormanesh N. Optimization of a modified double-turbine Kalina cycle by using Artificial Bee Colony algorithm. Applied Thermal Engineering, 2015, 91:19–32.
[18] Guo Z. Dual-pressure vaporization Kalina cycle for cascade reclaiming heat resource for power generation. Energy Conversion Management, 2015, 106:557–65.
[19] Li X, Zhang Q, Li X. A Kalina cycle with ejector. Energy, 2013, 54: 212-219.
[20] Rashidi J, Yoo C K. A novel Kalina power-cooling cycle with an ejector absorption refrigeration cycle: Thermodynamic modeling and pinch analysis. Energy Conversion and Management, 2018, 162: 225–238.
[21] Seckin C. Thermodynamic analysis of a combined power/refrigeration cycle: Combination of Kalina cycle and ejector refrigeration cycle. Energy Conversion and Management, 2018, 157: 631–643.

[22] Dubey A M, Agrawal G D, Kumar S. Performance evaluation and optimal configuration analysis of a transcritical carbon dioxide/propylene cascade system with vortex tube expander in high-temperature cycle. Clean Technology and Environmental Policy, 2016, 18:105-122.
[23] Wang E, Yu Z. A numerical analysis of a composition-adjustable Kalina cycle power plant for power generation from low-temperature geothermal sources, Applied Energy, 2013, 180: 834–848.
[24] Hua J, Chen Y, Wu J, Zhi Z, Dong C. Waste heat supply-side power regulation with variable concentration for turbine in Kalina cycle. Applied Thermal Engineering, 2015, 91:583–90.
[25] Nasruddin, Usvika R, Rifaldi M, Noor A. Energy and exergy analysis of Kalina cycle system (KCS) 34 with mass fraction ammonia-water mixture variation. Journal of Mechanical Sciences and Technolog, 2009, 23:1871–6.
[26] Hettiarachchi H D M, Golubovic M, Worek W M, Ikegami Y. The performance of the Kalina cycle system 11 (KCS-11) with low-temperature heat sources. ASME J Energy Resources Technology, 2007, 129:243–7.
[27] Sun F, Zhou W, Ikegami Y, Nakagami K, Su X. Energy exergy analysis and optimization of the solar-boosted Kalina cycle system 11 (KCS-11). Renewable Energy, 2014, 66:268–79.

AUTHOR INDEX

Institute of Physics
Dirac House, Temple Back
Bristol BS1 6BE UK

ISSN:1757-8981
ISBN 978-1-7138-3432-8

9 781713 834328